Neuronal Growth Cones

"I had the good fortune to behold for the first time that fantastic ending of the growing axon. In my sections of the spinal cord of the three day chick embryo, this ending appeared as a concentration of protoplasm of conical form, endowed with amoeboid movements. It could be compared with a living battering ram, soft and flexible, which advances, pushing aside mechanically the obstacles which it finds in its path, until it reaches the region of its peripheral termination. This curious terminal club, I christened the growth cone." (Santiago Ramón y Cajal, *Recollections of My Life*, 1937)

In *Neuronal Growth Cones*, Phillip Gordon-Weeks presents the molecular biology of the behaviour of growth cones. The book discusses the history of the discovery of growth cones and their importance in the development of a properly connected nervous system. This book is the first detailed, critical analysis of all aspects of growth cone biology.

Neuronal Growth Cones covers the basic morphology and behaviour of growth cones, motility and neurite extension via the growth cone cytoskeleton, pathfinding, intracellular signalling, and synaptogenesis.

This detailed treatment of our current knowledge of the growth cone is intended for advanced graduate students, postgrads, and researchers in cellular and molecular neuroscience, developmental biology, and anatomy.

Phillip R. Gordon-Weeks is in the Developmental Neurobiology Centre of King's College London.

T0276099

Developmental and cell biology series

SERIES EDITORS

Jonathan B. L. Bard, *Department of Anatomy, Edinburgh University*
Peter W. Barlow, *Long Ashton Research Station, University of Bristol*
David L. Kirk, *Department of Biology, Washington University*

The aim of the series is to present relatively short critical accounts of areas of developmental and cell biology where sufficient information has accumulated to allow a considered distillation of the subject. The fine structure of cells, embryology, morphology, physiology, genetics, biochemistry and biophysics are subjects within the scope of the series. The books are intended to interest and instruct advanced undergraduates and graduate students and to make an important contribution to teaching cell and developmental biology. At the same time, they should be of value to biologists who, while not working directly in the area of a particular volume's subject matter, wish to keep abreast of developments relevant to their particular interests.

RECENT BOOKS IN THE SERIES

Neuronal Growth Cones

PHILLIP R. GORDON-WEEKS
Developmental Neurobiology Centre, King's College London, UK

CAMBRIDGE
UNIVERSITY PRESS

CAMBRIDGE UNIVERSITY PRESS
Cambridge, New York, Melbourne, Madrid, Cape Town, Singapore, São Paulo

Cambridge University Press
The Edinburgh Building, Cambridge CB2 2RU, UK

Published in the United States of America by Cambridge University Press, New York

www.cambridge.org
Information on this title: www.cambridge.org/9780521444910

First published 2000
This digitally printed first paperback version 2005

A catalogue record for this publication is available from the British Library

Library of Congress Cataloguing in Publication data

Gordon-Weeks, Phillip R., 1952–
    Neuronal growth cones/Phillip R. Gordon-Weeks.
        p.   cm. – (Developmental and cell biology series : 37)
    Includes bibliographical references and index.
    1. Developmental neurophysiology.   2. Neurons – Growth.   I. Title.
  II. Series
  QP356.25.G67   2000
  573.8 – dc21   99-34186
                                         CIP

ISBN-13 978-0-521-44491-0 hardback
ISBN-10 0-521-44491-8 hardback

ISBN-13 978-0-521-01854-8 paperback
ISBN-10 0-521-01854-4 paperback

Contents

Preface

I had the good fortune to behold for the first time that fantastic ending of the growing axon. In my sections of *the spinal cord of** the three day chick embryo, this ending appeared as a concentration of protoplasm of conical form, endowed with amoeboid movements. It could be compared with a living battering ram, soft and flexible, which advances, pushing aside mechanically the obstacles which it finds in its path, until it reaches the region of its peripheral termination. This curious terminal club, I christened the **growth cone**.

Santiago Ramón y Cajal, *Recollections of My Life*, 1937 and 1989

If I had known how long it would take me to write this monograph on neuronal growth cones for Cambridge University Press, I would never have started it! I agreed to do so over seven years ago at a time when the task seemed considerably less daunting than it would be if I were starting now. However, as the years passed and the annual rate of publication of papers on growth cones became exponential, I began to feel I was facing a Herculean task. Of course, I did manage to get in a few games of tennis between writing chapters. I have tried to cover all topics concerning growth cones with the exception of regeneration, and I hope that the book is useful to both those entering the field and those who already work in it.

Many people have helped me to write this book, and I only have space to thank a few. David Tonge, Philip Beesley and Max Bush read large chunks and made many helpful and constructive suggestions for improvements; I owe them a debt of thanks. I would also like to thank Kate Kirwan for many of the line drawings and Leon Kelberman for photography, my family for tolerating my neglect, and the members of my lab for reading and commenting on earlier versions. Finally, I would also like to say a special thank you to Robin Smith, my commissioning editor at Cambridge University Press (now with Springer), for his elastic patience and seemingly limitless ability to tolerate my repeated breaking of deadlines and for his imaginative and practical suggestions on reading my efforts.

Phillip R. Gordon-Weeks

* As pointed out by Jacobson (1991, p. 195), the Craigie (Ramón y Cajal, 1937) English translation from the French omitted the words in italics.

Abbreviations

ADP	adenosine diphosphate
AMP	adenosine monophosphate
ATP	adenosine triphosphate
cAMP	cyclic adenosine monophosphate
DC	direct current
DRG	dorsal root ganglion
ECM	extracellular material
F-actin	filamentous actin
G-actin	globular actin
GABA	γ-amino butyric acid
GAP-43	growth-associated protein 43
GDP	guanosine diphosphate
GTP	guanosine triphosphate
MAP	microtubule-associated protein
micro-CALI	micro-chromophore-assisted laser inactivation
MTOC	microtubule-organising centre
MuSK	muscle-specific kinase
NCAM	neural cell adhesion molecule
NgCAM	neuron-glial cell adhesion molecule
NrCAM	neuronal cell adhesion molecule
NGF	nerve growth factor
PC12	phaeochromocytoma cells
RAGS	repulsive axon guidance signal
VASE	variable alternative spliced exon

1

Introduction

What are growth cones?

Of all animal cells, neurons have the most diverse and remarkable shapes. They vary in appearance from the simple, unipolar cells of invertebrate ganglia to the highly complex pyramidal cells of the human cerebral cortex with their profusion of processes and thousands of synapses. The basis for this diversity of shape is the ability of developing neurons to produce long and branching cellular processess, of which two distinct kinds are recognised: axons and dendrites. In mature neurons, axons convey action potentials away from the neuronal cell body, or soma, to the axon terminals, whereas dendrites transmit information toward the soma. Primary sensory neurons form an exception to this general rule since they have an axon which conveys action potentials *towards* the cell body. Axons and dendrites, collectively referred to as neurites when they are growing, are necessary for neurons to carry out their primary function, in which they have no equal, that of intercellular communication. In the central nervous system (brain and spinal cord), neurons usually communicate with each other, whereas in the peripheral nervous system they also communicate with a variety of effector cells such as muscle and secretory cells. In the adult, the distance over which the neuron must sustain its cellular processes to communicate with another cell may be considerable; it is many metres in the case of pyramidal motoneurons in large mammals such as the whale.

From an evolutionary standpoint, the diversity of neuronal shape has increased as the basic function of neurons to communicate with each other and with effector cells has become more sophisticated. Diversity of shape in neurons is essentially a matter of the distribution of axons and dendrites and their branches in three-dimensional space. How do developing neurons achieve this diversity? What mechanisms determine the final, adult morphology and the pattern of synaptic connections between cells? These questions are currently at the forefront of research in the neurosciences, not only because of their intrinsic interest but also because of the hope that answering them may help us to repair the damaged nervous system in adult humans, where the capacity to regenerate lost connections and damaged processes is impaired.

In the majority of animals, neurons are generated (neurogenesis) during embryogenesis or larval stages and the process is largely complete shortly after hatching, eclosion, or birth. In vertebrates, neurons originate during embryogenesis from cells that are produced in specialised regions of the ectoderm: the outer, epithelial cell layer of the three germ-layers of the vertebrate embryo. The origin of neurons in invertebrates is more complicated. In flies, such as *Drosophila*, neurons derive from the ectoderm, but in nematodes, such as *Caenorhabditis elegans*, neurons originate in a fragmented pattern throughout the worm, whereas in the leech all of the five embryonic stem cells (teloblasts) that derive from the egg, give rise to neurons (see Shankland & Macagno, 1992). In *Hydra*, neurons are generated throughout the life of the animal from multipotential interstitial stem cells.

In the embryo, neurons begin their postmitotic lives as unremarkable, non-process-bearing, rounded cells and must, therefore, extend neurites (Fig. 1.1). The extension of neurites, *neuritogenesis*, begins at the surface of the neuron as a localised, highly active membrane protrusion that develops into an entity called the growth cone (Fig. 1.1). In the simplest sense, these are motile enlargements at the tip of the extending neurite (Fig. 1.2) that were discovered and named over one hundred years ago by one of the founders of modern neurobiology, Ramón y Cajal (1890). Growth cones are present at the tips of both growing axons and dendrites. Growth cones are also formed by regenerating mature axons and dendrites when they are damaged by trauma such as crushing or cutting.

What do growth cones do and why have they attracted so much attention? Broadly speaking, growth cones have three crucial functions in the developing nervous system. First, they guide the growing neurite through the developing embryo to locate the cell with which the neuron will form a synapse. Secondly, they build the extending neurite behind them as they advance, like a spider spinning a web or the laying of a telephone cable by a cable-laying ship. Lastly, they form the pre- or postsynaptic element of the synapse. Their importance thus lies in the fact that they establish the process-bearing morphology characteristic of the neuron and that they form the correct connections between cells. In many cases, growth cones are required to navigate complicated routes through a variety of terrains – both hostile and friendly – to arrive at their destinations. Once they reach their target, their task is not over, for they must select an appropriate synaptic partner and make the transformation into a synapse.

Although growth cones have attracted considerable attention since their discovery, it is only in recent years that the appropriate technologies have

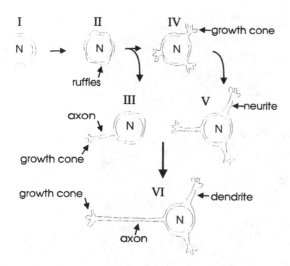

Figure 1.1 Drawings illustrating the development of a neuron. Immature neurons are small, unremarkable cells with a large nucleus (N) and a small amount of cytoplasm (I). After a period of surface membrane ruffling (II), neurons extend processes that end in growth cones (III, IV). *In vivo*, the first process to emerge is usually destined to become the axon (III) and later emerging processes become dentrites (VI), whereas *in vitro* the processes are bi-potential (neurites) and may develop into either an axon or a dendrite (V, VI).

become available to study them in detail and at a molecular level. When I first began my own research on growth cones in 1982 the number of original papers on the subject was relatively small. Now, little more than a decade and a half later, there are as many *reviews* on growth cones as those original research papers and the numbers of research papers are increasing at an astonishing rate, so much so that keeping abreast of the literature is a major undertaking.

This book attempts to bring together all that is currently known about growth cones that may help us to explain how they achieve these remarkable feats. Our ultimate aim is to describe growth cone behaviour completely. When we have achieved this, we shall have made a significant advance in understanding how the most important part of our bodies, our nervous systems, are properly formed. In addition, we may have clues as to how we can repair the damaged nervous system.

Growth cone morphology and behaviour

What do growth cones look like and how do they behave? Answering these questions has provided us with clues as to how growth cones function.

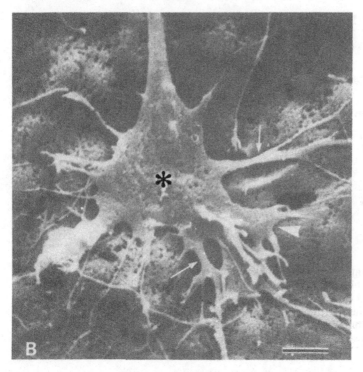

Figure 1.2 A scanning electron micrograph of a growth cone from a sensory neuron growing in the skin of a tadpole. The neurite, in this case an axon, enlarges at the growth cone (asterisk), which extends at its margins filopodia (arrows) and lamellipodia (arrowhead). In this location, the growth cone is largely two-dimensional, as growth cones in culture. The bar marker is 2 μm. (From Roberts, 1976, with permission of Company of Biologists Ltd.)

Appearance and behaviour under the light microscope

Growth cones have been observed with the light microscope in fixed and stained tissue sections of developing and regenerating neural tissue for over one hundred years. In the living state they have been observed in developing animals and in culture. The first observations of growth cones were made by Ramón y Cajal in sections of the embryonic chick spinal cord stained with the silver chromate method of Camillo Golgi, which entirely impregnates a small (about 5%) proportion of the cells in a tissue with a dense, black precipitate (Fig. 1.3; Ramón y Cajal, 1890). Although more detailed descriptions using more powerful techniques have been made since, the salient features of Ramón y Cajal's observations remain valid today. Growth cones are highly motile enlargements at the ends of growing neurites and are capable of extending two principal types of processes at their margins: fine, finger-like extensions called filopodia, or microspikes, and flat, sheet-like expansions called lamellipodia (literally sheet-foot), or veils, which are usually extended between adjacent filopodia (Figs 1.2 and 1.5). Although Ramón y Cajal only saw growth

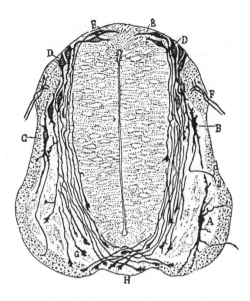

Figure 1.3 The first published picture of a growth cone. A drawing of a transverse section from the spinal cord of a 4-day-old chick embryo stained with the Golgi method. Commissural interneurons (D) have extended axons towards the floor-plate (H) and are tipped with growth cones (G). (From Ramón y Cajal, 1890.)

cones in static images, he recognised their significance and imagined them, with characteristic insight, to be highly motile – rapidly changing their shape as they insinuated or battered their way through the neuropil to their destinations or crawled through pre-existing spaces between cells (Ramón y Cajal, 1990, p.149). Ramón y Cajal's curiosity about growth cones led him to pose many questions about how they grow, how they navigate a path to their appropriate destinations, and how they then select a suitable synaptic partner. These questions, which are the central theme of this book, therefore have a long history. For many of these events he proposed molecular mechanisms, some of which, for instance chemotropism (see Chapter 3), are finding experimental support today.

Proof of the motile activity of growth cones was first demonstrated by Ross G. Harrison (1907, 1910), who developed tissue culture techniques for the express purpose of seeing them in the living state for the first time. In his first experiments he cultured explants of frog embryonic neural tissue in lymph clots and remarked on the rapid, amoeboid-like movements of the growth cones, which extended and retracted many fine, 'filamentous' processes, or pseudopodia as he called them, at their periphery (Fig. 1.4). Growth cones in these cultures could move with speeds of up to about 1 μm/min, although these were not sustained for long. Harrison observed growth cones bifurcating to form two new neurites and the retraction of side-branches (Fig. 1.4). He suggested that growth cones required a substratum in order to extend; they could not grow through a fluid medium (Harrison, 1914). This was an impor-

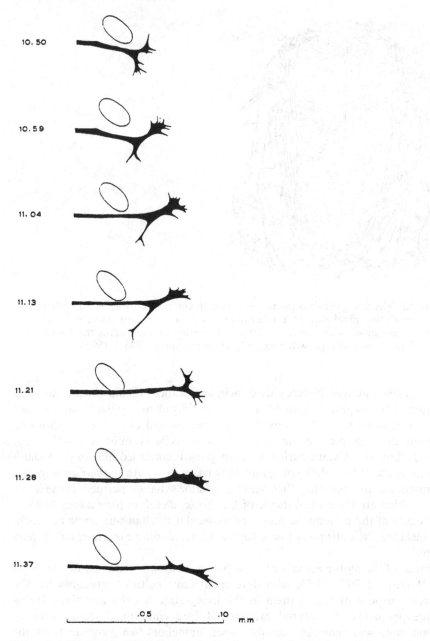

Figure 1.4 Harrison's (1910) camera lucida drawings of a neurite and its growth cone growing in a lymph clot from an explant of frog branchial ectoderm. The time of day is indicated on the left. The erythrocyte, shown in outline, marks a fixed point. Note how the morphology of the growth cone changes rapidly on a time-scale of minutes. The sequence also shows the extension and retraction of a side-branch.

tant idea from the point of view of understanding how growth cones advance since it implied that growth was not simply the extrusion of material from the neuronal cell body, like toothpaste from a tube, but that purchase on a substratum was necessary. It so happens that growth cones can develop pull or traction by attaching to a substratum, a property that has been measured directly (Lamoureux, Buxbaum & Heidemann, 1989). Harrison saw neurites arising directly from isolated neuronal cell bodies and proved, therefore, that neurites were not derived from the fusion of other cell processes, an idea prevalent at the time (for reviews, see Harrison, 1935; Hughes, 1968; Jacobson, 1991).

By using suitably shaped lymph clots as bridging-implants, Harrison induced growth cones, unaccompanied by other cell processes, to grow across a surgical gap in the frog embryonic spinal cord, and thereby demonstrated the motile independence of growth cones (Harrison, 1907).

Harrison's pioneering observations of growth cones in culture (Harrison, 1907, 1910) were quickly confirmed and extended by others (Burrows, 1911; Lewis & Lewis, 1912; Levi, 1917, 1926, 1934; Matsumoto, 1920; Lumsden, 1951; Hughes, 1953; Nakai, 1956; Nakai & Kawasaki, 1959). Burrows (1911), working in Harrison's laboratory, developed the culture technique further by growing neural tube from a warm-blooded animal (chick) in chicken plasma, which was more convenient to work with than lymph, and confirmed Harrison's main findings. Burrows obtained the first permanent histological preparations of neuronal cultures, staining them so that neurofilaments could be seen. He also mechanically dissociated neural tube fragments into small clumps of cells, so that neurons could be viewed in their entirety and confirmed that neurites originate directly from neuronal cell bodies (Burrows, 1911). He observed that elongation or growth of the neurite was not of uniform rate but was interspersed with frequent quiescent periods and that growth cones and neurites could retract as well as advance. Lewis and Lewis (1912) substituted simple salt solutions for plasma clots, one advantage of which is that neurites grow directly on the substratum, usually a glass coverslip, and therefore in one plane, rather than at different levels, as in a clot. This meant that neurites and growth cones could be followed continuously in the same focal plane of the microscope. Levi (1917) described in detail the behaviour of the growth cone during continuous observation. He, and others (Matsumoto, 1920; Lewis, 1950; Hughes, 1953; Nakai, 1956), noticed translucent vacuoles of up to 1 μm in diameter in growth cones. Hughes (1953) thought that the vacuoles formed by pinocytosis at the distal margin of the growth cone and that this allowed an intake of water for growth. Whilst we now know that biological membranes are generally freely permeable to water and our views on the mechanism of axon growth have changed somewhat (see Chapter 2), there is ample evidence for endocytotic uptake of extracellular fluid at the growth cone (Birks, Mackey & Weldon, 1972; Wessells & Ludueña, 1974; Weldon, 1975; Bunge, 1977; Cheng & Reese, 1987; Dailey & Bridgman, 1993). However, to what extent endocytosis at the growth cone underlies membrane recycling or pinocytosis, i.e. the bulk uptake of extracellular fluid, is not clear (Gordon-Weeks, 1988a; see Chapter 2).

Hughes found that some vacuoles were retrogradely transported and he measured their rate as 3.8–17 μm/min (see also Nakai, 1956). This is within the range of the rates of retrograde movement of vacuoles and actin filaments reported more recently (e.g. 1–6 μm/min; Forscher & Smith, 1988; Dailey & Bridgman, 1993). It would be interesting to know more about these organelles, especially their relation to microtubules or actin filaments. The ultrastructural correlate of the vacuoles has been established and they are a common feature of electron micrographs of growth cones (e.g. Tennyson, 1970; Dailey & Bridgman, 1993; and see below).

Hughes also noted that growth cones could retract on contacting other growth cones or neurites; this was the first observation of contact inhibition. This finding presaged an important topic of current research: the role of growth cone inhibition and collapse in pathfinding (see below and Chapter 3). Earlier, Levi had shown that neurites severed from their cell bodies in explant cultures continue to elongate (Levi, 1926; Levi & Meyer, 1945). Hughes (1953) also experimented on the effects of severing neurites in explants of chick brain and spinal cord using fine metal needles and found, remarkably, that the distal neurite and its growth cone survived for over 3 hours, elongating at a speed indistinguishable from the unsevered neurite. These observations suggested that growth cones possess considerable autonomy from the cell body and also, possibly, that the addition of new material for growth of the neurite occurred at the growth cone (Hughes, 1953).

The technique of isolating growth cones in culture from their parent neurite, and hence the cell body, has been used on a number of occasions in order to answer a variety of questions about growth cones (Shaw & Bray, 1977; Haydon, McCobb & Kater, 1984; Baas, White & Heidemann, 1987; Guthrie, Lee & Kater, 1989; Martenson et al., 1993). The technique has also been used in vivo to isolate retinal ganglion cell growth cones from their cell bodies, by removing the retina after the axons of the retinal ganglion cells had left the retina and entered the optic nerve (Harris, Holt & Bonhoeffer, 1987). They labelled growth cones with a fluorescent dye and filmed them, using video microscopy, growing in the optic tract and tectum in Xenopus embryos. Severed growth cones continued to grow for up to 3 hours and their behaviour was essentially indistinguishable from unsevered growth cones, demonstrating again that growth cones have considerable autonomy from their cell bodies.

Nakai and Kawasaki (1959) studied filopodial behaviour and in particular the consequence of filopodia contacting various cell types (Schwann, fibroblast and macrophage) and inanimate objects such as glass particles. They established that filopodia were indiscriminate in what they touched, but that they showed differential adhesion to structures and were possibly contractile. Filopodia can extend anywhere along the growing neurite in culture, but this is not true of lamellipodia (Lewis & Lewis, 1912; Nakai, 1956; Pomerat et al., 1967). Filopodia are extended and retracted regardless of whether the neurite itself is extending, retracting or not growing. Watching growth cones in culture one gains a sense of the independence of their highly motile behaviour from the overall stately advance of the neurite. The rate of extension of filopodia and their length vary between different neuronal types (Table 1.1). In a detailed study of growth cone filopodial extension in rat superior cervical ganglion neurons, Argiro, Bunge and Johnson (1985) found that filopodia could extend from the neurite shaft, the growth cone margin or from each other and that extension was preceded by the formation of a phase-dense nodule of cytoplasm at their origin. When branched filopodia extended, the position of the branch point remained fixed with respect to the base of the filopodium, suggesting that growth is by addition of material distally. The rate of extension declined as the filopodium increased in length and the initial rate of extension (maximally 0.12 $\pm$ 0.4 μm/sec) was positively correlated with the final length of the filopodium. Bray and Chapman (1985) also made a detailed analysis of filopodial behaviour of growth cones in culture, comparing growth on plain glass with growth on glass coated with polylysine, a highly adhesive substratum. Neurite extension rates were considerably slower on the more adhesive coated-glass and filopodial length and life-times increased (Table 1.1). Filopodia were extended at the leading edge of the growth cones, moved laterally and then shortened or became lateral filopodia on the axon shaft. Bray and Chapman (1985) interpreted the behaviour of filopodia in their cultured cells as supporting a role for filopodia in growth cone advance. This important topic, the function of filopodia, is discussed in detail elsewhere (see Chapters 2 and 3).

Growth cones were first observed in the living animal by Speidel (1933). Taking advantage of the transparency of the tail fin of the frog tadpole, Speidel was able to observe the growth cones of cutaneous sensory neurons, continuously for hours or intermittently over a period of weeks, establishing axonal arbors in the skin of anaesthetised animals. He established that the appearance and behaviour of growth cones in culture, as reported by Harrison (1907, 1910) and others (see above), was representative of their appearance and behaviour *in vivo* and thereby gave credibility to tissue culture work. However, culture systems do not completely mimic *in vivo* conditions and, therefore, individual growth cones may not display the full behavioural repertoire *in vitro* that they possess in the animal. Furthermore, cultures tend to be two dimensional, whereas most *in vivo* situations are three dimensional. In the skin of the tail fin, growth cones had a central light region, usually showing within it darker regions, which he called granules, and a peripheral region with up to eight filopodia. The rate of translocation of the growth cone changed continuously from a maximum of about 1 μm/min, or approximately one growth cone length every 10 minutes, which was similar to that seen previously in tissue culture, to an average of about 200 μm in 24 hours. Often the growth cone stopped temporarily, for no apparent reason, or retracted, and although this last behaviour may have been a consequence of morbidity, it had been seen previously in culture (e.g. Burrows, 1911). Growth cones, on encountering objects in their path, such as fibroblasts, formed an enlargement or varicosity, which was left behind at the obstruction. Larger obstructions caused "giant" growth cones to form, and often a change in the direction of growth. Growth

Table 1.1. *Growth cone characteristics*

Source	Substrate	Speed (μm/hr)	Filopodia Diameter (μm)	Length (μm)	Extension rate (μm/sec)	Reference
Aplysia PC12/SCG	Poly L-lysine Collagen	5–25 20 ± 2 (PC12) 18 ± 4 (SCG)	50	5–15 10–25		Goldberg & Burmeister, 1986 Aletta & Greene, 1988
Ti1 pioneers	Grasshopper limb bud	10 ± 2 (filopodial) 4 ± 1 (lamellipodial)		Longest 34 Average 16 ± 7		O'Connor *et al.*, 1990
Q1 commissural neuron	Grasshopper CNS			59 ± 3 (all forms) 21.2 ± 7.7 (turning) 20.7 ± 7.1 (turned)	0.025 ± 0.012 (turning) 0.028 ± 0.017 (turned)	Myers & Bastiani, 1993
SCG	Collagen	8–22 (embryonic) 4–13 (adult)				Argiro *et al.*, 1984
RGC (*Xenopus*)	*In vivo*	16 (filopodial) 52 (lamellipodial)				Harris *et al.*, 1987
RGC (zebrafish)	*In vivo*	10.7 (tectum) 17.8 (optic tract)				Kaethner & Stuermer, 1992
RGC (chick)	Poly L-lysine L1 N-Cadherin Laminin	10 (23)[a] 59 (122) 90 (151) 123 (188)				Lemmon *et al.*, 1992

						Reference
Leech	In vivo	5				Braun & Stent, 1989
DRG	Glass Poly L-lysine	64.8	18	6.89 ± 0.42	0.73	Bray & Chapman, 1985
SCG	Collagen	approx. 10			0.13 ± 0.04 (embryonic) 0.09 ± 0.04 (postnatal)	Argiro et al., 1985
SCG	Laminin	54–84 (embryonic) 18–30 (postnatal)		2.69 ± 1.88		Kleitman & Johnson, 1989
Forebrain (hamster)	In vitro	77.9 ± 7.0 (corpus callosum) 42.4 ± 4.0 (cerebral cortex)				Evans et al., 1997 Halloran & Kalil, 1994
Cerebral cortex (mouse)	Glass				0.24 ± 0.011 (extension) 0.21 ± 0.013 (retraction)	Sheetz et al., 1992
Spinal cord (Xenopus)	In vivo	31–73				Jacobson & Huang, 1985

[a] Number in parentheses is maximum growth rate.
Abbreviations: CNS, central nervous system; DRG, dorsal root ganglion; RGC, retinal ganglion cell; SCG, superior cervical ganglion.

cones readily grew along axons, which led to the bundling of neurites into fascicles. Whilst growing along an axon, a growth cone could pass another growth cone growing along the same axon in the opposite direction. Growth cones were observed to branch, as Harrison had seen them do in tissue culture (Harrison, 1910), although this is not the only way in which neurite branches can form. There are well-documented examples of secondary neurites, and therefore branches, emerging from the primary neurite – some distance from the growth cone (Harris et al., 1987; O'Leary & Terashima, 1988; Heffner, Lumsden, & O'Leary, 1990; McCaig, 1990; Kaethner & Stuermer, 1992; O'Rourke, Cline & Fraser, 1994; Bastmeyer & O Leary, 1996). This process is sometimes referred to as collateral sprouting or 'back-branching'.

Methodological advances

There have been a number of important technical advances in microscopy and methodological advances in cell culture which have played an incisive part in growth cone studies. The introduction of phase-contrast microscopy to tissue culture by Pomerat (Costero & Pomerat, 1951) allowed better, more contrasted images of growth cones to be obtained and hence a better appreciation of lamellipodia (Fig. 1.5), which are more difficult to see than filopodia under ordinary light. Time-lapse cinematography enabled more detailed analysis of growth cone movements to be made (Nakai, 1956, 1960; Goldstein & Pinkel, 1957; Nakai & Kawaski, 1959; Pomerat et al., 1967).

Other important developments in tissue culture techniques included the introduction of mechanical and enzymatic means to dissociate neural tissue into single cells (St. Amand & Tipton, 1954; Nakai, 1956; Levi-Montalcini & Angeletti, 1963; Varon & Raiborn, 1969; Chen & Levi-Montalcini, 1970). However, despite these technical advances, the experimental advantage of being able to observe single, dissociated neurons in their entirety in culture was not fully appreciated until the introduction of cultures of clonal lines of neuroblastoma cells (Augusti-Tocco & Sato, 1969; Schubert et al., 1969). This also made possible biochemical studies of homogeneous populations of neuronal cells uncontaminated by other cell types (see Chapter 4), and resulted in the publication of two papers, which showed how the growth cones of isolated cells could be analysed experimentally (Bray, 1970; Yamada, Spooner & Wessells, 1970). Yamada et al. (1970) cultured dissociated neural tissue from chick embryos and investigated the effects of agents that cause the disassembly of microtubules and microfilaments on neurite outgrowth and growth cone ultrastructure. They provided the first experimental evidence that neurite outgrowth depends on microtubules, while the motile activity of the growth cone, the extension of filopodia and lamellipodia, depends on actin filaments (see also Seeds et al., 1970; Daniels, 1972; Bray, Thomas & Shaw, 1978; and Chapter 2 for further discussion). Bray (1970) studied the surface movements of isolated rat sympathetic neurons growing in culture by filming the movements of glass or carmine particles adhering to growth cones and neurites. Such

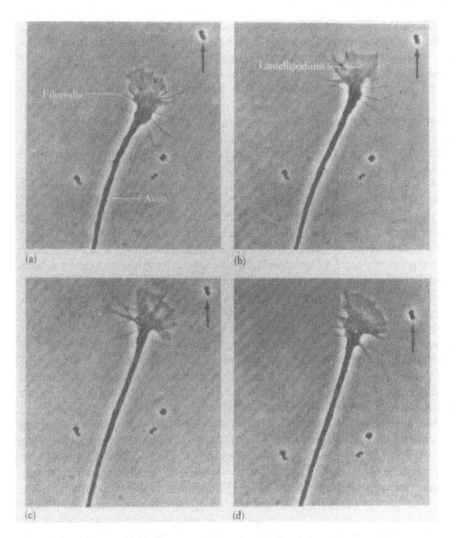

Figure 1.5 Phase-contrast micrographs of a growth cone from a chick dorsal root ganglion cell growing in culture. The interval between micrographs is 1 minute. The rapid extension and retraction of filopodia and lamellipodia are clearly visible. The progress of the neurite across the culture dish can be gauged with reference to stationary particles (arrow). (From Walbot & Holder, 1987, with permission of The McGraw-Hill Companies. Original by D. Bray.)

particles, when attached to filopodia, tended to move in a retrograde direction along the dorsal surface of the growth cone, finally coming to rest near the junction between the growth cone and the parent neurite. Particles on the neurite itself remained for long periods of time at the same distance from the cell body. Bray concluded from these observations that growth of the neurite surface membrane occurs by addition of surface membrane at the growth cone, an idea that has since found considerable experimental support (see Chapter 2). In further experiments, Bray (1973a) studied the branching pattern of individual sympathetic neurons growing on glass coverslips. He found that, under these conditions, the point of attachment of the neuron to the glass was mainly at the growth cone, as indicated by the observation that the cell body and the neurite, but not the growth cone, could be displaced mechanically using microelectrodes. Growth cones growing on glass moved at a fairly constant speed (40 μm/hr; see also Table 1.1) and branching was mainly produced by bifurcation of the growth cone rather than 'back-branching' along the neurite (see below) (Bray, 1973a).

Another way of looking at growth cones in the light microscope is by interference reflection microscopy. This method produces images, the light intensity of which depends on the proximity of the cell or its process to the substratum. Cell processes that are closely apposed to the substratum appear darker than those more distant. Bray (1973a) had shown previously that, when growing on uncoated glass coverslips, neurons are mainly attached by their growth cones. Using interference reflection microscopy, Letourneau (1979) confirmed that, when growing on glass, only the base or proximal region of the growth cone is close to the glass. However, when growing on glass coated with an adhesive molecule (polylysine), a greater area of the growth cone is closely apposed to the glass including the filopodia and the proximal regions of the lamellipodia (see also Gundersen, 1988). Despite the obvious usefulness of the technique, it has not found widespread application in assessing the effects of different substrata on growth cone contact (Wu & Goldberg, 1993; Gomez & Letourneau, 1994; Drazba et al., 1997).

More recently, video-enhanced contrast differential–interference contrast microscopy, introduced by Allen (1985), has spectacularly improved the light microscopic image of living growth cones in culture (Fig. 1.6). This technique can visualize objects of about 30–50 nm in diameter, such as microtubules ($\sim$ 25 nm diameter), i.e. nearly an order of magnitude smaller than the conventional limit of resolution of the light microscope ($\sim$200 nm). Observations can be made in real time and specimens can be viewed at different levels in the vertical dimension, that is 'optically sectioned'. Intracellular organelles within cells can be seen with great clarity and structures that are difficult to visualise by conventional techniques, such as lamellipodia, can be seen readily. The observations that have been made using video microscopy have had a considerable influence on our thinking about growth cone motility and behaviour (see Chapter 2). Using video microscopy, Goldberg and Burmeister (1986) and Forscher et al. (1987) defined two regions, that differed in their appearance, within the large (up to 50 μm; Table 1.1) growth cones of the giant neurons of the sea hare Aplysia californica growing in culture. A more proximal, central (C)-domain, which is relatively thick and contains abundant organelles, including mitochondria and vesicles of various kinds (see below), and a more distal, peripheral (P)-domain, which is thinner and relatively devoid of organelles and from which filopodia and lamellipodia emerge (Fig. 1.6). The existence of these two regions had been suspected for some time and, in fact, they are roughly equivalent to two regions identified by Ramón y Cajal (1960, page 45) that he distinguished on the basis of differences in their staining properties. This regionalization of the growth cone is a general phenomenon because similar regions have been seen in vertebrate (rat superior cervical ganglion) growth cones and phaeochromocytoma (PC12) cells (Aletta & Greene, 1988; Bridgman & Dailey, 1989). A distinction between two regions in the growth cone can also be made from electron microscope images (see below). These domains are also distinguished by differences in the distribution of cytoskeletal filaments within them (see Chapter 2). The C-domain is dominated by microtubules whereas the P-domain contains mainly microfilaments. Such a segregation between a cortical, actin-rich cytoskeleton and a central, microtubule-rich cytoskeleton is a feature of all animal cells (Bray, 1992). Occasionally, the C-domain is sufficiently thin for the movement of individual organelles to be seen, and they can invariably be seen in lamellipodia (e.g. Goldberg & Burmeister, 1986; Forscher *et al.*, 1987). These organelles undergo either Brownian motion or more directed movements, which may be associated with microtubule-dependent transport (see Chapter 2).

Time-lapse video microscopy is now being successfully applied to whole embryos or large explants of embryos in which growth cones have been supra-vitally labelled with fluorescent dyes (reviewed in Stirling & Dunlop, 1995). This approach has enabled observations of living growth cones to be made *in vivo* for hours or even days (Harris *et al.*, 1987; O'Connor *et al.*, 1990; O'Rourke & Fraser, 1990; Sabry *et al.*, 1991; Kaethner & Stuermer, 1992; Myers & Bastiani, 1993; Sretavan & Reichardt, 1993; Godement, Wang & Mason, 1994; Halloran & Kalil, 1994; O'Rourke, Cline & Fraser, 1994; Bastmeyer & O Leary, 1996). Living growth cones in these experiments behave, to a large extent, much as would be predicted from the observations made of growth cones *in vitro* and in fixed embryos. However, the picture that is emerging of the behaviour of *in vivo* growth cones suggests that they are far more dynamic than had previously been realised. Even when neurites are temporarily not elongating, their growth cones are, nevertheless, continuously extending and retracting filopodia and lamellipodia. Back-branching occurs more frequently than had been supposed and can lead both to collateral branches and neurite turning. Growth cone collapse and neurite retraction (up to 80 μm within 15 minutes), sometimes followed by recovery and re-extension, are relatively common events.

Factors influencing growth cone shape

Growth cones can have a wide spectrum of shapes depending on a variety of circumstances, including neuron type, location and age. They can vary from a simple paintbrush-like shape with a single filopodium to a complex, fan-like expansion bristling with many filopodia and numerous lamellipodia (Fig. 1.7). Although Harrison (1910, page 799) suspected that growth cones could exist in a variety of shapes and Hughes (1953) and Nakai (1956) alluded to the possi-

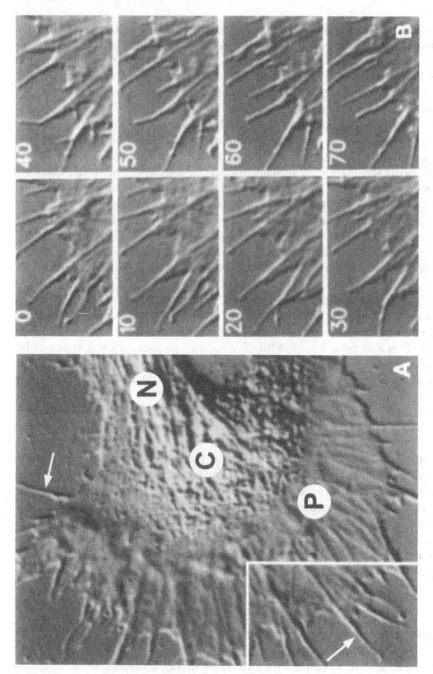

bility, variation in growth cone shape was first established in culture by Pomerat *et al.* (1967), who found differences in the morphology between growth cones of chick dorsal root ganglion cells and chick spinal cord neurons growing in similar culture conditions; the former were predominantly lamellipodial while the latter were more filopodial (see also Kapfhammer & Raper, 1987b). This observation suggests that characteristic growth cone shape can be an intrinsic property. Subsequent studies have supported this notion; for instance, certain adult invertebrate neurons in culture produce growth cones of such individual morphology that this feature can be used to identify the neuron type (Haydon *et al.*, 1985; Acklin & Nicholls, 1990; Kim & Wu, 1991).

Decision regions
Growth cone shape may also vary as a consequence of the local, microenvironment in the embryo. For example, when fasciculating on axons of other neurons, growth cones tend to be simple whereas when they are at so-called choice points or decision regions, a phrase coined by Tosney and Landmesser (1985a), where it is thought that decisions about the direction of growth are being made (see Chapter 3), growth cones tend to have elaborate shapes. There are many examples where growth cones of particular classes or populations of neurons, e.g. retinal ganglion cells, have been shown to assume very different morphologies depending on their exact location on the pathway that they navigate to their target site during development (Taghert *et al.*, 1982; Roberts & Taylor, 1983; Bentley & Caudy, 1984; Tosney & Landmesser, 1985a,b,c; Mason, 1986; Bovolenta & Mason, 1987; Kuwada, Bernhardt & Chitnis, 1990). During the formation of the lumbosacral plexus in the chick limb bud, the growth cones of motoneurons, while growing toward the plexus region in the spinal nerve or in the peripheral nerves, have relatively simple morphologies (Tosney & Landmesser, 1985a). In contrast, growth cones have complex shapes at the plexus itself and at other locations where they probably face decisions about the direction of growth. Similarly, retinal ganglion cell growth cones have simple morphologies while traversing the optic nerve and tract, but complex shapes at the optic nerve head and on first entering the tectum (Scalia & Matsumoto, 1985; Bovolenta & Mason, 1987; Harris, Holt & Bonhoeffer, 1987; Holt, 1989; Brittis & Silver, 1995; Mason & Wang, 1997).

Figure 1.6 Video images of a living growth cone from an *Aplysia* neuron growing in culture. A single image of the entire growth cone is shown in A. The neurite (N), C-domain (C) and P-domain (P) are clearly discernible. The neurite and C-domain appear thicker than the P-domain. In the P-domain, filopodia (arrows) and, between them, lamellipodia are visible. The rod-like appearance of the filopodia is due to the presence of an actin-filament bundle. Notice how this bundle extends proximally from the filopodia to the C/P border. B shows a series of images taken at 10-second intervals from the region within the box in A. The protrusion and retraction of filopodia are evident. (Reprinted with permission from Smith, 1988. Copyright 1988 American Association for the Advancement of Science.)

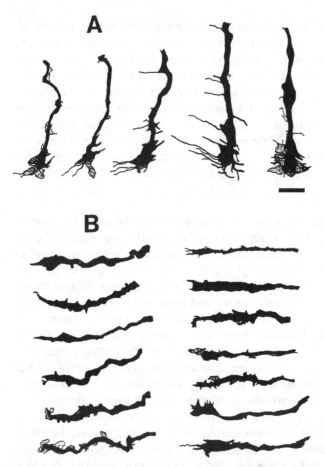

Figure 1.7 Drawings of horseradish peroxidase-labelled growth cones from the embryonic chick spinal cord. Examples of complex growth cones growing toward the floor-plate in the lateral region of the neural tube (transverse sections) are shown in A. These growth cones have many filopodia and lamellipodia (stippled regions). Medial is toward the right, ventral is toward the bottom. In contrast, simpler, less filopodial growth cones are found within the floor-plate itself as shown in B. Growth cones on the left are transverse views with ventral towards the bottom. Growth cones on the right are horizontal views with rostral towards the top. Bar in A = 10 μm. (From Yaginuma *et al.*, 1991. Reprinted by permission of Wiley-Liss, Inc., a subsidiary of John Wiley & Son, Inc.)

In the developing spinal cord, the axonal growth cones of commissural inter-neurons change their morphology as they grow through the floor-plate (Bovolenta & Dodd, 1990; Yaginuma *et al.*, 1991), as do the central projections of dorsal root ganglion growth cones in *Xenopus* tail as they navigate across the cord to the dorsolateral fascicle (Nordlander, Gazzerro & Cook, 1991). Interestingly, in the rat, commissural axon growth cones become *more* complex as they traverse the floor-plate, whereas in the chick they are *less* complicated

(Bovolenta & Dodd, 1990; Yaginuma *et al.*, 1991). These environmental effects are not generalised ones acting on all neurons since individual populations of neurons respond to the same pathway in different ways (Tosney & Landmesser, 1985a,b,c; Nordlander, 1987).

Fasciculation
Growth cones tend to have relatively simple shapes when fasciculating with or growing along other axons either *in vivo* (Roberts, 1976; Macagno, 1978; Raper, Bastiani & Goodman, 1983a,b,c; Shankland & Bentley, 1983; Bastiani, Raper & Goodman, 1984) or *in vitro* (Nakai, 1960; Rutishauser, Gall & Edelman, 1978; Wessells *et al.*, 1981; Tosney & Wessells, 1983). There are, however, some exceptions; for instance, in the developing corpus callosum, where axons are oriented in tightly packed, parallel bundles, growth cones have diverse and often complicated morphologies (Norris & Kalil, 1990); although this may be because some growth cones are changing to a topographically more appropriate axon on which to fasciculate. When they reach the region where they will form synapses, growth cones tend to become simple and small again (Mason, 1982, 1985a; Reh & Constantine-Paton, 1985; Sakaguchi & Murphey, 1985; Sachs, Jacobson & Caviness, 1986; Sretavan & Shatz, 1986). This is surprising since they would be expected to assume more elaborate forms when they are searching for the correct partners.

The complexity of growth cone shape correlates with the rate of growth: simpler, more lamellipodial growth cones grow faster than more complex, filopodial ones. This has been observed both *in vivo* (Harris *et al.*, 1987; but see O'Connor *et al.*, 1990; Table 1.1) and *in vitro* (Tosney & Wessells, 1983; Argiro, Bunge & Johnson, 1984; Aletta & Greene, 1988; Kleitman & Johnson, 1989). Exceptions to this correlation include the finding that, although chick dorsal root ganglion growth cones are lamellipodial when growing on laminin, they are very filopodial when growing on the extracellular matrix protein tenascin and move more rapidly (Taylor *et al.*, 1993). Abosch and Lagenaur (1993) found that mouse cerebellar granule cells were filopodial when growing on the cell adhesion molecules L1 and P84 (see Chapter 3), but lamellipodial on laminin and neural cell adhesion molecule (NCAM) and that they grew more rapidly on L1 and laminin than on P84 and NCAM.

Developmental age and type of substratum
In addition to the influence of neuron type and location on growth cone morphology, the developmental age of the neuron (Argiro, Bunge & Johnson, 1984) and, *in vitro*, the type of substratum on which the growth cone is moving (Luduena, 1973; Letourneau, 1975a,b, 1979; Helfand, Smith & Wessells, 1976; Hawrot, 1980; Hammarback *et al.*, 1985; Gundersen, 1987; Kleitman & Johnson, 1989) have been shown to affect growth cone shape and the rate of growth. For instance, Kleitman and Johnson (1989) found that the growth cones of *embryonic* rat superior cervical ganglion neurons growing in culture on the extracellular matrix protein laminin had relatively high speeds of extension (0.9–1.4 μm/min) and were largely lamellipodial whereas growth

cones from *postnatal* superior cervical ganglion neurons extended more slowly (0.3–0.5 μm/min) and were more filopodial. In embryonic cultures they found that growth cones had no filopodia and were entirely lamellipodial – these grew fastest. When growing on collagen rather than laminin, postnatal rat superior cervical ganglion growth cones do not extend lamellipodia (Argiro *et al.*, 1984). In marked contrast to the predominantly lamellipodial growth cones of rat embryonic superior cervical ganglion neurons growing on laminin, chick superior cervical ganglion growth cones are predominantly filopodial under these conditions (Kapfhammer & Raper, 1987a,b). These variations in growth cone morphology may depend on the types of receptors for extracellular matrix proteins carried by growth cones from different sources and the influence these have on the cytoskeleton of the growth cone (Payne, Burden & Lemmon, 1992; Abosch & Lagenaur, 1993; see Chapters 2 and 3).

The general view that has emerged from these studies is that when growth cones are travelling along pre-formed pathways such as bundled axons, they tend to be relatively small, simple and lamellipodial, and move swiftly, whereas, when they are required to select their direction of growth, they slow down, become larger and extend more filopodia. One explanation given for this change of behaviour is that by slowing down and assuming a broader distal margin, growth cones can 'sample' their environment more effectively. An alternative explanation, suggested to me by Dennis Bray, is that slowing down may cause an accumulation of material delivered to the growth cone by slow axoplasmic transport and thus an increase in growth cone size (see Speidel's observations of growth cones encountering obstructions above). This may also include an increase in filopodial length. The primary function of filopodia is thought to be detection of extrinsic signals or so-called 'guidance cues' that may influence the direction of growth by sampling or sensing their local environment (see Chapter 3). If this explanation is correct, then growth cone morphology can be used to assay for guidance cues.

Growth cone collapse/contact inhibition

Under certain circumstances in culture, growth cones collapse upon themselves and their neurites retract. As I have already mentioned, Hughes (1953) was the first to observe growth cone retraction on contact with another neurite. Since that time, growth cone collapse and neurite retraction have been observed in a number of situations in culture (Bray, Wood & Bunge, 1980; Kapfhammer, Grunewald & Raper, 1986; Kapfhammer & Raper, 1987a,b; Schwab & Caroni, 1988; Cox, Müller & Bonhoeffer, 1990; Davies *et al.*, 1990; reviewed in Patterson, 1988; Walter, Allsopp & Bonhoeffer, 1990). More recently, direct observation of living growth cones *in vivo* has revealed that growth cone collapse is not restricted to neurons in culture (Halloran & Kalil, 1994: reviewed in Stirling & Dunlop, 1995). In general, growth cones collapse following filopodial contact with the surfaces of certain cell types, for instance, when retinal ganglion cell growth cones encounter sympathetic neurons in culture. Collapse is frequently preceeded by an inhibition of motility and followed by neurite retraction. Often growth cones will recover and make additional attempts to cross the inhibitory surface, a behaviour that has not been demonstrated *in vivo* and is probably mediated by specific, cell surface receptor interactions. This important topic is discussed in detail in Chapter 3.

Leader (pioneer) and follower growth cones

In some situations a distinction can be made between the first growth cone(s) to advance along a pathway, often called 'pioneer' growth cones, a term introduced by Harrison (1935), and those that follow the same path at a later stage (followers). Pioneer growth cones tend to be morphologically distinct from followers: the former are larger and more complex, whereas the latter, generally fasciculating, are simpler and lamellipodial. This feature has been more thoroughly documented in invertebrates (LoPresti, Macagno & Levinthal, 1973; Bate, 1976; Bentley & Keshishian, 1982) and there is some disagreement about the extent to which it occurs in vertebrates (e.g. Roberts, 1988). However, clear examples have recently been described in the zebrafish forebrain (Wilson & Easter, 1991) and cat cerebral cortex (Kim, Shatz & McConnell, 1991).

Appearance under the electron microscope: organelles of the growth cone

Growth cones have been studied in the electron microscope using conventional transmission electron microscopy (see below), scanning electron microscopy (e.g. Ebendal, 1976; Roberts, 1976; Roberts & Taylor, 1983; Roberts & Patton, 1985; Payne, Burden & Lemmon, 1992; Fig. 1.2) and by various freezing techniques including freeze–fracture and freeze substitution (Rees & Reese, 1981; Cheng & Reese, 1985, 1987, 1988; Bridgman & Dailey, 1989).

The first electron microscope studies of growth cones were made in regenerating, peripheral nerves (Estable, Acosta-Ferreira & Sotelo, 1957; Wettstein & Sotelo, 1963; Lentz, 1967). For two reasons, this may not be an ideal system in which to examine growth cone ultrastructure; one concern is the difference between growth cones during development and regeneration, another is the difficulty in distinguishing in the electron microscope between structures peculiar to degeneration/regeneration, such as end bulbs and retraction bulbs on the one hand and growth cones on the other. The earliest observers reported the ultrastructure of terminal enlargements of axons, which have abundant, smooth endoplasmic reticula and mitochondria (Estable *et al.*, 1957; Wettstein & Sotelo, 1963; Lentz, 1967). Some of these may be growth cones, but the relationship between these structures and growing axons has not been established.

Early electron microscope studies of growth cones in the developing nervous system *in vivo* were disappointing, showing only 'watery' terminal enlargements of neurites largely devoid of organelles except for the occasional membrane sac or vesicle (Bodian, 1966; Bodian, Melby & Taylor, 1968; Del Cerro & Snider, 1968). As pointed out by Tennyson (1970), the absence of mitochondria was particularly surprising since electron microscope studies of the growth cones of regenerating peripheral nerves showed that they did have mitochondria (Estable *et al.*, 1957; Wettstein & Sotelo, 1963; Lampert, 1967; Lentz, 1967). Also mitochondria would be expected to be present since growth cones are highly motile and therefore require ATP. There is some doubt whether growth cones were properly identified in these early studies (and also some later ones, e.g. Fox, Pappas & Purpura, 1976). Serial section analysis to confirm that the structures under study were the anatomical ends of neurites was not done and no filopodia or lamellipodia were observed. The structures in these early reports were probably inadequately preserved and later studies following better fixation revealed a rich variety of organelles, including mitochondria, in growth cones (Table 1.2, and Figs 1.8 and 1.9).

The first convincing electron microscope report of growth cones *in vivo* was from Tennyson (1970) who studied, in serial sections, the entry of the central axons of dorsal root ganglion neurons into the spinal cord in rabbit embryos. She noticed that these growth cones had a central expanded region, corresponding to the C-domain, which she called a varicosity. This contained mitochondria, microtubules, neurofilaments, vesicles and smooth endoplasmic reticulum (Table 1.2). Peripheral to the varicosity were thin processes filled with a filamentous matrix and occasional vesicles. These were almost certainly the filopodia and lamellipodia of the P-domain.

Since Tennyson's work, there have been numerous electron microscope studies of growth cones in a wide range of species and regions of the developing and regenerating nervous system both *in vivo* and *in vitro* (Table 1.2). One important point that has emerged from this work is that no organelle is uniquely associated with, and therefore diagnostic of, growth cones. However, a set of organelles is almost invariably present, including a system of smooth endoplasmic reticulum, mitochondria and vesicles. These organelles tend to occupy the C-domain, as if somehow excluded from the peripheral regions of the growth cone, particularly the filopodia and lamellipodia. The P-domain and its extensions seem to be mainly composed of microfilaments (see Chapter 2).

Membrane-bounded organelles
A wide variety of membrane-bounded organelles are found in growth cones including membrane sacs, vacuoles, vesicles, tubules, reticula, and, of course, mitochondria, which are ubiquitous. In at least one case, a membrane structure is known to be an artefact of chemical fixation with glutaraldehyde. It takes the form of a mound or bulge under the plasma membrane of the growth cone containing collections of heterogeneously sized membrane-bounded vesicles (Fig. 1.9; Bunge, 1977; Nuttall & Wessells, 1979; Pfenninger, 1979; Rees and

Table 1.2. *Growth cone organelles*

Source	ser	ves/vac	mito	nf	mt	Reference
			Organelles			
Forelimb nerves[a] *Triturus*	+ +	+ +	+	+	+	Lentz, 1967
Dorsal root neurites Rabbit	+ +	+	+	+	+	Tennyson, 1970
Cerebellar cortex Rat and cat	+ +	+	+	−	−	Kawana *et al.*, 1971
Hippocampal pyramidal neurons Rat	+ +	+	+	−	+	Deitch & Banker, 1993
Dorsal root ganlion[b] Chick	+ +	+	+	+	+/−	Yamada *et al.*, 1971
Olfactory bulb Mouse	+	+	+/−	−	−	Hinds & Hinds, 1972
Superior cervical ganglion[b] Rat	+	+	+	+/−	+	Bunge, 1973
Spinal cord Chick						Skoff & Hamburger, 1974
Dendritic	+/−	+	−	−	−	
Axonal	+ +	+	+	−	−	
Medulla Cat	+	+	−	−	−	Fox *et al.*, 1976
Cerebral cortex Human	+ +	+	+	−	−	Povlishock, 1976
Interosseous nerve (wing) Chick	+ +	+	+/−	−	−	Al-Ghaith & Lewis, 1982
Spinal cord *Xenopus*	+	+	+	−	−	Nordlander & Singer, 1982b
Trochlear nerve[a] Goldfish	+ +	+	+	−	−	Scherer & Easter, 1984
Optic tectum Chick	+ +	+	+	−	+	Cheng & Reese, 1985
Optic nerve Cat	+ +	+	+	+	+	Williams *et al.*, 1986
Pyramidal tract Rat	+ +	+	+	−	+	Gorgels, 1991

[a]Regenerating.
[b]In culture.
Abbreviations: ser, smooth endoplasmic reticulum; ves/vac, vesicles/vacuoles; mito, mitochondria; nf, neurofilaments; mt, microtubules. + +, invariably present; +, common; −, absent; +/− variable.

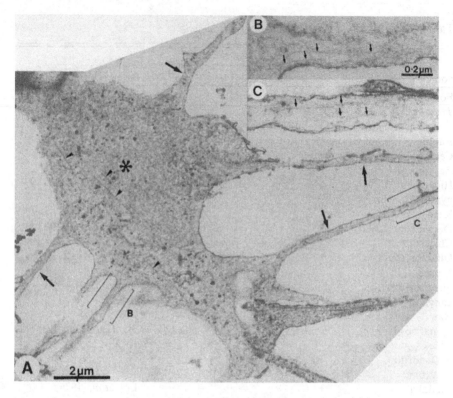

Figure 1.8 Electron micrograph of a chick dorsal root ganglion cell growth cone growing on a laminin substrate in culture. The C-domain (asterisk) of the growth cone is filled with a system of smooth endoplasmic reticulum (ser) and vesicles, and coursing through it toward the P-domain are single microtubules (arrowheads). The growth cone extends several filopodia (arrows). High-power views of the bracketed regions of two filopodia reveal fascicles of longitudinally oriented microfilaments (inset arrows, B and C).

Reese, 1981). Similar vesicle-containing mounds in highly motile regions of other cell types, such as fibroblasts, have been shown to be artefacts of glutaraldehyde fixation (Hasty & Hay, 1978; Shelton & Mowzcko, 1978). They are not seen in growth cones fixed with osmium tetroxide or with glutaraldehyde–osmium tetroxide–ferrocyanide mixtures (Pfenninger, 1979; Rees & Reese, 1981) or, more significantly, in rapidly frozen, freeze-substituted growth cones, unless previously exposed to glutaraldehyde (Rees & Reese, 1981; Cheng & Reese, 1985, 1987). Glutaraldehyde has been shown to induce mounds in growth cones of neurons in culture (Nuttall & Wessells, 1979).

Three types of smooth endoplasmic reticulum can be distinguished in growth cones on the basis of their morphology (reviewed in Gordon-Weeks, 1988a). The most characteristic membrane system within the growth cone, in conventionally fixed material, consists of highly branched cisternae whose lumens contain a slightly electron-opaque material (Figs 1.8 and 1.9)

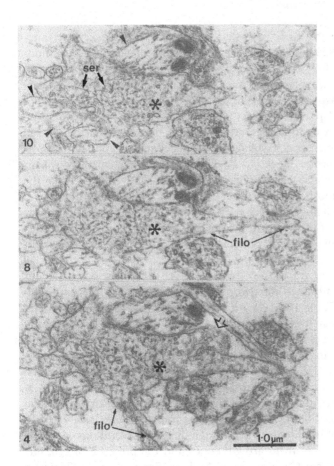

Figure 1.9 Electron micrographs of serial sections of a growth cone (asterisks) from the cerebral cortex of a 5-day-old rat. Characteristically, the C-domain of the growth cone is filled with a system of smooth endoplasmic reticulum (ser) and vesicles embedded in floccular material, presumably actin filaments. The growth cone is fasciculating with a small group of axons (arrowheads), with which it makes close contact, and extends two filopodia (filo) into the adjacent neuropil. Note the artefactual mound of vesicles (open arrow). The numbers in the bottom left-hand corner refer to the section numbers in the series.

(Yamada, Spooner & Wessells, 1971; Bunge, 1973; Rees & Reese, 1981; Gorgels, 1991). This form of reticulum may be present in copious amounts. A second type of smooth endoplasmic reticulum consists of short, blind-ended tubules with a wider lumen than the first type and a more electron-dense lumenal content (Yamada *et al.*, 1971; Bunge, 1973), and the third type, which is not always present, that consists of flat discs or sheets of cisternae occurring singly or, more commonly, stacked on top of each other forming a pile not unlike the arrangement of the cisternae of the Golgi apparatus (Rees & Reese, 1981; Gorgels, 1991). These membrane stacks were first described in rapidly frozen, freeze-substituted growth cones of rat superior cervical ganglion cultures (Rees & Reese, 1981) but they have also been seen in conventionally prepared material, for instance, in the growth cones of grasshopper central nervous system (CNS) neurons, where they were mistakenly identified as Golgi apparatus (Fig. 7 in Bastiani et al., 1985; see also Fig. 6 in Bunge, 1973), and in growth cones of the rat pyramidal tract (Gorgels, 1991). The sheet-like nature of this form of smooth endoplasmic reticulum has been confirmed by serial section analysis of growth cones isolated as a subcellular fraction from postnatal rat forebrain (Gordon-Weeks & Lockerbie, 1984) and, in freeze-substituted growth cones in chick tectum (Cheng & Reese, 1985, 1987). Unlike the branching form of smooth endoplasmic reticulum, the sheet-like form has a translucent lumen which appears to be partially lost after freeze-substitution by artefactual apposition of the adjacent membrane; only at the rim of the cisternae, does the lumen remain patent, possibly because of the greater curvature of the membrane here (Gordon-Weeks, 1988a).

In addition to smooth endoplasmic reticulum, growth cones may also contain other types of membrane-bounded organelles, such as vacuoles and vesicles of various kinds (Table 1.2). The difference between vacuoles and vesicles is not always stated, but it seems to be accepted that vacuoles are generally larger and always have a clear lumen. Vesicles may be clear or granule-containing and range in size from 50 nm, which is similar to that of synaptic vesicles, to about 100 nm. Landis (1978) has shown that the growth cones of sympathetic neurons in culture contain large numbers of small granular vesicles, which are diagnostic for noradrenergic synaptic vesicles in the peripheral nervous system (PNS) (Gordon-Weeks, 1988b). The presence of synaptic vesicles in other types of growth cones has not been reported, although there are several reports of the release of neurotransmitters from growth cones and the presence of synaptic vesicle-specific proteins (see Chapters 3 and 5). Vacuoles may attain diameters of up to 0.5 μm (e.g. Cheng & Reese, 1987) and may correspond to the pinocytotic vacuoles seen under the light microscope in growth cones in culture (see above).

Coated vesicles are also present in the cytoplasm of growth cones, and coated pits in the plasma membrane (Bastiani & Goodman, 1984; Cheng & Reese, 1985; Williams et al., 1986; Gordon-Weeks, 1987a). These have been encountered more frequently in the P-domain, particularly at the bases of filopodia (Cheng & Reese, 1985; Gordon-Weeks, 1987a). Lysosomes have been seen occasionally in growth cones (Rees et al., 1976).

The functional role played by these various forms of reticula and vesicles in growth cones is not entirely clear. At the outset, it would be anticipated that some of these organelles are involved in the growth and recycling of surface (plasma) membrane that must occur at a high rate in growth cones. This topic is discussed in detail in Chapter 2.

Cytoskeletal elements
The cytoskeletal elements of the growth cone observable in the electron microscope are microfilaments, microtubules and neurofilaments (Table 1.2).

Microfilaments are particularly concentrated in the P-domain whereas microtubules and, when present, neurofilaments, are mainly found in the C-domain (Fig. 1.8). In filopodia, microfilaments are organised into longitudinally oriented bundles that run the length of the structure and may extend across the P-domain and enter the C-domain (Figs 1.6 and 1.8). In the lamellipodia, microfilaments occur singly or in bundles, and they tend to be randomly oriented forming a dense space-filling meshwork (Isenberg, Rieske & Kreutzberg, 1977; Kuczmarski & Rosenbaum, 1979; Rees & Reese, 1981; Letourneau, 1983; Tosney & Wessells, 1983; Gordon-Weeks, 1987a; Lewis & Bridgman, 1992). The majority, although probably not all, of the microfilaments in growth cones are composed of actin (see Chapter 2).

In a growing neurite, microtubules are a dominant feature and are often observed in bundles running parallel to the longitudinal axis of the neurite. At the distal end of a growing neurite, where it enlarges into the growth cone, the microtubules in the bundle separate from each other, like the ribs of a fan, and extend individually into the C-domain of the growth cone (Yamada et al., 1971; Bunge, 1973; Letourneau, 1983; Tsui et al., 1984; Cheng & Reese, 1985, 1988; Bridgman & Dailey, 1989; Dailey & Bridgman, 1991; Sabry et al., 1991). As they diverge from each other on entering the C-domain, microtubules may have a winding course at their distal ends (Cheng & Reese, 1985, 1988) or loop back into the neurite shaft forming hairpin bends (Tsui et al., 1984; Sabry et al., 1991), whereas in the neurite they are usually straight. Furthermore, in the morphologically simple growth cones of the chick optic tectum, there is a marked preference for smooth endoplasmic reticulum and vesicles to be associated with the straight regions but not the winding regions (Cheng & Reese, 1988). This, however, is not the case in rat superior cervical ganglion neurons in culture, where reticulum in the growth cone is seen associated with microtubules throughout their length (Dailey & Bridgman, 1991).

Until recently (see Chapter 2), microtubules had rarely been seen to extend far beyond the C-domain of the growth cone, and had never been found in the filopodia at the edge of the growth cone and only very occasionally in lateral filopodia on neurite shafts (e.g. Fig. 10 in Yamada et al., 1971).

Axonal versus dendritic growth cones
Neurons are highly polarised cells in terms of the type of process that they produce. Most mature neurons have both an axon and dendrites. Axons are quite different from dendrites. Axons are characteristically long, cylindrical processes of fairly uniform cross-section that conduct action potentials away from the cell body (except in the case of primary sensory neurons) and terminate as pre-synaptic terminals at the synapse. Usually, the pre-synaptic terminal region consists of a number of enlargements, or varicosities, strung out along the axon, which contain the characteristic organelles of the synapse, such as synaptic vesicles. Dendrites, on the other hand, are tapered, highly branched processes, which function to receive synaptic input and transmit the information toward the cell body.

Neuritogenesis *in vivo* usually begins with the extension of a single neurite that is destined to become the axon (Roberts, 1988; Jacobson, 1991; Fig. 1.1). Later emerging neurites develop into dendrites. Clear-cut examples of this are found, for instance, in the retina. There, the principal projection neurons are the retinal ganglion cell neurons, whose axons form the optic nerve and tract. The first neurite that retinal ganglion cells produce is always the axon, which normally grows in an appropriate direction to find its target – radially towards the optic nerve head at the centre of the retina. Later-emerging neurites from retinal ganglion cells are the dendrites and they are oriented in the opposite direction to the axon. The temporal sequence of axon followed by dendrite is invariant. This programme of differentiation is thought to depend partly on signals impinging on the neuron from the local environment (see Chapter 3). When neurons are removed from the embryo, and thus deprived of extrinsic signals, their differentiation programme can become altered. For instance, when cultured as dissociated cells, hippocampal pyramidal neurons (Dotti, Sullivan & Banker, 1988) and cerebellar macroneurons (Ferreira, Busciglio & Caceres, 1989) differentiate by extending, more-or-less simultaneously, several neurites that appear to be bi-potential in terms of whether they will become axons or dendrites (Fig. 1.1). After these neurites have grown a short distance, one of them, randomly, starts to grow faster than the others, and this one will become the axon while the others differentiate into dendrites. Not all neurons in culture extend neurites which later differentiate into axons and dendrites. Cerebellar granule cells in dissociated, low-density cultures, i.e. such that the cells do not contact each other, first extend an axon, followed by a second and then, much later, dendrites (Powell *et al.*, 1997).

Although dendritic growth cones have been studied under the light microscope both *in vivo* and *in vitro*, there are few studies of dendritic growth cones in the electron microscope and it is not certain whether it is possible to distinguish these from axonal growth cones using ultrastructural criteria. We may expect the growth cones on dendrites to grow more slowly and their parent neurites to have different branching patterns, from those on axons. Furthermore, since dendrites have polyribosomes throughout their length both during growth and in the adult (reviewed in Steward & Banker, 1992), and polyribosomes can be identified in the electron microscope, this might form the basis for identification. The light microscopic appearance of dendritic growth cones shows that they are enlargements of the neurite and extend filopodia (Morest, 1969a,b; Miller & Peters, 1981). From the few electron microscope studies that have been conducted of dendritic growth cones it appears that their ultrastructure is similar to that of axonal growth cones (Fig. 1.10; Hinds & Hinds, 1972; Skoff & Hamburger, 1974; Vaughn, Henrikson & Grieshaber, 1974; Miller & Peters, 1981), with the exception of the presence of polyribosomes in dendritic growth cones (Skoff & Hamburger, 1974). Axonal synapses are frequently found on the surfaces of dendritic growth cones (Skoff & Hamburger, 1974; Vaughn, Henrikson & Grieshaber, 1974; Miller & Peters, 1981; Miller, 1988).

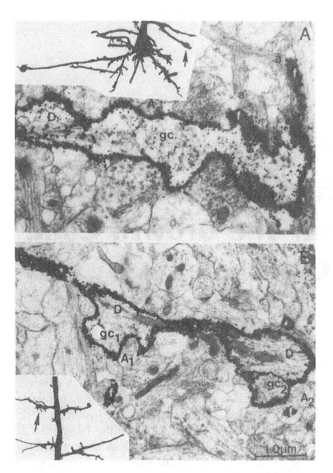

Figure 1.10 Examples of dendritic growth cones. The insets show drawings of Golgi-impregnated layer V pyramidal neurons from a postnatal day 6 rat cerebral cortex. The arrows indicate the growth cones shown in the accompanying electron micrographs. The growth cones (gc) extend filopodia (f) into the adjacent neuropil, which contains axons (a) some of which form synapses (A_1–A_2) on the body of the growth cone. (From Miller & Peters, 1981. Reprinted by permission of Wiley-Liss, Inc., a subsidiary of John Wiley & Sons, Inc.)

2

Motility and Neurite Extension: The Growth Cone Cytoskeleton

Introduction

Apart from our curiosity about the mechanisms underlying neurite extension and growth cone motility, understanding the cytoskeleton in growth cones is important because it is the 'final common path' of action of extrinsic guidance cues (reviewed in: Letourneau & Cypher, 1991; Strittmatter & Fishman, 1991; O'Connor & Bentley, 1993; Bentley & O'Connor, 1994; Lin, Thompson & Forscher, 1994; Tanaka & Sabry, 1995; Challacombe, Snow & Letourneau, 1996a; Letourneau, 1996). Broadly speaking, neurite extension depends upon the integrity of microtubules whereas, in common with motile events in other cell types, the motility of the growth cone, the extension and retraction of filopodia and lamellipodia, is supported by actin microfilaments. Pathfinding, the ability of the growth cone to navigate a route through the embryo to its target cell (see Chapter 3), depends on both of these parts of the cytoskeleton and their associated proteins.

Neurite extension is a special case of cellular motility. When a fibroblast or other motile cell in culture migrates across a substratum it extends a cellular process (lamellipodium) at the leading edge of the cell which attaches to the substratum so that, following traction, the entire cell including the nucleus, can move forward (Bray, 1992). Although there are similarities between neurite extension and cellular translocation, there are two important differences. First, neurite extension occurs by growth, in the sense of the addition of newly synthesised components of the neurite and second, neurite extension is not generally associated with the movement of the cell body, and therefore the nucleus, of the neuron. We know far less about growth cone motility than motile events in non-neuronal cells (reviewed in: Mitchison & Cramer, 1996; Welch *et al.*, 1997; Cramer, 1997).

Organisation of the growth cone cytoskeleton

The filamentous cytoskeleton of the growth cone observable in the electron microscope consists of microtubules, microfilaments and, in some growth cones, neurofilaments (see Table 1.2). In addition to this filamentous part of the cytoskeleton, there is a wide variety of actin-binding proteins, microtubule-associated proteins and other proteins that interact with the filamentous cytoskeleton. These proteins and their associated structures are thought to play a part in neurite elongation, growth cone shape, motility and pathfinding.

Microtubules

Microtubules are hollow, unbranched cylindrical filaments (24 nm in diameter), which in axons can attain lengths of over 100 μm. They are composed of heterodimers of the related proteins α- and β-tubulin, with molecular weights of about 55 kD, and a diverse group of accessory proteins collectively termed microtubule-associated proteins (MAPs). MAPs characteristically bind to the tubulin backbone of microtubules and can be grouped into several families according to their molecular structure and function (Wiche et al., 1991). Tubulin dimers self-associate to form non-covalent polymers or protofilaments, of which normally thirteen assemble to form a microtubule.

Microtubules within cells are in equilibrium with soluble α- and β-tubulin dimers, which are added to or lost from the filament ends. However, microtubules are not simple equilibrium polymers; by utilising the energy released from trinucleotide hydrolysis they achieve far greater excursions in length than would be possible under simple equilibrium, mass action dynamics. Microtubule assembly requires the presence of tubulin dimer-associated GTP, which is hydrolysed to GDP at some point after assembly. GDP-tubulin cannot polymerise and GDP-tubulin within the microtuble lattice destabilises the polymer and leads to depolymerisation. Microtubules are inherently polar, displaying different kinetics of tubulin addition and loss at each end. The end at which the kinetics favour net tubulin addition is known as the 'plus' end whereas the other end is the 'minus' end. Isolated microtubules, free of cellular influences, are in a state of dynamic instability because of GTP hydrolysis, undergoing phases of relatively slow elongation by polymerisation at their 'plus' ends alternating with rapid, brief depolymerisation (Mitchison & Kirschner, 1984a,b; Schulze & Kirschner, 1986, 1987; reviewed in Kirschner & Mitchison, 1986a,b). Within cells, microtubules exhibit less dramatic excursions in length change and thus their dynamic instability is said to be 'tempered', probably due to the actions of MAPs. Individual microtubules interconvert between these two phases apparently at random. It has been suggested that this behaviour allows microtubules continually to probe the cell periphery, or cortical cytoskeleton, and that this may form the basis for changes in cell morphology as a consequence of the action of extracellular signals (Mitchison & Kirschner, 1988). Such signals, acting locally at the cell surface, may enable the cortical cytoskeleton of the cell to 'capture' microtubules probing the cell periphery, and so stabilise them at appropriate sites within the cell. Microtubules are the structural substrates for the transport of membrane-bound organelles and so cortical capture of microtubules may result in the delivery of material along microtubules to that region, and therefore to the spatial regulation of cellular growth. This model could equally well apply, with minor modifications, to growth cones, as discussed below (Mitchison & Kirschner, 1988).

In cells in general, microtubules are formed, or nucleated, *de novo* and grow from microtubule-organising centres (MTOCs) (reviewed in: Tucker, 1992; Joshi, 1994). MTOCs are associated with distinct organelles of which the basal bodies at the base of each flagellum and the centriole-containing centrosomes are well-known examples. In neurons, the centrosome is usually close to the nucleus. It so happens that microtubules nucleated at MTOCs are organised with their 'minus' ends nearest or proximal to the MTOC. This arrangement imposes a polarity to the organisation of microtubules throughout the cell. Uniquely associated with MTOCs is a member of the tubulin superfamily of proteins, known as γ-tubulin, which appears to be essential for microtubule formation in eukaryotic cells (Oakley & Oakley, 1989; Joshi *et al.*, 1992; reviewed in Joshi, 1994).

In neurites, microtubules subserve two main functions: they provide a structural basis for the transport of membrane-bound organelles (i.e. fast axonal transport) and they contribute to the maintenance of the structural integrity of the neurite (reviewed in: Bershadsky & Vasiliev, 1988; Díaz-Nido, Hernandez & Avila, 1990). The function of microtubules in growth cones is less clearly defined, although they are probably important for transport processes, and perhaps for directional or vectorial growth (see below).

Microtubule organisation in neurites and growth cones

The organisation of microtubules in neurites and growth cones has been examined by immunofluorescence microscopy with antibodies to α- or β-tubulin, by electron microscopy and, to a limited extent, by video microscopy (Fig. 2.1). In most circumstances, tubulin antibodies label both microtubules and the soluble pool of tubulin, whereas the two other techniques reveal only microtubules. In the neurite, as in most mature axons and dendrites, microtubules are mainly oriented parallel to the longitudinal axis of the neurite and are often grouped into bundles or fascicles. In mature dendrites the inter-microtubule spacing in bundles is at least 90 nm, whereas in axons it is at least 40 nm. Neurites frequently exceed many millimetres in length but, although microtubules in neurites can be as long as 100 μm, they clearly cannot extend the full length of a neurite and are usually wholly contained within the neurite (Bray & Bunge, 1981; Jacobs & Stevens, 1986).

Although microtubules are abundant in neurites, they are not always seen by electron microscopy in growth cones *in vivo* (Table 1.2), perhaps because they are less stable in growth cones and therefore less well preserved by fixation (see below). In growth cones in culture, where microtubules have been more extensively studied, they seem to be more easily preserved. The microtubules in growth cones probably all have their proximal ends within the parent neurite. As they enter the C-domain from the neurite shaft, the microtubules tend to diverge from each other and cross the C-domain as single microtubules (Yamada et al., 1971; Bunge, 1973; Letourneau, 1983; Tsui et al., 1984; Cheng & Reese, 1985, 1988; Bridgman & Dailey, 1989; Dailey & Bridgman, 1991; Gordon-Weeks, 1991; Sabry et al., 1991; Tanaka & Kirschner, 1991; Tanaka, Ho & Kirschner, 1995). Their course across the C-domain can be straight or winding (Cheng & Reese, 1985, 1988) and, in some circumstances, they may form kinks or hairpin bends, looping back upon themselves to enter the neurite shaft (Tsui et al., 1984; Sabry et al., 1991; Tanaka & Kirschner, 1991; Fan et al., 1993; Lin & Forscher, 1993). The majority of the microtubules in the growth cone are confined to the C-domain but, occasionally, they can extend into the P-domain, where they may align with the longitudinal axis of a filopodium (Letourneau, 1983), and even gain entry into the proximal part of a filopodium (Gordon-Weeks, 1991; DiTella et al., 1996; Williamson et al., 1996). Single microtubules may enter the lateral filopodia of neurite shafts (Yamada et al., 1971). Occasionally, microtubules are also found bundled in the C-domain (Sabry et al., 1991; Tanaka & Kirschner, 1991).

Microtubules in growing axons and their growth cones are all oriented with their 'plus' ends distally (Heidemann, Landers & Hamborg, 1981; Baas et al., 1987). In growing dendrites, however, microtubules oriented with their 'plus' ends proximal are also encountered; except at the growth cone where they are all oriented as in axonal growth cones (Baas et al., 1988).

Neurite extension and microtubule dynamics

Microtubules have long been suspected of playing an important role in neurite growth. Indirect evidence for this idea is that agents that depolymerise microtubules, such as colchicine and colcemid, inhibit neurite elongation by neurons in culture and, when substrate adhesion is low, may cause neurite retraction (Seeds et al., 1970; Yamada et al., 1970; Daniels, 1972, 1975; Bray et al., 1978). Furthermore, antisense oligodeoxynucleotides to tubulin mRNA (Teichman-Weinberg et al., 1988) and microinjection of colchicine-tubulin (Keith, 1990), a synthetic derivative of tubulin which is unable to polymerise, block neurite extension in culture. Interestingly, when substrate adhesion is high, microtubule depolymerisation is associated not with neurite retraction but with the formation of growth cones along the neurite shaft – where they do not normally appear (except during 'back-branching', see Chapter 1), as if microtubules suppress an underlying potential for growth cone formation in the neurite (Bray et al., 1978; Joshi et al., 1986; Matus et al., 1986).

These experiments indicate that microtubules are important for neurite growth but they do not shed light on exactly what that role is. It is clear,

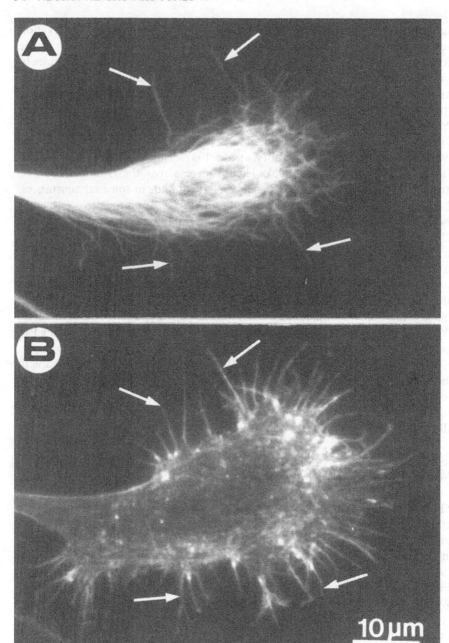

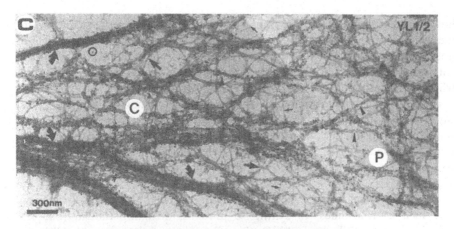

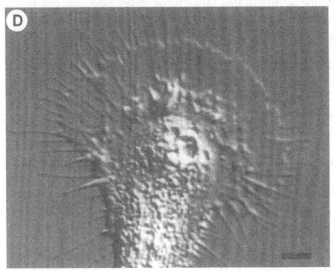

Figure 2.1 Organisation of microtubules and actin filaments in growth cones.

A/B: A growth cone from a chick dorsal root ganglion cell in culture fixed with fixatives containing detergent, to reveal the cytoskeleton, and immunostained with an antibody to tyrosinated α-tubulin (A), which labels all microtubules in growth cones, and rhodamine-phalloidin, which labels F-actin (B). In the C-domain of the growth cone the microtubules are bundled but extend singly across the P-domain, and occasionally enter filopodia (arrows in A and B) as indicated by co-localisation with filopodial F-actin bundles (B).

C: A growth cone from a rat dorsal root ganglion cell prepared as a whole mount for electron microscopy and immunogold labelled with antibodies to tyrosinated α-tubulin. The gold particles (black dots) decorate bundled microtubules (curved arrows) in the C-domain (C) and also single microtubules (arrowheads) advancing through the P-domain (P). Actin filament bundles (arrows) are unlabelled.

D: A growth cone from an *Aplysia* neuron growing in culture and viewed by high-resolution video microscopy. The long, rod-shaped elements visible in the P-domain (P) are actin filament bundles. Bar = 5 μm.

(A, B: from Gordon-Weeks, 1991, with permission; C: from Bush *et al.*, 1996a, with permission of Blackwell Science Ltd.; D: from Forscher & Smith, 1988, by copyright permission of the Rockefeller University Press.)

however, that microtubules are important in maintaining neurite shape and in the transport of material along the neurite. Furthermore, these experiments do not provide information about the location of those microtubules in the differentiating neuron that are important for neurite extension, and, in particular, whether microtubules can originate *de novo* in the neurite as well as the cell body, or whether microtubules or tubulin subunits (dimers or oligomers) are transported into the neurite. A considerable amount of experimental data, some of it contradictory, has accumulated in an attempt to answer these questions (reviewed in: Hollenbeck, 1989; Solomon, 1992; Gordon-Weeks, 1993; Joshi & Baas, 1993; Baas, 1997; Baas & Brown, 1997; Bray, 1997; Hirokawa *et al.*, 1997).

It is a widely held dogma that axons do not possess the biochemical machinery, such as ribosomes, to synthesise proteins, including tubulin (but see Olinck-Coux & Hollenbeck, 1996); this takes place in the neuronal cell body (perikaryon). Although dendrites do have ribosomes along their entire length, both when growing and when fully formed (reviewed in: Craig, Jareb & Banker, 1992; Steward, 1997), mRNA for tubulin is confined to the cell body (Garner, Tucker & Matus, 1988). Therefore, microtubules are either assembled in the cell body and transported into the axon and dendrite, or are formed *de novo* in these processes from tubulin subunits (dimers or oligomers) that have been transported into the axon or dendrite. Of course these two mechanisms are not mutually exclusive. The evidence for microtubule formation (nucleation) at the cell body is incontrovertible, but there is considerable doubt about the existence and location of *de novo* microtubule formation in growing axons and growth cones, although it is fairly clear that extensive polymerisation of existing microtubules does occur in growing axons and growth cones. Exactly how tubulin is transported along axons and dendrites and in what form, as polymer, i.e. microtubules, or as subunits, is a matter or considerable current debate (Baas & Brown, 1997; Hirokawa *et al.*, 1997).

Early biochemical analysis of the incorporation and transport of metabolically radiolabelled tubulin in neurons suggested that tubulin is transported as a polymer in axons after assembly at the cell body (reviewed in Lasek, 1982, 1988). In these experiments, radiolabelled amino acids, usually ^{35}S-methionine, are introduced *in vivo* into the vicinity of the cell body where they become incorporated into newly synthesised tubulin, among other proteins. The movement of these proteins into and along the axon is then analysed biochemically by cutting the axon into segments at various time intervals and separating soluble (dimeric) from insoluble (presumed microtubule-incorporated) tubulin. Separation is usually achieved with detergents; the cytoskeleton is operationally defined as detergent insoluble. These experiments showed that radiolabelled tubulin moves down the axon in a polymerised form, i.e. as microtubules, in a coherent wave that does not spread out appreciably or decrease in amplitude as it progresses down the axon, and therefore does not exchange extensively with soluble tubulin. The rate of movement was about 40 μm/hr; a speed not too dissimilar from the fastest rate of growth cone migration (Table 1.1).

On the basis of these findings, Lasek and co-workers have proposed that microtubules are pre-assembled in the neuronal cell body and translocated as such down the axon by an active, i.e. energy-dependent, transport mechanism. Although the transport mechanism has not been identified, an active mechanism was postulated because it was assumed that diffusion was insufficient to account for the distances involved. While this may certainly be true for large polymers such as microtubules, it is probably not true for tubulin dimers. Calculations based on diffusion coefficients for molecules the size of tubulin dimers in cytoplasm (Jacobson & Wojcieszyn, 1984) indicate that the $t_{\frac{1}{2}}$ for diffusion along a 100 μm length of axon would be about 55 seconds (Keith, 1990). Furthermore, when microinjected into PC12 cells, colchicine-tubulin, which cannot polymerise, can travel distances of several 100 μm along growing neurites within a few hours (Keith, 1990).

Black, Keyser and Sobel (1986) used ^{35}S-methionine to label newly synthesised tubulin metabolically in cultured sympathetic neurons and then used detergents to separate soluble tubulin from insoluble tubulin. They showed that newly synthesised tubulin remains soluble for up to 10 minutes following a pulse-chase with unlabelled methionine, whereas by 120 minutes it is mainly associated with the detergent-insoluble fraction (presumably microtubules). This result might be taken to imply that tubulin is assembled into microtubules near its site of synthesis, the cell body, since, if tubulin was moving along the axon solely by slow axoplasmic transport, it would only have travelled a maximum distance of about 80 μm. Such an interpretation supports Lasek's hypothesis. However, the interpretation hinges entirely on the assumption that tubulin can only travel along axons by slow axonal transport – a point not yet established (see Campenot, Lund & Senger, 1996). Indeed, fluorescently labelled dextrans and proteins such as bovine serum albumen (60 kD), when microinjected into the soma of *Xenopus* spinal cord neurons in culture, diffuse along neurites at a rate only five times less than they do in aqueous solution (Popov & Poo, 1992; see also, Sabry, O'Connor & Kirschner, 1995). Based on such experiments, these workers estimated that a typical globular protein of 70 kD takes about 30 minutes to diffuse a distance of 200 μm along a neurite. Biotinylated actin (approx. 42 kDa) takes less than 5 min to diffuse 50 μm along the neurites of PC12 cells and mouse dorsal root ganglion neurons in culture (Okabe & Hirokawa, 1991). Whilst diffusion may be sufficient to supply tubulin to growth cones of short neurites, less than a few hundred micrometres in length, it is unlikely to be an adequate mechanism for long neurites or axons in the adult (Sabry et al., 1995).

Lasek's model of microtubule formation in the cell body and subsequent axonal transport implies that, at some stage after MTOC nucleation, microtubules must detach and move into the axon. If the MTOC in the cell body is the only microtubule nucleation site in the neuron, then this must be true since microtubules are found entirely within neurites (see above). Experiments from Baas' group provide indirect evidence that microtubule detachment from centrosomes occurs at a high rate in neurons in culture (Yu *et al.*, 1993; Ahmad *et al.*, 1994; Yu, Schwei & Baas, 1996). Using cultured sympathetic neurons they showed that very few (< 10) microtubules are normally attached to the centrosome. However, within 5 minutes after recovery from extensive microtubule depolymerisation with nocodazole, a specific and reversible microtubule depolymerising drug, many hundreds of short microtubules were attached to the centrosome. As time passed, this number declined and a wave of detached microtubules moved away from the centrosome. More recent experiments from the same group have shown that the wave of microtubules travels to the periphery of the neuronal cell body away from the centrosome and that some of them are destined to enter the neurite (Ahmad & Baas, 1995; see also Slaughter, Wang & Black, 1997). The behaviour of microtubules in these experiments was inferred from electron micrograph images. Confirmation of the release of microtubules nucleated at the centrosome and their dispersal throughout the cytoplasm has now come from direct observation of fluorescently labelled microtubules by time-lapse video microscopy in living epithelial cells (Keating *et al.*, 1997). Cultured cells were microinjected with fluorescently labelled tubulin which became incorporated into newly nucleated microtubules at the centrosome. Microtubules were seen nucleating at the centrosome and detaching from it. Microtubules leaving the centrosome moved radially away from it and individually, rather than in bundles. They had their 'plus' ends directed away from the centrosome, suggesting that they were transported by a 'minus' end-directed microtubule motor. By photobleaching a segment of fluorescently labelled microtubule with a laser and monitoring the positions of the microtubule ends relative to the photobleached mark over time, Keating *et al.* (1997) showed that both microtubule translocation and 'minus' end depolymerisation occurred. Microtubules often buckled and formed loops as if encountering obstacles as they moved away from the centrosome.

Collectively, these experiments demonstrate that the neuronal centrosome is highly efficient at microtubule nucleation and subsequent detachment, a result which supports one feature of Lasek's model. Clearly, such a mechanism can contribute to the population of non-centrosomal associated microtubules found in neurites. But can microtubules form in other regions of the growing neuron, for instance in the neurite itself or at the growth cone? There is no structural evidence for MTOCs in either the neurite or the growth cone. Furthermore, γ-tubulin, which seems to be essential for microtubule nucleation at MTOCs (Oakley & Oakley, 1989; Joshi *et al.*, 1992; Oakley, 1992), is not present in growing axons (Baas & Joshi, 1992). Such negative findings do not, however, rule out novel MTOCs.

There is a subset of microtubules within axons that are resistant to depolymerisation by drugs. After maximal microtubule depolymerisation, repolymerisation following drug washout takes place only at the distal ends of those microtubules that resisted drug-induced depolymerisation; no new microtubules are formed (Baas & Ahmad, 1992). This finding also suggests that there are no MTOCs within the axon.

Location of microtubule assembly

If microtubules are formed *de novo* in the growth cone and become incorporated into the growing neurite, they would become stationary within the neurite as the growth cone advances and therefore their behaviour would be very different from microtubules entering the neurite at the cell body and migrating down the axon towards the growth cone. Furthermore, the site of action of those microtubule depolymerising drugs that block neurite outgrowth would be different.

To test the role of *local* microtubule dynamics in axon growth, Bamburg and co-workers, in a provocative set of experiments, applied various microtubule depolymerising or stabilising drugs to discrete sites on chick dorsal root ganglion neurons in culture (Bamburg, Bray & Chapman, 1986). Drugs were applied by micropipette to cell bodies, axons or growth cones in the presence of a continuous flow of culture medium to ensure discrete drug localisation (Fig. 2.2). As expected, drug application to all sites blocked axon elongation. The important observation, however, was that the growth cone was more sensitive than the cell body by two orders of magnitude; the neurite was intermediate in sensitivity. Assuming that these drugs only affect microtubule dynamics, the authors suggested that microtubule dynamics at the growth cone are important in axon elongation. These results were also interpreted to indicate that tubulin reaches the growth cone by transport down the axon in an unassembled form, either as dimer or small oligomer. A re-examination of the effects of these drugs on growth cone microtubules showed that they caused depolymerisation, and hence loss of microtubules, from growth cones, rather than prevented assembly (Yu & Baas, 1995).

Is neurite elongation affected if tubulin assembly/disassembly is blocked *without* affecting the numbers of existing microtubules? Microtubule dynamic behaviour is suppressed by substoichiometric (nanomolar) concentrations of microtubule binding drugs, such as nocodazole, vinblastine and Taxol (Jordan & Wilson, 1990; reviewed in Wilson & Jordan, 1994). When such drugs are applied to growing neurons in culture at substoichiometric concentrations, although the rate of axon elongation is reduced – sometimes by as much as one half – it is generally not blocked (Letourneau & Ressler, 1984; Letourneau, Shattuck & Ressler, 1986; Zheng, Buxbaum & Heidemann, 1993; Yu & Baas, 1995; Rochlin, Wickline & Bridgman, 1996; Williamson *et al.*, 1996; Challacombe, Snow & Letourneau, 1997; but see Tanaka, Ho & Kirschner, 1995). These results suggest that microtubule dynamics contribute to axon elongation because, in their absence, growth is reduced, but they also imply that microtubule translocation is important (Joshi & Baas, 1993).

Microinjection of colchicine-tubulin, a derivatised form of tubulin that is unable to polymerise, blocks neurite elongation in PC12 cells only when it has diffused to the growth cone, but not when restricted to the cell body or proximal neurite (Keith, 1990). Also, when the critical concentration for tubulin polymerisation is lowered in growth cones using high concentrations of Taxol (micromolar, see below), the soluble pool of tubulin is polymerised on to the

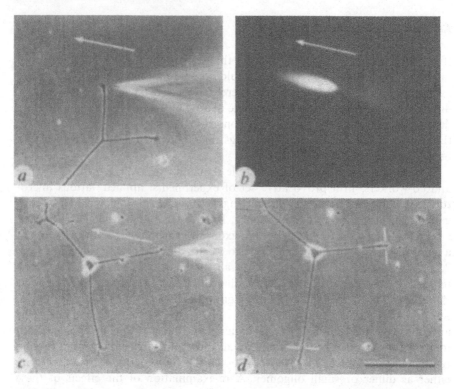

Figure 2.2 The effects of microtubule depolymerising agents on axon growth in chick dorsal root ganglion neurons. A micropipette filled with colcemid, a drug which depolymerises microtubules, and a fluorescent dye (fluorescein), was placed over a growth cone (a), and the flow of the drug solution from the pipette monitored by the fluorescein fluorescence (b). The effects of colcemid application are shown in (c), at the time of application, and (d), 30 minutes after application. Only the growth cone to which the drug had been applied stopped advancing. The arrows indicate the direction of flow of the culture medium. (From Bamburg *et al.*, 1986, with permission.)

'plus' ends of the existing microtubules in the growth cone (Letourneau & Ressler, 1984; Gordon-Weeks, Mansfield & Curran 1989a). This changes the microtubule organisation dramatically and, interestingly, results in blockade of axon growth (see below).

Direct visualisation of microtubule dynamics

The experiments described so far are all indirect. Ideally, we would like to be able to observe individual microtubules directly in living neurites and growth cones. The crowding of microtubules in axons and the unfavourable geometry compared with the spread and thin regions of fibroblasts in culture, for example, make this difficult. However, attempts to visualise bulk microtubule movement directly within growing axons has met with some success but has, confusingly, produced diametrically opposite results in different species. Two methods have been used to visualise microtubule movements within axons. Originally, fluorescent derivatives of tubulin were microinjected into cells in culture. After the fluorescent tubulin had distributed throughout the cell and become incorporated into microtubules, a segment of the axon was marked by photobleaching using an intense light source. The recovery of the bleached segment, owing to tubulin subunit exchange, and its movements, were then monitored. The first experiments of this kind were done using PC12 cells by Keith, who found that the bleached patch moved slowly distally, implying that microtubules are translocated along neurites toward the growth cone (Keith, 1987). In contrast, similar experiments using PC12 cells and cultured chick and mouse dorsal root ganglion neurons failed to detect movement of the bleached zone, which instead recovered its fluorescence, suggesting that microtubules are stationary and exchanging polymeric tubulin for soluble dimers (Lim, Sammak & Borisy, 1989; Okabe & Hirokawa, 1989; Lim et al., 1990). It was presumed, therefore, that tubulin dimers were translocated down the axon and not microtubules, as in the Lasek model.

More recently, caged-fluorescein derivatives of tubulin have been used. Such derivatives become fluorescent by photoactivation with ultraviolet light and require less intense light than photobleaching (for a methodological review, see Sawin, Theriot & Mitchison, 1993). The fluorescent zone appears bright against a dark background. Also, caged-fluorescein derivatives are not thought to produce cell-damaging free radicals, unlike the earlier method. Experiments using microinjected caged-fluorescein-labelled tubulin provide support for both microtubule transport along growing axons *and* assembly at growth cones. In *Xenopus* spinal cord motoneurons, microtubules were not stationary, as indicated by the distal movement of fluorescence during neurite elongation (Reinsch, Mitchison & Kirschner, 1991; Tanaka & Kirschner, 1991; Okabe & Hirokawa, 1992), whereas in mammalian (mouse) dorsal root ganglion neurons there was no translocation (Okabe & Hirokawa, 1992). Curiously, in *Xenopus* neurons, fluorescent segments near the cell body moved more slowly than those more distally and even distal segments moved at a slower rate than that of neurite elongation. This apparent species difference is not attributable to differences in the two techniques, because, although the caged-fluorescein tubulin method is more sensitive, photobleaching in *Xenopus* neurons also shows segment movement (Okabe & Hirokawa, 1993). A re-examination of PC12 cells by Keith and Farmer (1993) confirmed the original observation of microtubule movement by Keith (1987), but showed considerable variation in fluorescence recovery after photobleaching and local loss of microtubules in some cases, suggesting that photobleaching can cause local damage. It may also be important to bear in mind that PC12 neurites have mixed properties between axons and dendrites, although they most closely resemble sympathetic axons (Greene & Tischler, 1976), whereas all the other experiments were done with axons. These *in vitro* experiments on microtubule translocation in growing axons have now been extended to neurons growing in their natural environment in the embryo (Sabry, O'Connor & Kirschner, 1995; Takeda et al., 1995).

In living zebrafish embryos in which spinal cord neurons had been labelled by microinjecting fluorescein-conjugated tubulin into blastomeres, photobleaching of growing axon segments failed to uncover movement of the bleached segment, although fluorescence did recover in the bleached patch (Takeda et al., 1995). Similarly, in whole limb-bud explants from grasshopper embryos in which the Ti1 pioneer sensory neurons had been labelled with caged-fluorescein tubulin, photoactivated segments of growing axon failed to translocate during axon extension and growth-cone pathfinding, although the fluorescence of the mark did fade in time, suggesting subunit exchange (Sabry et al., 1995). At present this entirely unsatisfactory paradox, and whether microtubules or tubulin dimers are transported down growing axons (see Campenot et al., 1996), remains unresolved (Baas, 1997; Baas & Brown, 1997; Bray, 1997; Hirokawa et al., 1997).

Post-translational modifications of tubulin

Additional evidence that microtubules in neurites and growth cones are elongating distally by subunit addition comes from experiments examining the distribution of post-translationally modified forms of tubulin in neurites and growth cones. Several post-translational modifications of tubulin have been identified including phosphorylation, acetylation, glutamylation and a reversible removal of the C-terminal tyrosine of α-tubulin by specific enzymes (reviewed in Serrano & Avila, 1990).

With a few minor exceptions, most of the genes for α-tubulin code for a protein with a C-terminal tyrosine (Serrano & Avila, 1990). This tyrosine can be selectively removed by a specific tubulin tyrosine carboxypeptidase (Argarana, Barra & Caputto, 1978) or added by a specific tubulin tyrosine ligase (Barra et al., 1973; Raybin & Flavin, 1977). These post-translational modifications are unique to tubulin. Most, if not all, of the soluble α-tubulin in cells is C-terminally tyrosinated whereas polymerised tubulin contains both forms of α-tubulin. In non-neuronal cells, where the existence of biochemically distinct but overlapping subpopulations of microtubules is well established (Gundersen, Kalnoski & Bulinski, 1984; Piperno, Ledizet & Chang, 1987), α-tubulin becomes de-tyrosinated and acetylated some time after assembly into microtubules (Kreis, 1986; Schulze et al., 1987). This phenomenon produces microtubules with variations in α-tubulin isoforms along their length. There is a correlation between microtubule populations that are relatively stable to depolymerisation by cold shock and microtubule depolymerising agents, such as nocodazole, and acetylation and detyrosination on the one hand, and labile microtubules and tyrosination on the other (Kreis, 1986; Khawaja, Gundersen & Bulinski, 1988). Although these post-translational modifications of tubulin correlate with an increase in the stability of the microtubules to depolymerisation, they are not causal to microtubule stability (Piperno et al., 1987; Schulze et al., 1987; Khawaja et al., 1988). What cellular factors are involved in microtubule stability in vivo has yet to be determined, although it seems likely that cross-linking of adjacent microtubules, i.e. micro-tubule bundling, would increase stability. Microtubules are cross-linked to each other, to form bundles, and to other cytoskeletal elements by MAPs (see below). The effects of transfection experiments with MAPs are consistent with the idea that they modify microtubule stability (e.g. Knops *et al.*, 1991; Takemura *et al.*, 1992).

The relative age of a microtubule population can be gauged by the distribu-tion within it of the post-translationally modified forms of tubulin. Immunofluorescence and immunoelectron microscopy, using antibodies speci-fic for acetylated, tyrosinated or de-tyrosinated α-tubulin, have revealed differ-ences in post-translationally modified forms of tubulin in microtubule populations in both neurites and growth cones in neurons in culture. Immunofluorescence has shown that, in proximal regions of growing neurites, de-tyrosinated and acetylated forms of α-tubulin predominate over tyrosinated and unacetylated forms, whereas in distal neurites, and particularly in growth cones, the reverse is true (Black *et al.*, 1989; Lim *et al.*, 1989; Robson & Burgoyne, 1989a; Baas & Black, 1990; Arregui *et al.*, 1991; Mansfield & Gordon-Weeks, 1991). For instance, in dorsal root ganglion neurons in cul-ture, the majority of the axonal growth cones (these cells do not have dendrites) stain with antibodies specific for tyrosinated α-tubulin, but not for de-tyrosi-nated α-tubulin (Robson & Burgoyne, 1989a). This is also the case with growth cones of PC12 cells (Lim *et al.*, 1989), and the axonal and dendritic growth cones of cerebral cortical neurons in culture (Mansfield & Gordon-Weeks, 1991). Immunoelectron microscopy has revealed that individual microtubules in neurites have distinct domains of post-translationally modified α-tubulin (Black *et al.*, 1989; Baas & Black, 1990; Baas & Ahmad, 1992; Brown *et al.*, 1993). There is a proximal domain which has predominantly de-tyrosinated and acetylated α-tubulin, and a distal domain at the 'plus' end composed of unacetylated, tyrosinated α-tubulin. The border between the two domains tends to be quite sharp. Furthermore, microinjection of biotinylated tubulin subunits into neurons in culture shows that the tubulin can be incorporated into axonal microtubules but only at their distal ends, i.e. microtubules in growing axons elongate distally (Okabe & Hirokawa, 1988; Baas & Ahmad, 1992). Unfortunately, in these experiments the growth cone was not investi-gated directly. Thus, while formation of microtubules *de novo* in neurites remains controversial, the weight of evidence favours growth of pre-existing microtubules that have been transported into the neurite and growth cone from the cell body (reviewed in: Solomon, 1992; Gordon-Weeks, 1993; Joshi & Baas, 1993).

These observations support the idea that microtubules in growing neurites are elongating by distal polymerisation and that this process is most active in the growth cone and distal neurite. As the growth cone advances and the microtubules elongated in the growth cone become incorporated into the neur-ite cytoskeleton, the α-tubulin is de-tyrosinated at its carboxy-terminus and acetylated (Lim *et al.*, 1989; Robson & Burgoyne, 1989a; Mansfield & Gordon-Weeks, 1991). These changes are associated with a disto-proximal gradient of increasing microtubule stability. For instance, in chick dorsal root ganglion neurons, recovery of a photobleached segment is slower in more proximal regions of the neurite and fastest in the growth cone, and may be incomplete near the cell body (Edson *et al.*, 1993b). Also, quantitative analysis of the ratio of tyrosinated to acetylated forms of α-tubulin along individual microtubules in neurites of sympathetic neurons in culture showed that more proximally located individual microtubules have a higher proportion of acetylated and a lower proportion of tyrosinated α-tubulin, indicating their enhanced stability (see below) (Brown *et al.*, 1993). How this is brought about remains unclear.

Assembly competent tubulin in growth cones

Growth cones contain a large, soluble pool of assembly competent tubulin. This was first shown by observing the effects of the agent Taxol on growth cones in culture (Letourneau & Ressler, 1984). Taxol, a compound obtained from the bark of the Pacific yew (*Taxus brevifolia*), lowers the critical concentration point for tubulin assembly within cells and thus forces the soluble pool of tubulin to polymerise on to existing microtubules or to form microtubules *de novo*, presumably without the involvement of MTOCs (Schiff, Fant & Horwitz, 1979; Schiff & Horwitz, 1980; reviewed in Horwitz, 1992). The effect of Taxol treatment on growth cones is to assemble the soluble pool of tubulin on to the 'plus' ends of the microtubules that enter the C-domain from the neurite shaft (Letourneau & Ressler, 1984; Letourneau, Shattuck & Ressler, 1986, 1987; Gordon-Weeks, 1987a; Gordon-Weeks *et al.*, 1989a; Mansfield & Gordon-Weeks, 1991). At high, stoichiometric concentrations of Taxol (low micromolar), when this effect goes to completion, microtubule loops appear in the C-domain of the growth cone because of the large size of the soluble tubulin pool. One of the consequences of this artificial hyper-polymerisation of the tubulin in growth cones is to block neurite elongation, further supporting a role for microtubules in neurite growth (Peterson & Crain, 1982; Letourneau & Ressler, 1984; Mansfield & Gordon-Weeks, 1991). At sub-stoichiometric concentrations of Taxol (nanomolar), microtubule dynamic instability is suppressed without a change in polymer mass; in other words the microtubule population is effectively 'frozen' (reviewed in Wilson & Jordan, 1994). These low concentrations of Taxol have dramatic effects on the distribution of microtubules in growth cones, essentially restricting them to the C-domain and increasing their bundling (Williamson *et al.*, 1996; Challacombe *et al.*, 1997).

Biochemical analysis of growth cones isolated as a subcellular fraction from developing mammalian brain (reviewed in Lockerbie, 1990) also supports the existence of a large soluble tubulin pool (Gordon-Weeks & Lang, 1988; Gordon-Weeks *et al.*, 1989a). Growth cones are isolated under conditions (4°C) in which the microtubules within them are depolymerised. Paradoxically, despite the large pool of soluble tubulin, microtubules are not present in isolated growth cones, even after incubation at 37°C in physiological buffers (Gordon-Weeks, 1987a,b). However, after treatment with Taxol, microtubules are formed *de novo* in isolated growth cones, thereby demonstrating that the soluble tubulin pool is assembly competent and supporting the suggestion (see above) that there are no MTOC in growth cones (Gordon-Weeks, 1987a,b). Presumably, microtubules do not form in isolated growth cones because the tubulin concentration is below the critical concentration for *de novo* microtubule assembly.

How the soluble pool of tubulin in the growth cone is maintained is not yet clear. It could be sustained by delivery of tubulin in a soluble form down the axon or by depolymerisation of microtubules entering the growth cone. The factors controlling the size of the soluble tubulin pool in growth cones are also unknown, although these might include the number of microtubule 'plus' ends, the rate of supply of tubulin to the pool and post-translational modifications of tubulin. Most of the α-tubulin in growth cones is C-terminal tyrosinated, as indicated by biochemical experiments with isolated growth cones (Gordon-Weeks & Lang, 1988; Gordon-Weeks *et al.*, 1989a), and immunofluorescence studies of cultured neurons using antibodies specific for either tyrosinated or de-tyrosinated α-tubulin (Lim *et al.*, 1989; Robson & Burgoyne, 1989a; Arregui *et al.*, 1991; Mansfield & Gordon-Weeks, 1991). Tyrosinated α-tubulin is no less able to polymerise than de-tyrosinated α-tubulin. However, if the C-terminal tyrosine is phosphorylated, then assembly is markedly impaired (Wandosell, Serrano & Avila, 1994). Even if assembly occurs, the binding of MAPs may be altered by microtubule phosphorylation and this in turn may lead to less stable microtubules (see below). Experiments with isolated growth cones have shown that tubulin can be phosphorylated on tyrosine residues, but the location of the tyrosine within the molecule is not known (Cheng & Sahyoun, 1988; Lockerbie, Edde & Prochiantz, 1989). The proto-oncogene tyrosine kinase pp60^{c-src} is present in an active form in growth cones (Maness *et al.*, 1988) and phosphorylates tubulin in them, although probably not at the C-terminal tyrosine (Matten *et al.*, 1990). Tubulin phosphorylation in growth cones can apparently be influenced by cell adhesion molecules (Atashi *et al.*, 1992; see Chapter 4).

A model for the organisation of microtubules in neurites and growth cones

Although some of the observations on microtubule dynamics in neurites and growth cones remain contradictory, models have been proposed that combine most of the data and which predict testable outcomes (Joshi & Baas, 1993). In differentiating neurons, microtubules are nucleated at the centrosome in the perikaryon. The centrosome is often conveniently positioned subjacent to the site of origin of the future growth cone of the first neurite to emerge from the cell body. For instance, in the pioneer sensory neurons of grasshopper limb buds, where the first neurite becomes the axon (Lefcourt & Bentley, 1989) and in PC12 cells (Stevens, Trogadis & Jacobs, 1988). Microtubules, capped at their 'minus' ends, can detach from the centrosome and enter the neurite, with their 'plus' ends distal, where they are transported to the growth cone. The mechanisms underlying these microtubule movements from the centrosome and into the neurite are unknown, although likely candidates include 'minus' end-directed motors attached to stationary elements of the axon. In growing axons, microtubules enter the neurite with their 'plus' ends oriented distally; this may not be true for all microtubules in growing dendrites (Sharp, Yu & Baas, 1995). As they migrate down the neurite, microtubules are able to exchange tubulin subunits at their 'plus' (distal) ends with a soluble pool of tubulin dimers. This exchange is detectable with fluorescence photobleaching techniques but not after radiolabelling. Microtubules become bundled and incorporated into a stationary phase of the neurite cytoskeleton near or within the growth cone where they may also be involved in pathfinding (see below).

The clearest images of the behaviour of microtubules in growth cones have come from experiments in which fluorescently labelled microtubules are viewed in living growth cones in culture using video cameras that are sensitive to low light (e.g. Sabry *et al.*, 1991; Tanaka & Kirschner, 1991; Tanaka, Ho & Kirschner, 1995). Several kinds of behaviour have been seen using this method: microtubules extend forward individually across the P-domain; microtubules become bundled in the C-domain and subsequently the plasma membrane collapses around the bundle to form new neurite; individual microtubules buckle and form loops, or splay out across the C-domain, and bundles of microtubules extend forward into lamellipodia. It is not possible to say whether any or all of these behaviours are due to microtubule assembly and/or movement of existing microtubules because of the difficulty of seeing individual microtubules in their entirety, particularly the proximal ends of microtubules, since they are always embedded in the neurite shaft along with a large number of microtubules.

A role for microtubules in growth cone pathfinding

In addition to their well-established role in neurite elongation, microtubules may be involved in growth cone pathfinding, specifically in growth cone turning (reviewed in Gordon-Weeks, 1993; Bentley & O'Connor, 1994; Lin *et al.*, 1994a; Bush *et al.*, 1996c). In this model it is proposed that those filopodia that interact with extrinsic guidance cues become stabilised and do not retract. Stabilised filopodia are then assumed to be able to 'capture' microtubules undergoing dynamic instability and randomly extending into the P-domain. Captured microtubules are in turn stabilised against depolymerisation and thereby provide routes within the growth cone for the differential delivery of organelles by fast axonal transport and thus vectorial growth (Gordon-Weeks, 1991; Sabry *et al.*, 1991). Such growth and microtubule stability may underlie growth cone turning (Fig. 2.3).

Although there is no direct evidence that microtubules in growth cones are dynamically unstable, in a study of *Aplysia* neurons observed in culture with

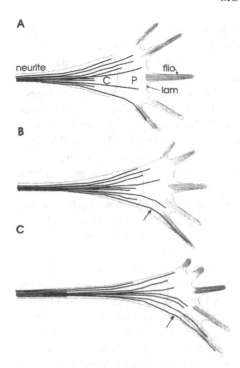

Figure 2.3 Diagrams showing the hypothetical reorganisation of microtubules (thick lines) and microfilaments (thin lines) in a growth cone undergoing a turning manoeuvre. Time elapses from A to C. It is imagined that the lower filopodia (filo) and lamellipodia (lam) encounter an extrinsic guidance signal that stabilises these structures. As a consequence, microtubules that are dynamically unstable and randomly probing the peripheral domain are 'captured' by filopodial actin filaments and stabilised (arrows in B and C). Dashed lines indicate the border between the central (C) and peripheral (P) domains. (From Bush *et al.*, 1996a, with permission of the Biochemical Society and Portland Press Ltd.)

video-enhanced microscopy, microtubules showing elongation at their distal ends were observed in growth cones (Forscher & Smith, 1988). However, the authors were not able to rule out forward sliding of the microtubules because the proximal ends of the microtubules were not visible. Similarly, Tanaka & Kirschner (1991) and Tanaka, Ho & Kirschner (1995) saw growth and shrinkage of fluorescently labelled microtubules in *Xenopus* neuronal growth cones. Also, microtubules advance further into the P-domain when microfilaments are disassembled with cytochalasins, suggesting forward sliding or distal polymerisation (Forscher & Smith, 1988; Lin & Forscher, 1993). The proposal that growth cone microtubules are undergoing dynamic instability is also consistent with the observation that these microtubules are turning over more rapidly than in other regions of the growing neuron (Lim *et al.*, 1989), and that these microtubules are largely composed of tyrosinated α-tubulin, which correlates with microtubule instability (see above).

Some of the other features of this model have also been observed. For example, there is some evidence that the microtubule cytoskeleton in growth cones is in some way coupled to microfilaments. For instance, the formation of microtubule loops following Taxol treatment is associated with the collapse of the P-domain of the growth cone (Letourneau & Ressler, 1984; Gordon-Weeks et al., 1989a; Mansfield & Gordon-Weeks, 1991). Also, microtubules in growth cones of neurons in culture have occasionally been seen to insert into filopodia (Gordon-Weeks, 1991). A protein has been identified in growth cones that may mechanically link microtubules to actin filaments (Goslin et al., 1989).

This model for vectorial growth predicts a correlation between microtubule capture by filopodia and the direction of growth (Fig. 2.3). This prediction has been tested in the Ti1 pioneer growth cones of grasshopper limb buds growing in explant culture (Sabry et al., 1991). These neurons were microinjected with fluorescent tubulin, which became incorporated into microtubules in the growth cone and could then be visualised with video cameras that are sensitive to low light. Although microtubule invasion into filopodia was not observed, microtubules were seen to invade selectively or be retained selectively in branches that developed from filopodia and became, eventually, new neurite. These authors suggested that differential organisation of microtubules across the growth cone is an important component of vectorial growth. A similar conclusion was also reached by Lin & Forscher (1993) studying Aplysia bag-cell growth cones interacting with each other in culture and stained by tubulin antibodies. They found that, when filopodia from one growth cone contacted another growth cone, F-actin accumulated at the contact site and subsequently microtubules selectively extended into the contact region in both growth cones and became aligned along the interaction axis. This microtubule reorganisation was associated with an advance of the C-domain along the direction of orientation of the microtubules. This sequence of events is consistent with the idea that stabilised filopodial actin filaments can recruit microtubules to support neurite growth.

In elegant experiments with Xenopus spinal cord neurons in which the microtubules had been labelled by microinjecting fluorescent tubulin into blastula stage embryos, Tanaka and colleagues visualised microtubule behaviour in growth cones turning at sharp substrate borders in vitro (Tanaka et al., 1995). They found clear-cut evidence that microtubules become re-oriented in the direction of the turn during growth cone turning. To test for a requirement of microtubule reorganisation in the growth cone for growth cone turning, Williamson et al. (1996) studied the effects of substoichiometric concentrations of Taxol on chick dorsal root ganglion cell growth cones making right-angled turns at sharp substrate borders between permissive and non-permissive substrates in vitro (Taylor et al., 1993). Substoichiometric concentrations of Taxol restrict the extension of dynamic microtubules to the C-domain in these growth cones and increase their bundling (see also Rochlin et al., 1996). This is associated with a failure of the growth cones to make a right-angled turn at the border. Instead, they remain at the border unable to move until the Taxol is removed. This is not due to a blockade of axon extension, although extension is reduced (see above). A similar result was found by Challacombe *et al.* (1997), also with chick dorsal root ganglion neurons but using a different non-permissive substrate. These experiments support the model for growth cone turning outlined in Fig. 2.3. A major question concerning the nature of the molecules that mediate the interaction of microtubules and actin filaments in growth cones remains unanswered.

Microtubule-associated proteins
Associated with microtubules are a structurally and functionally diverse group of proteins known as microtubule-associated proteins (MAPs) that profoundly modify the properties of microtubules (Olmsted, 1986; Wiche *et al.*, 1991; Lee, 1993; Hirokawa, 1994). Two broad categories of MAPs are recognised: structural MAPs and microtubule-motor MAPs. In the former category are tau, MAP4, MAP2, MAP1A, and MAP1B; also known as MAP1.2 (Asai *et al.*, 1985; Aletta *et al.*, 1988), MAP5 (Riederer *et al.*, 1986; Garner *et al.*, 1989, 1990) and MAP1x (Calvert & Anderton, 1985; Garner *et al.*, 1989). The microtubule-motor MAPs include kinesin and the cytoplasmic dyneins (Kreis & Vale, 1993). Mature neurons are particularly rich in MAPs, including the high-molecular-weight MAPs, MAP1A and MAP2, and the low-molecular-weight MAP tau (Matus, 1988). There are differences in the complement of MAPs found in developing neurons compared to mature neurons, suggesting that they play a role in neuronal development.

Many of the structural MAPs have the ability to promote microtubule assembly *in vitro* and have been shown to affect microtubule dynamic instability both *in vivo* and *in vitro* (Hirokawa, 1994). Microtubules are generally less dynamic in cells than they are in cell-free solution *in vitro* and because there are important differences in the behaviour of microtubules in different cell types (Pepperkok *et al.*, 1990; Shelden & Wadsworth, 1993) and in neurons during their development (reviewed in: Tucker, 1990; Hirokawa, 1991; Avila, Domínguez & Díaz-Nido, 1994; Bush, Eagles & Gordon-Weeks, 1996c), it seems likely that such differences are due to differences in the cellular complement of MAPs. Although there is no direct evidence, as well as stabilising microtubules to depolymerisation, structural MAPs are thought to cross-link microtubules and thus form microtubule bundles, such as those seen in neurites, and to cross-link microtubules to other cytoskeletal filaments (Olmsted, 1986; Matus, 1988). However, the claim that the structural MAPs produce microtubule bundling directly has been challenged (Lee & Brandt, 1992). Bundling of microtubules implies a cross-linking function of MAPs and this has not been demonstrated directly. Although transfection of some MAPs into cells can induce microtubule bundling, it is not clear if this is due to increased microtubule stability – which allows unidentified endogenous factors to cause bundling – or to a direct effect of the MAP. Furthermore, domains within MAPs that are associated with microtubule bundling but are distinct from binding domains have not been detected. However, transfection of insect cells with baculovirus expressing tau or MAP2 produces microtubule bundles with different inter-filament spacing characteristic of the MAP (Chen *et al.*,

1992). Also, filaments linking adjacent microtubules have been seen in quick-frozen, deep-etched electron micrographs of pelleted microtubules to which MAPs have been added (Hirokawa, 1991).

The structural MAPs appear at a very early stage in neural development, as soon as neurons become postmitotic (reviewed in: Nunez, 1986; Matus, 1988; Tucker, 1990; Bush, Eagles & Gordon-Weeks, 1996c). The first structural MAP to be expressed in developing neurons is MAP1B, which appears in axons at the time that they emerge from the cell body (Tucker, Binder & Matus, 1988; Tucker et al., 1988).

It has often been suggested that MAP1B may be important for neurite outgrowth (Matus, 1988; Tucker, 1990; Díaz-Nido et al., 1991; Gordon-Weeks & Mansfield, 1992; Gordon-Weeks, 1993; Avila et al., 1994). Indirect evidence for this is the correlation between neurite outgrowth and MAP1B expression. For instance, the molecule, particularly phosphorylated forms and its mRNA are rapidly up-regulated during neurite outgrowth in neuroblastoma cells (Gard & Kirschner, 1985; Díaz-Nido et al., 1988, 1991), PC12 cells (Greene et al., 1983; Drubin et al., 1985; Aletta et al., 1988; Brugg & Matus, 1988; Zauner et al., 1992) and P19 embryonal carcinoma cells (Tanaka et al., 1992). Furthermore, MAP1B is expressed in growing CNS axons in vivo (Riederer et al., 1986; Tucker et al., 1988; Schoenfeld et al., 1989; Díaz-Nido et al., 1990; Viereck & Matus, 1990) where it is strongly developmentally regulated, particularly phosphorylated forms, which decline rapidly during synaptogenesis in the CNS (Fischer & Romano-Clarke, 1990; Riederer et al., 1990; Viereck & Matus, 1990; Gordon-Weeks et al., 1993), except in the olfactory system, where the formation of axons continues throughout life (Viereck et al., 1989; Gordon-Weeks et al., 1993), and other restricted regions of the nervous system (Fawcett et al., 1995; Bush et al., 1996b; Nothias et al., 1996). There is also a decline in the levels of mRNA for MAP1B during brain development (Safaei & Fischer, 1989; Garner et al., 1990; Oblinger & Kost, 1994). More direct evidence for a role for MAP1B in axonogenesis comes from experiments in which MAP1B expression is suppressed in cells in culture. In PC12 cells (Brugg, Reddy & Matus, 1993) and in cerebellar macroneurons (DiTella et al., 1996), antisense oligodeoxynucleotides to MAP1B reduce or abolish MAP1B expression, and concurrently inhibit neurite outgrowth, effects which are reversed on antisense removal.

Two, independent, mouse mutants have been made in which the MAP1B gene has been deleted by homologous recombination (Edelmann et al., 1996; Takei et al., 1997). The phenotypes of these two MAP1B 'knockout' mice are very different. The 'knockout' produced by Edelmann et al. (1996) shows widespread neurological abnormalities in the heterozygotes, including limb spasticity, malformed cerebella, including Purkinje cells with stunted dendrites, and blindness; the homozygotes die in utero and have not been analysed. In contrast, in the Takei et al. (1997) 'knockout', both heterozygotes and homozygotes survive and are relatively normal. The homozygotes have low birth brain weights and show temporal delays in the development of their nervous systems but recover by adulthood. However, these animals may not be com-

Table 2.1. *Microtubule-associated proteins in growth cones*

Protein	Function	Reference
Ezrin	Actin filament/microtubule cross-linker	Goslin *et al.*, 1989; DiTella *et al.*, 1994
MAP1B	Microtubule stability	Mansfield *et al.*, 1991; Black *et al.*, 1994; Boyne *et al.*, 1995; Rocha & Avila, 1995; Ulloa *et al.*, 1997
MAP2	Microtubule bundling Microtubule stability	Cáceres *et al.*, 1984; Dotti, Banker & Binder, 1987; Gordon-Weeks *et al.*, 1989
SCG10	Microtubule destabiliser	DiPaolo *et al.*, 1997
Tau	Microtubule stability	Gordon-Weeks *et al.*, 1989a; DiTella *et al.*, 1994; Rocha & Avila, 1995

plete 'knockouts' since there is evidence that they express a truncated isoform of MAP1B, at low levels (Takei *et al.*, 1997). The tau 'knockout' mouse appears relatively normal (see below) and, since there is *in vitro* evidence that tau can functionally substitute for MAP1B (DiTella *et al.*, 1994), it will be interesting to see the phenotype of a tau and MAP1B double 'knockout'.

There have been several direct studies of the expression and distribution of MAPs in growth cones (see Table 2.1). Immunostaining suggests that MAP2, tau and MAP1B are present in growth cones of neurons in culture (Kosik & Finch, 1987; Gordon-Weeks *et al.*, 1989a; Mansfield *et al.*, 1991; Gordon-Weeks *et al.*, 1993; Black, Slaughter & Fischer, 1994; DiTella *et al.*, 1994; Black *et al.*, 1996; Kempf *et al.*, 1996; Mandell & Banker, 1996). Electron microscopy using immunogold labelled antibodies shows that MAP1B is bound to growth cone microtubules, including the individual microtubules that traverse the P-domain and insert into the proximal part of filopodia (Bush *et al.*, 1996a). In neurons in culture (Mansfield *et al.*, 1991; Fischer & Romano-Clarke, 1991; Black *et al.* 1994; Bush *et al.*, 1996a) and *in vivo* (Bush & Gordon-Weeks, 1994), some phosphorylated isoforms of MAP1B are present in axons in a concentration gradient that is highest at the growth cone, suggesting that these isoforms of MAP1B may have a role in neurite growth (Gordon-Weeks, 1993, 1997). Tau also exhibits a phosphorylation gradient in growing axons but, in contrast to MAP1B, the gradient is highest at the cell body (Mandell & Banker, 1996). Biochemical studies have confirmed the presence of phosphorylated MAP1B in growth cones and shown that growth cones contain a kinase(s) capable of phosphorylating MAP1B (Mansfield *et al.*, 1991; Rocha & Avila, 1995). The function of MAP1B in axon growth or the role of phosphorylation is unknown.

Experiments in culture with antisense oligodeoxynucleotides against tau (Cáceres & Kosik, 1990; Cáceres, Potrebic & Kosik, 1991; Hanemaaijer & Ginzburg, 1991; Shea *et al.*, 1992; DiTella *et al.*, 1994) and MAP2 (Dinsmore & Solomon, 1991; Solomon, 1991; Cáceres, Mautino & Kosik, 1992; Sharma,

Kress & Shafit-Zagardo, 1994) suggest that these structural MAPs are also important for neurite outgrowth. Embryonic rat cerebellar macroneurons, when grown in dissociated culture, initially extend neurites that are not differentiated into either dendrites or axons (Ferreira & Cáceres, 1989). Eventually, one neurite elongates more rapidly than its siblings and becomes the axon, while the other neurites differentiate into dendrites. A similar sequence of events has been documented in dissociated hippocampal neurons in culture (Dotti et al., 1988; Goslin & Banker, 1989). When antisense oligodeoxynucleotides to tau are introduced into cerebellar macroneuron cultures, suppression of tau synthesis occurs and the neurons fail to differentiate an axon, although neurite formation appears unaffected (Cáceres & Kosik, 1990; Cáceres et al, 1991). PC12 cells, genetically engineered to overproduce tau, extend neurites more rapidly than normal cells while the reverse is true of cells under-expressing tau (Esmaeli-Azad, McCarty & Feinstein, 1994). When tau or MAP2 is expressed in insect Sf9 cells, which do not normally express these MAPs, the cells extend long, neurite-like processes that contain bundled microtubules (Baas, Pienkowski & Kosik, 1991; Knops et al., 1991; LeClerc et al., 1996). These results suggest that tau and MAP2 may play a role in neurite growth or in the differentiation of a neurite into an axon or a dendrite. Expression of tau or MAP2 in vertebrate non-neuronal cells promotes dramatic microtubule bundling but does not lead to process outgrowth (Kanai et al., 1989; Lee & Rook, 1992; Weisshaar, Doll & Matus, 1992). However, if the cortical actin cytoskeleton in MAP2 transfected cells is disrupted, microtubule containing processes are produced (Edson, Weisshaar & Matus, 1993). A role for tau in axon elongation has recently been questioned by observations made on transgenic mice lacking the tau gene (Harada et al., 1994). These mice have grossly normal nervous systems and show only subtle abnormalities in the organisation of microtubules in their small calibre axons. Also, cultured neurons from these animals differentiate axons and dendrites normally. Tau may have a role in growth cone function unrelated to its role in axons (DiTella et al., 1994). Growth cones contain an actin binding protein related to ezrin that co-localises with actin filaments in growth cones (Goslin et al., 1989). When tau expression in growth cones of embryonic cerebellar neurons in culture is inhibited using tau antisense oligodeoxynucleotides, the ezrin-like protein is lost from the growth cones and accumulates in axons (DiTella et al., 1994). This suggests that the localization of the ezrin-like protein to actin filaments is dependent upon tau.

Microfilaments

Microfilaments are 5–7 nm in diameter and may be several micrometres in length. Microfilaments are composed of actin. Actin is a highly conserved, myosin-binding globular protein consisting of a single polypeptide chain of 374 or 375 amino acids with a molecular weight of about 42 kD. Actin exists in cells either as a monomer, called globular or G-actin, or as a filament – F-actin. Actin has two globular domains divided by a cleft, within which binds one molecule of ATP or ADP, and one molecule of Ca^{2+} or Mg^{2+}. One domain is slightly smaller than the other and contains both the N- and C-termini and the myosin binding site. Actin is an adenosine triphosphatase.

Higher eukaryotes have several isoforms of actin encoded by a family of genes. In warm-blooded vertebrates there are six isoforms. Three actin classes have been defined differing in their isoelectric point: α (the most acidic), β and γ. All embryonic cells and cells in the adult nervous system contain β and γ actin isoforms. These actin isoforms differ from each other only by four conservative amino acid substitutions at the amino terminus, a region involved in interactions with actin-binding proteins. There do not seem to be mechanisms to differentiate between the two isoforms within neurons (Bamburg & Bernstein, 1991), although in non-neuronal cells the different isoforms are differentially distributed (Gimona et al., 1994).

The main function of actin is to form microfilaments in cells. Microfilaments have a role in a wide range of cellular functions from exocytosis to mitosis (Bray, 1992). A molecule of G-actin can bind non-covalently to two other monomers and form a helical polymer or filament. Two filaments associate in parallel to form a helical coiled microfilament. Like tubulin, actin is inherently polarised and, because of a head-to-tail type of assembly, this imposes a polarity on microfilaments. Actin filament polarity is most readily observed by decorating filaments with the proteolytic fragments of myosin containing the globular head domain, either heavy meromyosin or myosin subfragment-1. In the electron microscope this produces an arrowhead pattern along the filament such that one end is barbed (the 'plus' end) and the other is pointed (the 'minus' end). G-Actin addition takes place preferentially at the barbed end, whereas at the pointed end, actin monomer disassembly predominates. Heavy meromyosin or myosin subfragment-1 decoration of actin filaments has been used as a means to determine filament orientation in cells. Normally, actin assembly in vitro requires the presence of monomer-associated ATP, which is hydrolysed to ADP at some point after monomer addition to the 'plus' end of an existing filament (Carlier, 1989). At steady state, the net assembly of G-actin at the 'plus' end and the net disassembly of G-actin at the 'minus' end creates a treadmilling effect in which individual subunits pass through the filament from 'plus' to 'minus' ends. ADP-actin can still polymerise but at a very much slower rate than ATP-actin, and treadmilling does not occur in filaments composed of only ADP-actin. What function this in vitro effect may have in the cell is unknown, for instance, how this might affect interactions with actin-binding proteins (see below) such as myosin. Actin filaments have not been seen to undergo dynamic instability, unlike microtubules; however, microtubules sometimes exhibit treadmilling.

Microfilament organisation in neurites and growth cones

In neurites, microfilaments are concentrated in the cortical (peripheral) cytoskeleton (Metuzals & Tasaki, 1978; Tsukita, Kobayashi & Matsumoto, 1986).

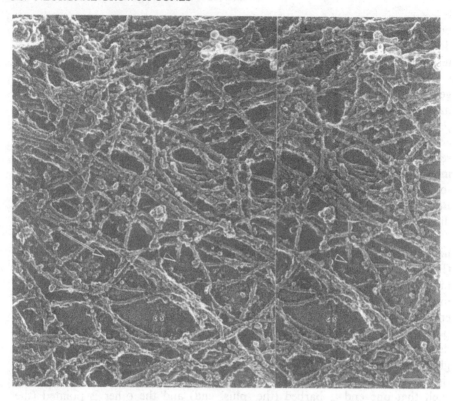

Figure 2.4 A freeze-etch stereo view of a superior cervical ganglion cell growth cone in culture after detergent extraction. Two actin filament populations can be seen: one is composed of bundles of filaments radiating from the leading edge of the growth cone (large arrow), while the second intersects the first in an orthogonal orientation. The relationship between actin filaments and the cytoplasmic surface (cs) of the ventral plasma membrane of the growth cone can be seen. As filaments approach the membrane surface, they either bend and follow it or make contacts with their ends on filaments that appear to lay upon the membrane surface (arrowheads). Bar = 0.1 μm. (From Lewis & Bridgman, 1992, by copyright permission of The Rockefeller University Press.)

In growth cones, the cytoskeleton of the P-domain is dominated by microfilaments, in contrast to the C-domain, where microtubules predominate (Fig. 2.1; Yamada *et al.*, 1971; Bunge, 1973; Isenberg *et al.*, 1977; Kuczmarski & Rosenbaum, 1979; Rees & Reese, 1981; Letourneau, 1983; Tosney & Wessells, 1983; Tsui, Ris & Klein, 1983; Gordon-Weeks, 1987a, 1991; Bridgman & Dailey, 1989; Lewis & Bridgman, 1992). Electron microscopy has shown that microfilaments in lamellipodia occur singly or in bundles forming a dense meshwork (Fig. 2.4). Single microfilaments are relatively short (< 1 μm), may branch and are randomly oriented with respect to the plane of the plasma membrane, as is the case in the cortical (peripheral) region of most cells (Bray, 1992). Microfilament bundles in lamellipodia, however, tend to radiate from the leading edge of the growth cone and are predominantly ventrally located and tend to be found only in expanding lamellipodia (Lewis &

Bridgman, 1992). Each bundle contains 6–12 parallel filaments, which are generally longer (3–6 μm) than the singly occurring filaments. Bundles can extend the length of the lamellipodium, terminating near the junction between the P- and C-domain, but exactly where is not known.

In filopodia, the microfilaments are arranged in parallel bundles reminiscent of the core-bundle of actin filaments in the microvilli of epithelial cells (Mooseker & Tilney, 1975; Mooseker, 1985). The number of filaments in these bundles (> 15) tends to be greater than that in lamellipodial bundles and the bundles are cylindrical, whereas those in the lamellipodia tend to be planar (Lewis & Bridgman, 1992). Filopodial actin bundles may extend rearwards across the P-domain and enter the C-domain, although the location of their proximal ends is not known (Rees & Reese, 1981; Letourneau, 1983; Gordon-Weeks, 1987a; Lewis & Bridgman, 1992). These bundles are seen particularly well in video-enhanced images of growth cones in culture (Figs 1.5 and 2.5) (Goldberg & Burmeister, 1986; Aletta & Greene, 1988; Forscher & Smith, 1988).

Bundling of actin filaments requires cross-linking proteins and indeed filaments cross-linking actin bundles have been seen in electron micrographs of microvilli (Mooseker & Tilney, 1975; Mooseker, 1985). Similar cross-linking filaments of about 7 nm in diameter have also been seen in growth cone filopodia (Rees & Reese, 1981; Gordon-Weeks, 1987a). These have a regular periodicity along the longitudinal axis of the filopodium of 19 nm (Rees & Reese, 1981) or 30 nm (Gordon-Weeks, 1987a). The latter figure corresponds closely to the periodicity of cross-linking proteins of microvilli actin bundles (Mooseker & Tilney, 1975). In detergent-extracted growth cones, a bar of proteinaceous material is seen in the electron microscope at the distal ends of the filopodial actin bundles, presumably the tips of filopodia (Lewis & Bridgman, 1992). This material may be related to proteins required to anchor actin filaments to the membrane. Although the proteins involved have not been identified, actin-binding proteins that also interact with membranes have been found in growth cones (e.g. fodrin and vinculin; see Table 2.2 and below). Such proteins may control actin assembly, in analogy with MTOCs, since the filopodial actin filaments are all oriented with their 'plus' ends distal (see below).

The immunofluorescence staining pattern seen in growth cones with fluorescently labelled mycotoxins, which bind specifically to F-actin, such as phalloidin or phallacidin, is similar to that of the distribution of microfilaments, consistent with the idea that the microfilaments are composed of actin (Fig. 2.1; Letourneau, 1981, 1983; Forscher & Smith, 1988; Koenig et al., 1985; Bridgman & Dailey, 1989). Filopodia are particularly strongly labelled with these probes. The situation with antibodies to actin is complicated by the fact that these do not distinguish between G- and F-actin. In general, the immunofluorescence staining pattern seen with antibodies to actin is diffusely spread throughout the growth cone (Jockusch, Jockusch & Burger, 1979; Sotelo et al., 1979; Letourneau, 1981; Shaw, Osborn & Weber, 1981; Spooner & Holladay, 1981; Koenig et al., 1985). However, filopodia tend to stain more brightly than other regions of the growth cone and, on antibody dilution, the P-domain, and particularly the filopodia, are the last structures to remain stained (Letourneau,

Table 2.2. *Actin-binding proteins in growth cones*

Protein	Function	Present in filopodia	Reference
α-Actinin	Bundling	+	Jockusch & Jockusch, 1981; Shaw, Osborn & Weber, 1981; Letourneau & Shattuck, 1989; Sobue & Kanda, 1989
ADF/cofilin	Depolymerising G-actin binding	+	Bamburg & Bernstein, 1991
Caldesmon	Side-binding	+	Kira, Tanaka & Sobue, 1995
Ezrin, radixin, moesin (ERM proteins)	Tau binding	+	Goslin et al., 1989; DiTella et al., 1994; Gonzalez-Agosti & Solomon, 1996
Fascin	Bundling	+	Edwards & Bryan, 1995; Sasaki et al., 1996
Filamin	Cross-linking	+	Letourneau & Shattuck, 1989
Fimbrin	Bundling	+	Shaw, et al., 1981
Fodrin (calspectrin)	Membrane linkage	–	Koenig et al., 1985, Sobue & Kanda, 1989
GAP-43	Filament length regulation	+	Moss et al., 1990; Strittmatter et al., 1992; Hens et al., 1993; He et al., 1997
Gelsolin	Severing/barbed end capping	+/–	Tanaka, Kira & Sobue, 1993
	Monomer sequestering		
MAP1B	Microtubule cross-linking	+	Mansfield et al., 1991
MAP2	Microtubule cross-linking	+	Gordon-Weeks et al., 1989a; Cáceres et al., 1984

Myosin I	Membrane linkage	+	Roisen et al., 1978; Barylko et al., 1992; Wagner et al., 1992; Lewis & Bridgman, 1996; Wang et al., 1996a
Myosin II	Filament translocation Tension development	+/−	Roisen et al., 1978; Kuczmarski & Rosenbaum, 1979; Letourneau, 1981; Shaw et al., 1981; Bridgman & Dailey, 1989; Letourneau & Shattuck, 1989; Wagner et al., 1992; Cheng, Murakami & Elizinga, 1993
Myosin V	Vesicle transport, filopodial extension	+	Wang et al., 1996a; Evans et al., 1997
Profilin	Depolymerising/polymerising monomer sequestering	−	Faivre-Sarrailh et al., 1993
Tau	Microtubule binding	+	Gordon-Weeks et al., 1989a; DiTella et al., 1994
Tropomyosin	Side-binding	+	Shaw et al., 1981; Letourneau & Shattuck, 1989
Vinculin	Membrane linkage	+	Halegoua, 1987; Letourneau & Shattuck, 1989; Igarashi & Komiya, 1991; Varnum-Finney & Reichardt, 1994

Abbreviations: ADF, actin depolymerising factor; MAP, microtubule-associated protein; GAP-43, growth-associated protein. +, present; −, absent; +/− variable.

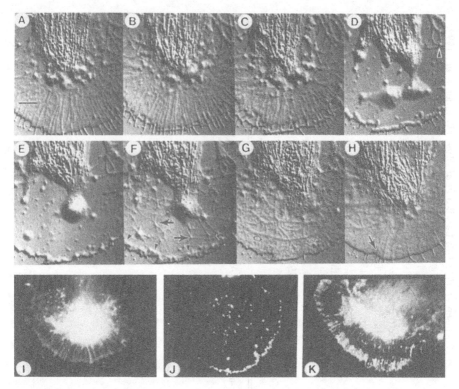

Figure 2.5 The effects of the actin-depolymerising drug cytochalasin on actin filaments in the growth cones of bag cell neurons of the marine mollusc *Aplysia*: (A) Video image of a living growth cone before cytochalasin addition. Notice the filopodia and the rods of actin filaments in the P-domain (lower half of figure). (B/C) Thirty seconds and 1 minute, respectively, after the addition of cytochalasin. A wave of phase contrast is separating from the distal margin of the growth cone and moving retrogradely (arrowheads). (D/E) Three minutes and 9 minutes, respectively, after cytochalasin addition. Actin filament rods in filopodia have disappeared concurrently with filopodial extension. Arrowhead indicates microtubule bundles. (F) Sixty seconds after cytochalasin removal. Actin filaments are re-forming (arrows) and a retrograde wave commences (arrowheads). (G) Three minutes after cytochalasin removal. Retrograde waves have now spread across the P-domain (arrows) and filopodia are emerging. (H) Seventeen minutes after cytochalasin removal. Recovery is almost complete and filopodia with associated actin filament rods are visible (arrow).

Rhodamine–phalloidin staining of growth cones showing the distribution of F-actin: (I) Early phase of cytochalasin action (80 seconds, corresponds to C above) showing that the distal ends of the actin filaments in filopodia have begun to depolymerise. (J) Late phase of cytochalasin action (30 minutes, corresponding to E above) showing complete loss of actin-organised filaments. (K) Early phase of recovery from cytochalasin effects (65 seconds after removal, corresponding to F above) showing re-assembly of the actin filaments within filopodia. (From Forscher & Smith, 1988, by copyright permission of The Rockefeller University Press.)

1983), suggesting higher concentrations of actin in these regions. Quantitative analysis of the concentration of F-actin in chick dorsal root ganglion growth cones in culture also shows that the highest concentrations are found in the P-domain and at the membrane skeleton (Fan *et al.*, 1993).

Direct evidence for the presence of actin in microfilaments comes from ultrastructural studies with heavy meromyosin or myosin subfragment-1, which bind to microfilaments in growth cones, thus confirming their actin composition (Chang & Goldman, 1973; Isenberg & Small, 1978; Letourneau, 1983; Markham & Fifková, 1986; Gordon-Weeks, 1987a; Lewis & Bridgman, 1992). These experiments also show that the microfilaments in filopodia are all oriented with their barbed or 'plus' ends, i.e. the end at which G-actin is preferentially added, distally, toward the tip of the filopodia (Chang & Goldman, 1973; Gordon-Weeks, 1987a; Lewis & Bridgman, 1992). The organisation of actin filaments in lamellipodia is more complicated (Lewis & Bridgman, 1992). These authors identified two populations of actin filaments in lamellipodia of embryonic rat sympathetic neurons in culture: bundles of long filaments that were ventrally located and oriented perpendicular to the leading edge of advancing lamellipodia with their barbed ends distal, and shorter, randomly oriented filaments that filled the lamellipodial space between dorsal and ventral membranes and whose barbed ends were randomly directed.

Growth cone motility and actin filament dynamics

Further confirmation of the actin composition of microfilaments in the growth cone and direct evidence for their role in growth cone motility derives from experiments with the cytochalasins (Yamada *et al.*, 1970, 1971; Wessells *et al.*, 1971; Marsh & Letourneau, 1984; Letourneau *et al.*, 1986; Cooper, 1987; Forscher & Smith, 1988; reviewed in: Smith, 1988; Sobue, 1993). These agents are mycotoxins that inhibit actin filament growth by capping their 'plus' ends (Cooper, 1987) and, in some circumstances, may depolymerise actin filaments. Cytochalasins disrupt microfilaments in the growth cone as seen in the electron microscope (Yamada *et al.*, 1971; Gordon-Weeks, 1987a) and with video microscopy (Forscher & Smith, 1988), and simultaneously cause a reversible loss of filopodial and lamellipodial extension (Yamada *et al.*, 1971; Forscher & Smith, 1988). Yamada *et al.* (1971) found that cytochalasins blocked neurite extension in neurons in culture, as well as causing growth cone collapse. However, this has not been confirmed by later studies, a discrepancy probably explained by differences in the adhesivity of the culture substrata. Neurite extension is not blocked by cytochalasins when neurons are grown on highly adhesive substrata such as poly-ornithine (Marsh & Letourneau, 1984; Letourneau *et al.*, 1987), or in explant cultures (Bentley & Toroian-Raymond, 1986; Chien *et al.*, 1993). Variations in the effect of cytochalasins on axon growth have been seen with different substrata (Abosch & Lagenaur, 1993).

In their detailed study of the effects of cytochalasins on growth cones, Forscher and Smith (1988) took advantage of the large (20–70 μm in width) growth cones of the bag cell neurons of the marine mollusc *Aplysia* (Figs 1.5 and 2.5). These neurons were grown on a highly adhesive substratum of poly-D-lysine, which prevented growth cone collapse consequent upon cytochalasin addition and so enabled clearer views of actin filament dynamics. Growth cones were viewed using high resolution video microscopy coupled with corre-lative immunofluorescence. In untreated growth cones, in addition to filopodial and lamellipodial extension and retraction, they noticed waves of phase con-trast that originated at the distal margins of the lamellipodia and moved toward the C-domain, i.e. retrogradely, at rates of 3–6 μm/min. Similar waves or 'ruffles' have been seen in a variety of motile cells in culture (reviewed in Bray & White, 1988). Retrograde waves are probably due to the co-ordi-nated movement of F-actin filaments from their site of assembly (see below) in the distal margin of the growth cone to the C-domain in the upper (dorsal) cortical cytoskeleton of the growth cone. Such an interpretation is supported by the effects of cytochalasins. In addition to reversibly blocking lamellipodial and filopodial extension, as anticipated, cytochalasins also reversibly inhibited the formation of retrograde waves but, most importantly, did not inhibit the movement of retrograde waves that had formed before drug application. On cytochalasin addition, a clear zone appeared at the distal margin of lamellipo-dia and, at the same time, the distal ends of the filopodial actin bundles retracted (Fig. 2.5). The clear zone, which was interpreted to be due to actin filament disassembly or retraction, an interpretation supported by the lack of phalloidin staining in the zone, progressed rearward until the entire P-domain became essentially featureless, apart from small aggregates of actin filaments (Fig. 2.5). The rate of progression of the clear zone (3.5 μm/min) was similar to the rate of retrograde waves seen in untreated growth cones, consistent with the phenomenon being due to actin filament movement. Such dorsal, retrograde waves in the cortical (membrane) skeleton probably explain the movement of exogenous particles that adhere to the dorsal surface of the growth cone (Bray, 1970; Koda & Partlow, 1976; Feldman *et al.*, 1981; Forscher, Lin & Thompson, 1992; Lin & Forscher, 1993) and the retrograde movement of neurites and glass fibres over the dorsal surface of growth cones (Heidemann *et al.*, 1990), particularly since their rate of translocation is similar. For instance, charged polystyrene beads moving retrogradely on filopodia and lamellipodia of *Aplysia* bag cell neurons have a translocation rate of about 3 μm/min (Lin & Forscher, 1993). Beads that attach to lamellipodia in which F-actin has been depolymerised with cytochalasin move randomly rather than retrogradely (centripetally) (Lin & Forscher, 1995). These observations suggest that F-actin, and not lipid as was once supposed, interacts with protein com-ponents in the dorsal plasma membrane of growth cones and moves them retrogradely (see also Sheetz *et al.*, 1989).

Forscher and Smith (1988) noticed that cytochalasin treatment, in addition to causing actin filament changes, resulted in an extension of the microtubules into more distal regions of the growth cone. This was associated with bi-directional transport of organelles along the microtubules. They attributed this microtubule extension to removal of a physical barrier to the microtubule ends which depended on the integrity of actin filaments. This explanation also accounts for observations made in hepatoma cells transfected with MAP2c (Weisshaar *et al.* 1992). These cells form circumferential bundles of microtubules when transfected. However, when the actin filaments in transfected cells are depolymerised with cytochalasin, the bundles of microtubules become re-organised and extend out from the cell surface in long processes (Edson *et al.*, 1993a). If such a barrier exists, however, it is very selective since fluorescently labelled dextran with a diameter of 38 nm, i.e. larger than the width of a microtubule (~ 24 nm), can penetrate into the tips of filopodia when microinjected into neurons in culture (Popov & Poo, 1992).

All of the effects of cytochalasin treatment were reversed on removal of the toxin. Filopodial and lamellipodial extension returned and concurrently there was a recovery of actin filaments, as judged both by video microscopy and phalloidin staining (Fig. 2.5). Actin filament assembly during recovery occurred predominantly at the distal margin of the growth cone. The first visible sign was a thickening of the distal edge of the lamellipodium, owing to the initiation of actin filament assembly at that location, which progressed rearward filling the entire lamellipodium (Fig. 2.5). This was followed by the recovery of filopodial actin filaments, which also appeared distally at first and then extended rearward.

These experiments firmly established the basic elements of actin filament dynamics in growth cones and provided strong evidence for the distal assembly of actin filaments and an associated rearward cortical (peripheral) flow of filaments, similar to that seen in other motile cells (Bray & White, 1988). What is the mechanism responsible for the retrograde flow of actin filaments? Since the retrograde flow is maintained in the early stages of cytochalasin action, even during the formation of a filament-free zone in the lamellipodia, it is unlikely to be due to a force generated by actin filament polymerisation. The most likely explanation is that myosins are involved and indeed there is evidence to support such a role for myosins in both the rearward flow of actin filaments and filopodial extension in growth cones. The role of these proteins and other actin-binding proteins is discussed in detail below.

What is the mechanism of lamellipodial and filopodial extension? Since they have similar kinetics of retraction and extension during normal growth cone advance (Sheetz, Wayne & Pearlman, 1992) and during cytochalasin poisoning and recovery, the same mechanism may underlie both events. The uniform organisation of filopodial actin filaments with their 'plus' ends oriented distally and the effects of cytochalasin discussed above suggest that, at least for filopodia, actin filament polymerisation occurs distally, i.e. at the tip of the filopodium (Fig. 2.6). If the kinetics of filopodial extension depend on the rate of diffusion of G-actin from the base of the extending filopodium to its tip then, since the tip is moving further away from the base during filopodial extension, the time for G-actin to diffuse to the tip should increase with time. Consequently, the rate of filopodial extension should decrease with time.

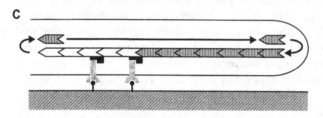

Figure 2.6 Schematic diagram of a growth cone filopodium showing how an actin filament might move retrogradely (A, B) and extend a filopodium (C). (A) All actin filaments in filopodia have their barbed ends, at which assembly occurs preferentially, distally. Monomeric actin (hatched chevrons) diffuses to the distal end of the growth cone filopodium where it assembles on to the barbed or 'plus' end of an actin filament. A 'plus' or barbed-end directed motor, such as a myosin molecule (ellipses with tails), attached either to the membrane cyctoskeleton on the dorsal surface of the filopodium, or directly to the membrane, drives the actin filament retrogradely, to the left of the diagram (B). This mechanism underlies the regrograde flow of actin filaments on the dorsal surface of the growth cone. Under these circumstances, there is no extension of the filopodium. (C) Extension of the filopodium will occur if the actin filament is immobilised relative to the substratum. This may occur by cross-linking actin filaments to membrane receptors (inverted 'Y's) in the ventral membrane of the filopodium that have engaged ligands (lollipops) anchored to the substratum. Here also a myosin molecule may be involved in actin filament translocation.

This has in fact been observed (Argiro *et al.*, 1985). The appearance of a retrograde wave of actin depolymerisation at the distal margin of lamellipodia in *Aplysia* bag-cell growth cones during cytochalasin treatment is also consistent with distal actin filament assembly (Forscher & Smith, 1988), although actin filament organisation is not uniform in lamellipodia as it is in filopodia (Lewis & Bridgman, 1992). In *Aplysia* growth cones, the rearward flow of actin filaments appears to be coupled to filopodial and lamellipodial extension (Lin & Forscher, 1995). Changes in the rate of each process are inversely related. Mechanistically, this can be explained by having a 'clutch' that reduces rearward actin filament flow by mechanically coupling filaments to the substratum, for instance, via membrane receptors (Fig. 2.6), and thereby allowing barbed end polymerisation or motor proteins to drive filopodial extension.

More direct evidence for distal assembly of actin filaments in filopodia and lamellipodia comes from experiments with biotinylated and fluorescently labelled actin (Okabe & Hirokawa, 1989, 1991; Sanders & Wang, 1991). When biotinylated actin was microinjected into the cell bodies of PC12 cells or mouse dorsal root ganglion neurons in culture, the labelled actin became incorporated into the actin filament cytoskeleton in these cells, including the growth cone cytoskeleton. At early time points after microinjection, biotinylated actin incorporated into the most distal regions of the actin cytoskeleton, including the filopodia and distal margins of lamellipodia in growth cones. At later time points, biotinylated actin became incorporated throughout the cytoskeleton. These results are consistent with actin polymerisation distally. Laser photobleaching experiments with fluorescently labelled actin microinjected into PC12 cells and dorsal root ganglion neurons provided further support for distal assembly and, in addition, showed that actin filaments are probably moving retrogradely (centripetally) after assembly (Okabe & Hirokawa, 1989, 1991). When a zone at the distal margin of a lamellipodium was photobleached, recovery of fluorescence occurred first at the most distal margin of the bleached zone. Full recovery of fluorescence occurred by a rearward advance of the fluorescence – presumably due to rearward movement of actin filaments away from the distal edge of the lamellipodium. This front of fluorescence moved retrogradely at an average rate of 1.6 μm/min. This rate is about half as fast as the retrograde movement of actin filaments seen in *Aplysia* bag-cell growth cones, but such differences may relate to differences in the temperature of the experiments (see above, Forscher & Smith, 1988). When filopodia were photobleached, a similar polarised pattern to the recovery of fluorescence was seen. Fluorescence first appeared at the distal end of the filopodium and advanced backward. Recovery of fluorescence in growth cones after photobleaching was about twice as rapid as that seen in neurites, consistent with a higher rate of actin filament turnover in growth cones (Okabe & Hirokawa, 1989, 1991).

Collectively, these experiments suggest that actin is either added to the distal ends of pre-existing filaments in lamellipodia and filopodia, or that *de novo* actin assembly occurs distally. Consistent with this notion is the finding that actin filaments accumulate in filopodia contacting guidance cues during growth cone pathfinding (O'Connor & Bentley, 1993, see Chapter 3). They also show that there is a continuous retrograde flow of actin filaments, formed at the distal margin, over the top (dorsal) surface of the growth cone (Fig. 2.6). For an individual growth cone, the rate of the rearward translocation of actin filaments is faster than the rate of growth cone advance (Okabe & Hirokawa, 1991), which may explain the fact that extension of filopodia and lamellipodia are usually followed by their retraction (Nakai & Kawasaki, 1959; Bray & Chapman, 1985). What happens to actin filaments once they reach the rear of the growth cone? Clearly they do not accumulate at this location, since the concentration of F-actin in the C-domain is low compared to the P-domain (see above). A similar sharp fall-off in actin filament density at the rear of lamellipodia has been seen in other motile cells, e.g. keratinocytes, but how this occurs is not known (reviewed in Theriot & Mitchison, 1992). One possibility is that there is a localised actin filament severing or depolymerising activity in the C-domain, giving rise to G-actin, which is then free to diffuse to the site of assembly, or to bind to actin-sequestering proteins, which transport actin to sites of assembly. Several actin-depolymerising factors have been detected in growth cones but their roles have yet to be established (see Table 2.2).

Do growth cones 'push' or 'pull'?

Do filopodia or lamellipodia pull the neurite forward by exerting tension on the growth cone or is advance produced by a push from behind? The idea that filopodia are contractile and can exert force on the rest of the growth cone, thereby pulling the neurite along, initially gained widespread acceptance (reviewed in: Johnston & Wessells, 1980; Bray, 1982; Letourneau, 1983b; Landis, 1983). The presence of a bundle of actin filaments within filopodia (see above) is consistent with this notion. The attraction of this idea was enhanced by experiments showing that growth cones *in vitro* can discriminate between different substrata, showing increasing preference for substrates of increasing adhesivity (Letourneau, 1975a,b). This led to the idea that growth cones are guided by differentially adhesive substrata and that individual filopodia compete with each other to pull the growth cone in a particular direction. Successful filopodia are those that attach to the most adhesive substratum and thus exert the highest tension on the growth cone. Pathfinding was therefore seen as the outcome of a competition between individual filopodia to pull the growth cone in a particular direction (Bray, 1982; Letourneau, 1983b; Bunge, Johnson & Argiro, 1983; Lockerbie, 1987).

The 'push–pull' controversy opened up a lively debate (Bray, 1987; Letourneau *et al.*, 1987; Goldberg & Burmeister, 1988; Turner & Flier, 1989). It has been clear for some time that the growth cone can exert tension on the neurite and that such tension can contribute to neurite elongation (Bray, 1979, 1984). By pulling growth cones of chick dorsal root ganglion cells in culture attached to a micropulling device, Bray (1984) showed that neurites can elongate by generating new neurite (rather than the neurite being elastically stretched) in response to tension applied at the growth cone. In this case, tension was not only affecting the growth cone locally because tension was presumably transmitted along the neurite to the cell body. Tension also exists in PC12 neurites, as indicated by the fact that they retract to the cell body if detached from the substratum, an effect antagonised by cytochalasin (Joshi *et al.*, 1985). The existence of tension in growing axons has been indirectly demonstrated in experiments on embryonic grasshopper limb sensory neurons (Ti1 pioneers) in explant culture (Condic & Bentley, 1989a,b). These neurons extend growth cones between the limb epithelium and its basal lamina. Enzymatic removal of the basal lamina whilst growth cones were extending on it caused them to retract back to the cell body, suggesting that extension is dependent upon adhesion to the basal lamina. In some cases retraction could be antagonised with cytochalasin D, implying that retraction was due to actin-dependent tension in the extending axon.

Growth cone tension has been measured in technically heroic experiments by Heidemann's group (Lamoureux, Buxbaum & Heidemann, 1989; Heidemann et al., 1990; reviewed in: Heidemann & Buxbaum, 1994) in which chick dorsal root ganglion neurons in culture were attached by their cell bodies to glass-rod tension transducers and then lifted away from the substratum, leaving the growth cone still attached. As the growth cone advanced, its pull – up to about 200 μdyn (2 nN) – on the neurite and hence the cell body could then be measured. These workers found that there was a good correlation between neurite tension and growth cone advance (Fig. 2.7). When growth cones advanced rapidly, the tension in the neurite increased rapidly. Conversely, during periods of growth cone quiescence, tension remained constant. Heidemann's initial experiments were done with uncoated substrata, which are not very adhesive (Lamoureux, Buxbaum & Heidemann, 1989; Heidemann et al., 1990). Later experiments with highly adhesive substrata (polylysine-coated plastic) confirmed these results (Heidemann et al., 1991). These experiments do not distinguish, however, between a pull originating from the growth cone filopodia or from some other structure within the growth cone such as the cortical actin cytoskeleton. That the pull can originate from filopodia is indicated by the fact that filopodia can elastically deform neurites in their path while shortening (Nakai, 1960; Wessells et al., 1980; Heidemann et al., 1990), and is consistent with the finding that, when filopodia are mechanically detached from the substrate, they retract into the growth cone (Wessells & Nuttall, 1978); although such an effect would be produced if the filopodium is pulling away from the growth cone or towards it. However, when filopodia are mechanically severed from the growth cone, they can be induced to contract independently (Davenport et al., 1993). Growth cones can be induced to turn by lifting filopodia on one side or to branch when filopodia are lifted at the front end (Wessells & Nuttall, 1978). A corollary of the idea that neurite extension occurs by growth cone pull is that external application of tension on the neurite should influence neurite elongation. This has, in fact, been found to be the case (Bray, 1984; Lamoureux et al., 1992).

Differential adhesivity and growth cone pathfinding
More recently, the role of differential adhesivity in pathfinding has been challenged by experiments showing that growth cones do not adhere particularly strongly to substrata consisting of naturally occurring guidance molecules, such as laminin (see Chapter 3). Doubt about the role of filopodia in neurite extension came initially from the striking observation that neurites lacking filopodia as a result of the action of cytochalasins still continue to advance

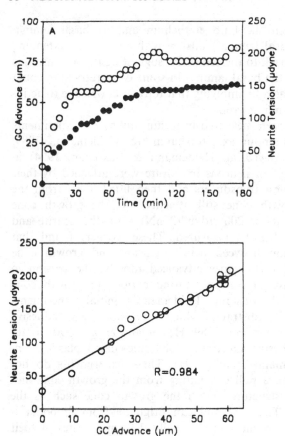

Figure 2.7 Axon tension and growth cone (GC) advance in a chick dorsal root ganglion neuron growing on polylysine. (A) Graph showing axon tension (open circles) and growth cone advance (filled circles) as a function of elapsed time. (B) Axon tension as a function of growth cone advance from the data in (A). R is the correlation coefficient between advance and tension. (From Heidemann et al., 1991a, with permission of the Company of Biologists Ltd.)

when the substratum is highly adhesive, albeit rather slowly (Marsh & Letourneau, 1984; Letourneau et al., 1987; McCaig, 1989). A similar lack of effect of cytochalasins on neurite extension has also been seen in short-term explant cultures (Bentley & Toroian-Raymond, 1986; Chien et al., 1993). In the experiments from Bentley's group, pioneer sensory neurons in explant cultures of embryonic grasshopper limb buds were studied. The growth cones of these neurons, which have many, extremely long filopodia, navigate stereotypical routes through the limb bud mesenchyme to the CNS that are invariant from animal to animal. In the presence of cytochalasins, growth cone filopodia (and lamellipodia) were lost but axons still continued to advance. Most importantly, however, about 40% of growth cones completely failed to navigate the appropriate route to the CNS and over 80% had disoriented segments along their trajectory. About 25% of axons would be expected to navigate an appropriate route by chance and, since more than this proportion did, it is assumed that cytochalasins do not completely destroy pathfinding ability in all cases. Nonetheless, these experiments directly demonstrate a role for filopodia in growth cone pathfinding. This important topic is considered more fully in Chapter 3. In the experiments reported by Chien *et al.* (1993), explants of embryonic *Xenopus* brain, which included the retino-tectal system, were studied. The effects of cytochalasins on retinal ganglion cell growth cones growing in the optic tract and chiasm in these explants were similar to those observed by Bentley's group. Growth cones lost their filopodia, but continued to advance and made pathfinding errors.

The lack of effect of cytochalasins on undirected neurite extension may be because extension can be sustained by the continuous, but non-vectorial, flow of material into the distal tip of the growth cone by axoplasmic transport; the fast component of which is microtubule based and therefore not affected by cytochalasins. The mechanism underlying slow axonal transport is unknown. Alternatively, cytochalasin treatment may not completely eradicate all actin filaments in the growth cone. For instance, although Chien *et al.* (1993) found that cytochalasin-treated growth cones lacked filopodia, they retained lamellipodia. Interestingly, the pathfinding ability of these growth cones was still impaired.

These doubts about the role of filopodia in growth cone pulling were compounded by close examination of the behaviour of growth cones in culture using high-resolution video microscopy (Goldberg & Burmeister, 1986; Aletta & Green, 1988; reviewed in Goldberg & Burmeister, 1989). This showed that growth cone advance was closely correlated with protrusive events within the growth cone, particularly to extension of lamellipodia, suggesting that forward pushing underlies neurite extension. Furthermore, at a border between two different substrata where growth cones turn, turning was not correlated with an increase in the number and length of filopodia in the direction of turning, but with a selective loss of filopodia on the side away from the direction of turning (Burmeister & Goldberg, 1988). Neither group saw any indication of isotonic tension, such as filopodial shortening, although, of course, tension may be developed isometrically. Burmeister & Goldberg (1988) suggested that the function of filopodia, in addition to their role in pathfinding (Mason, 1985b; Goldberg & Burmeister, 1986), is to provide support for the advancing lamellipodium and that it is these latter structures that are the important entities in neurite advance, since their engorgement by organelles and cytoplasm is the first step in the conversion of the P-domain into the C-domain, and thus into nascent neurite.

If, in addition to their role in pathfinding, filopodia are capable of exerting tension on the growth cone, then there should be a mechanical (molecular) link between the actin filament bundle of filopodia and other parts of the growth cone cytoskeleton. From this standpoint, an understanding of the actin binding proteins in the growth cone is important.

Actin binding proteins

A wide variety of actin binding proteins that modify actin filament dynamics have been identified (Weeds, 1982). Several of these have now been detected in growth cones, mainly by immunofluorescence with specific antibodies (Table 2.2). Of these, among the most interesting are the myosins, because of their ability to interact with actin to develop tension, and actin binding proteins such as profilin that are involved in phosphatidylinositol signalling pathways. Myosins move actin filaments in the direction of their pointed ('minus') ends while moving toward the barbed ('plus') end, i.e. they are 'plus' end-directed motors (Fig. 2.6). Myosins belong to a large superfamily of mechanoenzymes that utilise ATP as an energy source. Myosin II, or conventional myosin, exemplified by the type found in skeletal muscle fibres, has two globular heads that bind actin and a long tail that interacts with other myosin type II tails to form a filament (thick filament). Unconventional myosins, such as myosin I (mini-myosin), mainly have a single globular head and a tail that can, in some classes (I and VII), interact with membranes, but does not form filaments (Hammer, 1991, 1994; Pollard *et al.*, 1991; Mooseker & Cheney, 1995; Maciver, 1996; Titus, 1997). Myosins are actin-stimulated Mg^{2+}-ATPases and this activity is located in the globular head domain. Myosins share sequence homology, particularly in the head domain, but the sequences in their tail domains diverge.

Studies of the distribution of conventional myosin (type II) in growth cones, as revealed by immunostaining, were originally disappointing. In neuroblastoma cells, there was some indication that myosin type II may be distributed in a regular, banded pattern along filopodia (Sotelo *et al.*, 1979), whereas most studies of neurons showed a patchy distribution, mainly confined to the C- and proximal part of the P-domain (Roisen *et al.*, 1978; Kuczmarski & Rosenbaum, 1979; Letourneau, 1981; Shaw *et al.*, 1981; Bridgman & Dailey, 1989; Letourneau & Shattuck, 1989). Such a pattern did not readily suggest how myosin type II might interact with actin microfilaments in the growth cone to develop tension. However, more recently, Miller *et al.* (1992), using highly purified antibodies to chromaffin cell myosin II, found labelling of filopodia in neurons. Similarly, Cheng, Murakami and Elizinga (1993), using antibodies to a brain form of conventional myosin called type IIB, labelled filopodia and lamellipodia of the growth cones of rat dorsal root ganglion neurons in culture. Interestingly, myosin IIA, a non-neuronal form of non-muscle myosin II, although present in the neuronal cell body, was excluded from these regions of the growth cone, suggesting a role for myosin IIB in growth cone motility. The one immunoelectron microscopical study published so far indicated that myosin type II is predominantly located at the junction of the P- and C-domain in clusters that were sometimes associated with microfilaments (Bridgman & Dailey, 1989). A more recent report by this group confirmed the presence of myosin IIB at the C/P-domain junction and also found a correlation between regions of the growth cone where lamellipodia were retracting and antibody staining for myosin IIB, suggesting that this form of myosin is involved in retraction (Rochlin *et al.*, 1995). Myosin type II labelling was also occasionally seen at the base of and along filopodia (Bridgman & Dailey, 1989).

Myosin I, a single-headed unconventional myosin is mainly concentrated in the C-domain but is also clearly present along the entire length of filopodia (Barylko *et al.*, 1992; Wagner, Barylko & Albanesi, 1992; Lewis & Bridgman, 1996; Wang *et al.*, 1996a). Myosin I is enriched in isolated growth cones (Ruppert, Kroschewski & Bähler, 1993). Immunogold electron microscopy with antibodies to myosin Iα suggest that this myosin isotype is mainly associated with the growth cone plasma membrane, both at the dorsal and ventral membrane surfaces (Lewis & Bridgman, 1996). Myosin V is a two-headed myosin with a long tail that is present in axonal and dendritic growth cones where it co-localises with actin filaments, particularly in filopodia (Espreafico *et al.*, 1992; Cheney *et al.*, 1993; Wang *et al.*, 1996a). Thus, several different classes of myosins are present in growth cones and their distribution differs from each other, suggesting specific roles in growth cone motility and axon growth (reviewed in Bridgman *et al.*, 1994).

Although it has been suggested that myosins may be involved in filopodial and lamellipodial extension, there has been no direct test of this proposal until recently. Wang *et al.* (1996a) used micro-chromophore-assisted laser inactivation (micro-CALI) to probe the function of myosin V in growth cone motility in dorsal root ganglion cells *in vitro*. In the technique of micro-CALI, a laser beam is used to inactivate a protein by highly localised production of hydroxyl free radicals generated by laser irradiation of a chromophore (malachite green) covalently coupled to an antibody bound to the protein (Jay, 1988). Photochemical damage produced by micro-CALI has a half-maximal radius of about 1.5 nm, which makes it possible to inactivate discrete functional domains in a protein. Usually, the malachite green coupled antibody is micro-injected into the cell where it binds to its epitope on the protein of interest. The cell can then be irradiated in a spatially highly restricted region (about 10 μm in diameter) with a laser beam under the light microscope. Obviously only antibodies that do not interfere with the function of the protein can be used with this technique. By comparing the behaviour of the irradiated region of the growth cone with the non-irradiated part, the function of a number of proteins in growth cones have now been explored using this method (reviewed in Jay, 1996). Wang *et al.* (1996a) found that when they inactivated myosin V in dorsal root ganglion growth cones in culture using micro-CALI, the rate of filopodial extension was reduced, by a factor of two, in the irradiated zone, whereas the rate of filopodial retraction was unaffected. Since filopodial extension was not entirely abolished, myosin V may not be the only motor protein involved. Alternatively, an additional mechanism, such as barbed end polymerisation of actin filaments, may contribute to filopodial extension (Fig. 2.6). The effect was assumed to be specific because micro-CALI of other cytoskeletal molecules in growth cones, such as talin and vinculin, produced different effects on filopodia and lamellipodia (Sydor *et al.* 1995; Wang *et al.*, 1996a; reviewed in Jay, 1996). Furthermore, micro-CALI of myosin Iβ in dorsal root ganglion cell growth cones induced lamellipodial expansion but did not affect filopodial dynamics (Wang *et al.*, 1996a). These results suggests that myosin V is involved in filopodial extension and that filopodial retraction is independently regulated, possibly by another myosin, whereas myosin Iβ is involved in lamellipodia protrusion. Exactly how myosin V contributes to filopodial extension is not known. There is good evidence that myosin V is a vesicle motor that translocates a vesicle cargo along actin filaments (Titus, 1997). This may be important for membrane growth in filopodia. Consistent with this idea is the finding that myosin V is associated with a sub-population of small (50–100 nm) vesicles, as well as with microtubules and actin filaments, in the growth cones of superior cervical ganglion neurons (Evans, Hammer & Bridgman, 1997). Furthermore, in the *dilute-lethal* mutant mouse, which lacks myosin V, there are no abnormalities in axon growth, growth cone morphology or organisation of the cytoskeleton of superior cervical neurons (Evans *et al.*, 1997). These findings are consistent with a role for myosin V in vesicle movements along microtubules and actin filaments in growth cones and that this function supports plasma membrane addition and hence filopodial extension.

Alternatively, myosin V may exert force directly on actin filament bundles in filopodia (Fig. 2.6). Evidence for a differential regulation of filopodial extension and retrograde flow of actin filaments and the involvement of myosins in these processes has come from experiments in which myosins are inhibited directly (Lin *et al.*, 1996, see also: Jian, Hidaka & Schmidt, 1994; Ruchhoeft & Harris, 1997). In these studies, two independent approaches were used to inhibit myosin action in growth cones: a biochemical approach using *N*-ethyl-maleimide-inactivated myosin subfragment-1, which binds specifically to actin filaments but is incapable of generating force, and a pharmacological approach using 2,3-butanedione-2-monoxime, which inhibits myosin ATPases. These inhibitors had the same effects on growth cone behaviour when injected into *Aplysia* neurons in culture; they caused a dose-dependent inhibition of retrograde actin filament flow and an enhanced growth cone extension, especially of filopodia (Lin *et al.*, 1996). The latter effect was blocked by cytochalasin treatment, indicating that it was dependent on actin polymerisation and not mobilisation of pre-existing actin filaments. These findings strongly suggest that a myosin molecular motor generates rearward flow of actin filaments in growth cones.

Neurofilaments

Antibodies to neurofilament proteins do not usually stain growth cones (e.g. Shaw *et al.*, 1981), and neurofilaments have only occasionally been seen in growth cones by electron microscopy (Table 1.2). Several lines of evidence suggest that neurofilaments are involved in the radial growth of axons but their function in those growth cones where they occur is unknown. In higher vertebrates (birds, mammals), neurofilaments are composed of a triplet of proteins known as neurofilament high (mol. wt. 200 kD), neurofilament medium (mol. wt. 160 kD) and neurofilament low (mol. wt. 70 kD), or, in some classes of neurons such as cerebellar granule cells, they are composed of α-internexin or peripherin. Neurofilament low cannot form filaments *in vivo*, but can do so with the addition of either neurofilament medium or neurofilament high, i.e. neurofilaments are obligate heteropolymers (reviewed in Lee & Cleveland, 1994). In mature axons the number of neurofilaments correlates with the cross-sectional area of the axon and in a Japanese quail mutant called *quiver*, which lacks neurofilaments, radial growth of axons does not occur (Yamasaki, Itakura & Mizutani, 1991; Yamasaki *et al.*, 1992). In most other respects the nervous system of these mutants develops comparatively normally. Similarly, in transgenic mice completely lacking axonal neurofilaments, neural connections appear to form normally and the mice are grossly indistinguishable from wild-type mice (Eyer & Peterson, 1994). In hippocampal neurons in culture, neurite extension initially occurs in the absence of neurofilaments (Shaw, Banker & Weber, 1985). These observations, and the fact that arthropods do not have neurofilaments but do develop nervous systems (Phillips *et al.*, 1983), suggest that neurofilaments are not required for longitudinal axon growth or growth cone function. The prevailing view is that the function of neurofilaments is to support the radial growth of axons, an event which occurs largely after the cessation of neurite outgrowth, i.e. after synaptogenesis.

Neurite extension and surface membrane growth

How and where does growth of the surface membrane occur in the extending neurite? Are membrane components inserted into the plasma membrane at the cell body subsequently to diffuse down to the growth cone in the plane of the membrane, or are they transported down to the growth cone within the cytoplasm and incorporated at the growth cone (reviewed in: Bray, 1996; Futerman & Banker, 1996)? Growing axons are generally thought not to possess the biochemical machinery for the synthesis of surface membrane components (but see Davis *et al.*, 1992); these have to be manufactured in the cell body and delivered to the axon. While dendrites, both when growing and fully mature, do have rough endoplasmic reticulum (Nissl bodies) and free ribosomes, the latter along their entire length, the mRNA so far identified in dendrites codes for non-membrane proteins (reviewed in Craig *et al.*, 1992). It seems likely, therefore, that all surface membrane components of the neurite, glycoproteins, glycolipids, etc., are synthesised in the cell body. Furthermore, since there is no evidence for a Golgi apparatus located in the neurite or growth cone, terminal glycosylation must also take place in the perikaryon.

The first observation that shed light on the question of where in the neurite membrane components are incorporated came from the work of Speidel (1942). By observing the individual growing axons of living sensory neurons in frog tadpole skin over a period of weeks he was able to document the branching (arborisation) patterns as the axons established their innervation territory. Speidel noticed that, during arborisation, the distance between branch points remained fairly constant, despite the overall increase in the length of the arborisation. This suggests that there is no intercalated growth between branch points and therefore that addition of new membrane probably takes place at the growing tips or growth cones. Speidel's observations of axons *in situ* were later confirmed for chick dorsal root ganglion (DRG) neurons in dissociated cell culture (Bray, 1973a). Further indirect evidence for incorporation of membrane components at the growth cone came from observations made of the movements of glass or carmine particles picked up by the growth cones of sympathetic neurons growing in culture (Bray, 1970). Such particles, when attached to filopodia, tended to move in a retrograde direction along the dorsal surface of the growth cone, finally coming to rest near the junction between the growth cone, and the parent neurite. Particles on the neurite itself remained for long periods of time at the same distance from the cell body. Bray concluded from these observations that growth of the neurite surface membrane occurs by addition of surface membrane at the growth cone (Bray, 1973b). This conclusion has since found confirmation from direct studies of the incorporation of specific membrane components in growing or regenerating axons (Tessler, Autilio-Gambetti & Gambetti, 1980; Feldman et al., 1981, Griffin et al. 1981; Pfenninger & Maylié-Pfenninger, 1981). Griffin et al. (1981) studied the incoporation of radiolabelled fucose into fucosyl glycoproteins in regenerating adult rat sciatic nerve following a crush injury. Crushing a peripheral nerve severs many of the axons within it. Growth cones subsequently form at the severed proximal stump while the distal segment of axon, separated from its cell body, dies. The growth cone then grows a new distal axon segment in the nerve, sometimes called a 'sprout', thus providing an ideal experimental opportunity to study axonal membrane biosynthesis in growing axons. Griffin et al. (1981) found that there was a peak of fucose incorporation into the axon surface membrane in the region of the nerve where growth cones and sprouts were most abundant and that this newly inserted glycoprotein did not appreciably redistribute in the plane of the membrane after incorporation. They concluded that new membrane proteins are inserted at or near the growth cone in regenerating axons and that, once incorporated, the proteins are essentially stationary. Fucose incorporation also occurred along the parent axon but at a much lower level; such incoporation probably contributes to intercalated growth. Expression of exogenous membrane glycoprotein also leads to preferential addition at the growth cone in hippocampal neurons in culture (Craig, Wyborski & Banker, 1995). These authors induced expression of the lymphocyte transmembrane glycoprotein CD8α in embryonic hippocampal neurons in dissociated cell culture using a defective herpes virus vector. They found that CD8α first appeared in the plasma membrane at the growth cone and that in time a gradient of membrane expression was established that was highest in concentration at the growth cone and tailed off towards the cell body. The main conclusion from all of these findings is that surface membrane components in growing neurites are incorporated at the growth cone and only later become stably associated with the proximal membrane. In contradistinction to these observations are those of

Small et al. (1984), Small and Pfenninger (1984) and Popov et al. (1993), whose experimental work provides evidence that some membrane components are inserted at the cell body and diffuse into the neurite in the plane of the plasma membrane. Pfenninger's group (Small & Pfenninger, 1984; Small et al., 1984) found that the intramembranous particles seen in the neurite plasma membrane by freeze–fracture were distributed along growing primary olfactory neuron axons in a gradient that was highest at the cell body, while Popov et al. (1993) showed that local incoporation of a fluorescently labelled membrane lipid into cultured Xenopus spinal cord neurons led to bulk anterograde diffusion of the lipid at a rate that was faster the more distally along the axon was the incorporation of the lipid.

Experiments in chick dorsal root ganglion neurons in culture have shown directly that axon extension is dependent upon a supply of new membrane-bound organelles to the growth cone by fast axonal transport (Martenson et al., 1993). In these experiments the anterograde movement of membrane-bound organelles in growing neurites was physically blocked by local application of a laser optical trap or 'tweezers'. This caused a neuritic swelling to appear proximal to the site of action of the optical tweezers because of the build-up of anterogradely transported organelles. Neurite extension ceased after a time delay that was directly proportional to the distance between the site of application of the optical tweezers and the growth cone. Interestingly, when neurites were severed with optical tweezers, the neurite continued to elongate for a longer period of time than would have been predicted from the transport blocking experiments. In both cases neurite extension eventually ceased altogether. These experiments directly demonstrate that neurite extension is dependent on a supply of organelles to the growth cone or distal neurite. The time discrepancy between the two manipulations may relate to a restraining effect of the stationary cytoskeleton in the neurite on the growth cone which is relieved when the neurite is cut. Neurites are known to be under tension and filopodia can exert tension on the growth cone (see above) (Dennerll et al., 1988, 1989; Lamoureux et al., 1989; Heidemann et al. 1990; reviewed in Heidemann et al., 1991).

Additional indirect evidence for the insertion of new membrane components at the growth cone comes from in vitro experiments in which laser optical tweezers were used to apply small (sub-micrometre) polystyrene beads, coated with antibodies to either protein or lipid components of the neuronal plasma membrane, to chick dorsal root ganglion neurons in culture (Dai & Sheetz, 1995a,b). Polystyrene beads attached to surface membrane components exhibit two kinds of movement over the surface of growing axons in culture: a random, diffusive movement and a net retrograde movement towards the cell body of about 7 μm/min, which was five times greater than the rate of axon elongation (Dai & Sheetz, 1995a). The latter movement was interpreted as supportive evidence for the addition of plasma membrane components at the growth cone. It was assumed that membrane components added in excess at the growth cone were recovered by their removal at the cell body following retrograde flow back along the axon. One corollary of this idea is that the plasma membrane tension should be higher at the cell body than at the growth cone – with intermediate levels along the axon – and that this incremental gradient of membrane tension is the driving force for the retrograde flow of membrane components. To measure plasma membrane tension, Dai and Sheetz (1995a) measured the force required to pull out a tubular strand of membrane attached to a bead at different sites along the axon and at the growth cone. This force is proportional to membrane tension. This approach showed that the membrane tension at the growth cone is very low but that it increases progressively along the axon toward the cell body, a finding consistent with addition of membrane components at the growth cone (Dai & Sheetz, 1995b)

Membrane recycling in growth cones

Although a net increase in the area of the plasma membrane of the growth cone contributes to axon elongation, there is clear evidence that growth cones are also actively engaged in endocytotic activity. Extracellular tracers such as ferritin and horseradish peroxidase are readily taken up into endocytotic vesicles in growth cones (Birks, Mackey & Weldon, 1972; Wessells & Ludueña, 1974; Weldon, 1975; Bunge, 1977; Cheng & Reese, 1987; Dailey & Bridgman, 1993). Vacuoles that can be labelled with extracellular tracers have been shown to form transient openings with the plasma membrane – although whether they were undergoing endocytosis or exocytosis is not clear – and, in living growth cones in culture, these vacuoles can be seen appearing and disappearing in a manner consistent with local recycling through the plasma membrane (Dailey & Bridgman, 1993). Endocytosis appears to occur throughout the growth cone, both in the C- and P-domains, and following endocytosis, most vacuoles move retrogradely (centripetally) at between 1 and 6 μm/min (Hughes, 1953; Nakai, 1956; Goldberg & Burmeister, 1986; Forscher & Smith, 1988; Dailey & Bridgman, 1993). This is a similar rate to that of the retrograde flow of actin filaments, suggesting that vacuoles may be physically linked to actin filament retrograde flow. Orthograde or centrifugal movement of vacuoles in growth cones is rarely seen. If local recycling only recovers a portion of the membrane added by exocytosis, then net growth will occur. There is indirect, biochemical evidence for exocytosis in growth cones. Using isolated growth cone particles from embryonic rat brain, Pfenninger and colleagues have devised an assay, based on lectin binding of membrane proteins, that provides evidence for exocytosis in growth cones (Lockerbie, Miller & Pfenninger, 1991; Pfenninger et al., 1992). However, to what extent endocytosis at the growth cone underlies membrane recycling or pinocytosis, i.e. the bulk uptake of extracellular fluid, is not clear (Gordon-Weeks, 1988a). How membrane dynamics are co-ordinated spatially and temporally in the growth cone is also unknown.

3

Pathfinding

While studying the development of the retina and nervous system, we often asked ourselves this question: What are the mechanical causes of nerve fiber outgrowth and the sources of the marvellous power by which nerve expansions make direct contact with far-off neural, mesodermal, or epithelial cells?
Santiago Ramón y Cajal, *Studies on Vertebrate Neurogenesis*, 1960.

Introduction

It has been realised for over a century that growth cones can navigate precise routes through the developing embryo to locate an appropriate cell with which to form a synapse (Ramón y Cajal, 1892). Growth cone navigation is generally known as 'pathfinding'. Growth cone pathfinding has attracted considerable attention, both because of its importance in the formation of a properly connected, and therefore properly functioning, nervous system and because it is one of the most remarkable examples of cellular morphogenesis (for reviews see: Dodd & Jessell, 1988; Bixby & Harris, 1991; Hynes & Lander, 1992; Goodman & Shatz, 1993; Culotti, 1994; Tessier-Lavigne, 1994; Goodman, 1996; Goodman & Tessier-Lavigne, 1997). The distances over which growth cones must navigate in the embryo are usually relatively short, of the order of hundreds of micrometres (see 'Demonstration of chemotropic factors *in vitro*' below). Despite this, the navigational task facing growth cones is rarely simply a matter of following a straight path of least resistance; there may be many places along the route at which the growth cone must make large steering manoeuvres to locate the correct path. Right-angled turns are not uncommon, for instance at the mammalian optic chiasm and the floor-plate of the neural tube, and there are instances of growth cones having to change direction by as much as 180° (see below). Despite its long history, it is only in the last two decades that significant progress has been made in understanding the mechanisms underlying growth cone pathfinding and the molecules involved. How growth cones successfully pathfind is the subject of this chapter.

Decision regions and pathfinding

The locations where growth cones are required to make navigational decisions are known as 'decision regions' or 'choice points' and were first so named by Tosney and Landmesser (1985a,b,c) in the context of their pioneering studies of the lumbosacral plexus in the chick limb bud, where motor and sensory axons sort out into their appropriate nerves (reviewed in Landmesser, 1984, 1991). Other well-studied examples of decision regions include the mammalian optic chiasm, where retinal ganglion cell axons decide whether to cross the midline or not (reviewed in Godement, 1994; see Chapter 1), the sub-plate of the developing cerebral cortex, where thalamocortical afferents decide where and when to grow into the cortical plate (reviewed in: McConnell, 1992; Molnár & Blakemore, 1995) and the horizontal septum in the trunk of zebrafish embryos, where segmental motoneurons pathfind their way to their appropriate muscles (reviewed in Eisen, 1992).

At decision regions, growth cones exhibit a variety of behaviours as they navigate through the region (Taghert et al., 1982; Roberts & Taylor, 1983; Bentley & Caudy, 1984; Scalia & Matsumoto, 1985; Tosney & Landmesser, 1985a,b; Caudy & Bentley, 1986; Bovolenta & Mason, 1987; Harris et al., 1987; Holt, 1989; Kuwada et al., 1990; Nordlander et al., 1991). Generally, they become more filopodial and larger and may slow down, although this last change does not always happen (e.g. Sretavan & Reichardt, 1993). The local effects on growth cone behaviour at decision regions are not generalised ones affecting all classes of growth cones passing through the region, since individual populations of growth cones respond in different ways (Tosney & Landmesser, 1985a,b; Nordlander, 1987; Godement, Salaun & Mason, 1990; Godement, Wang & Mason, 1994). At decision regions, dramatic changes in direction are often seen and commonly include turns of 90° or more. For instance, in the retino-tectal system of the chick, the growth cones of retinal ganglion cell axons make right-angled turns in the superficial layer of the optic tectum as they descend to deeper levels (Vanselow et al., 1989). Similarly, the axonal growth cones of commissural interneurons turn rostrally through a right-angle when they reach the contralateral side of the floor-plate (Bovolenta & Dodd, 1990). In insect embryos, the growth cones of limb bud sensory neurons growing toward the central nervous system contact a series of so-called 'guide-post' cells along their route that constitute decision regions. Guide-post cells are individually identifiable from embryo to embryo and, after contacting particular guide-post cells, sensory neuron growth cones may make sharp turns that are remarkably similar from individual to individual (Bate,

1976). Such changes in behaviour are thought to reflect the fact that the growth cone is exploring the local extracellular microenvironment and responding to local influences. Although decision regions are clearly important for correct growth cone navigation, it is likely that growth cones are also making navigational choices outside these regions, but these are of a more minor nature.

Despite the requirement of growth cones to steer complicated routes through decision regions, the degree of precision of growth cone pathfinding is remarkable and, although mistakes are sometimes made, they are often corrected. Ramón y Cajal found many examples of incorrectly located (ectopic) neurons in embryos, for instance, lying within the central canal or within the pia, that nonetheless still managed to send their axons in the appropriate direction (Ramón y Cajal, 1892). This observation caused Ramón y Cajal to change his mind about the mechanisms of pathfinding. Previously, he had thought that the growth cone pushed aside cells in its way like a battering ram. He revised this view in the light of his observations on ectopic neurons and suggested that growth cones could respond to a factor diffusing out from the target tissue (see the section entitled 'Chemotropic (diffusible) factors'). Another striking example of corrective behaviour has been seen in the rabbit cerebral cortex (Van der Loos, 1965; for other examples, see: Jacobson, 1991; Purves & Lichtman, 1985). In the rabbit and other mammals, pyramidal neurons in the cerebral cortex normally have an axon that grows away from and orthogonal to the pial surface. Occasionally, pyramidal neurons become incorrectly oriented, so that they are inverted with respect to the normal orientation. In many of these inverted neurons, despite the initial outgrowth of the axon toward the pial surface, corrective events can produce complete 180° axon turns, so that the appropriate orientation is achieved. These axons then make normal connections.

Experimental alteration of the position or orientation of neurons often fails to prevent axon growth in the appropriate direction. Rotation or transposition of the giant Mauthner neurons in the brain stems of fish and amphibians often produces initial axon outgrowth in an inappropriate direction, but this is usually corrected, even to the extent of axons forming hairpin bends, and the correct path is then found (Stefanelli, 1951; Hibbard, 1965). When small regions of the embryonic chick spinal cord are translocated, the motoneurons developing within them can still find their correct muscle cells despite extending their axons along novel pathways (Lance-Jones & Landmesser, 1980, 1981b). When eye primordia containing precursor cells of the retinal ganglion neurons are transplanted to ectopic sites in the *Xenopus* embryo, the axons of the retinal ganglion neurons that subsequently develop can still correctly find the tectum (e.g. Harris, 1986). Like the translocated spinal cord motoneurons, they also take abnormal routes to their normal target. When portions of the rhombencephalon are rotated by 180° along their rostro-caudal axis and transplanted into a host embryo, the motoneurons that subsequently develop in the transplant still manage to find the appropriate exit point from the neural tube into the periphery (Guthrie & Lumsden, 1992). These examples illustrate the remarkable ability of the developing nervous system to correct mistakes that would otherwise lead to errors in connectivity. In theory, the mechanisms underlying this ability to correct mistakes could operate either through positive or negative influences on the growth cone. As we shall see in this chapter, both kinds of influences are employed in developing nervous systems and corrective re-orientation of growth cones can be achieved both through the effects of a growth inhibiting zone and a gradient of an attractive molecule.

Suitable animals and appropriate techniques

Considerable progress has been made in recent years in elucidating the mechanisms underlying pathfinding, although we are far from a complete understanding of how pathfinding is achieved for any particular axonal pathway. Much of this work has been done in lower vertebrates and invertebrates because of the experimental advantages offered by such organisms. These advantages include: nervous systems with small numbers of identifiable cells, e.g. the roundworm *Caernorhabditis elegans* – 302 adult neurons – and the fruit fly *Drosophila melanogaster* – 250 neurons per hemisegment; short reproductive cycles, as in *C. elegans* – 3-day life cycle; large number of offspring, e.g. *Drosophila*; transparent embryos, such as *C. elegans* and, more recently, the zebrafish *Brachydanio rerio*; accessible genetics and readily generated mutants, e.g. *Drosophila*. Work with invertebrates was originally criticised because its relevance to vertebrates, and thus to humans, was not clear. Recent molecular studies, however, have shown that many of the molecules involved in pathfinding are conserved across species as phylogenetically far apart as *Drosophila* and chick, and therefore have lent justification to invertebrate studies (see below). However, conservation of molecules or, more precisely, families of molecules, does not necessarily mean conservation of mechanisms.

Although the idea that growth cones grow along distinct and specific pathways, rather than growing out randomly to find their targets, derived originally from the early Golgi studies of Ramón y Cajal and others, it was not until the advent of anterograde axonal tracers for visualising individual or small subsets of axons, such as horseradish peroxidase, lectins and fluorescent molecules, that the generality of this idea became obvious (Crossland *et al.*, 1974; Lance-Jones & Landmesser, 1981a,b; Holt & Harris, 1983; Raper *et al.*, 1983a,b,c; Jacobson & Huang, 1985). Anterograde tracing also revealed unsuspected features of the behaviour of growing axons and their growth cones. For instance, tracer studies revealed that many axons undergo a waiting period at certain locations along their pathway before resuming their journey (Hollyday, 1983; Myers, Eisen & Westerfield, 1986; Shatz & Luskin, 1986; Shatz *et al.*, 1988, 1995). These observations were largely made on fixed material, an approach which restricts the experimenter to 'snapshot' views of dynamic processes. The introduction of supra-vital staining, that is staining of living tissues, with fluorescent dyes and their imaging with low-light-sensitive video cameras has made possible time-lapse observations on living growth cones *in vivo* (Harris *et al.*, 1987; Myers & Bastiani, 1993; Sretavan & Reichardt, 1993;

Godement *et al.*, 1994; Halloran & Kalil, 1994). Supra-vital staining with fluorescent dyes, such as the lipophilic dicarbocyanine dyes DiI and DiO, are non-specific in the sense that they stain a cellular compartment, e.g. cell membranes, rather than a specific molecule (Terasaki, 1993). These dyes have been used to examine the endoplasmic reticulum in living growth cones *in vitro* (Dailey & Bridgman, 1989). The discovery of the green fluorescent protein from the jellyfish *Aequorea victoria* and the cloning of the gene encoding the protein has made it possible to visualise specific molecules in living cells by expressing the protein of interest as a protein fused with green fluorescent protein (Wang & Hazelrigg, 1994). It is likely that these supra-vital techniques will make an increasingly important contribution to our understanding of growth cone pathfinding.

Many other techniques have contributed to our understanding of growth cone pathfinding, and some are discussed later in the text (see the section entitled 'How can guidance cues be identified?').

Pathfinding is a property of growth cones

Pathfinding is an autonomous property of growth cones – the cell body is not necessary, although it can clearly influence pathfinding, for instance, when changes in gene expression are involved. Growth cone autonomy during pathfinding has been strikingly demonstrated in the retino-tectal system of *Xenopus* embryos (Harris *et al.*, 1987). In these experiments, retinal ganglion cell growth cones were separated from their cell bodies by removing the eye in embryos after the retinal ganglion cells had extended their axons into the optic nerve. Under these circumstances, growth cones continued to grow towards the tectum as normal, presumably relying on the supply of new material by axoplasmic transport from that part of the axon still attached to the growth cone. Despite being severed from their cell bodies, retinal ganglion cell growth cones, on arriving at the tectum, could still make appropriate directional choices.

Not surprisingly, it is the filopodia of the growth cone that are involved in the initial detection and response to pathfinding cues, since they are the most distal structures of the growth cone and therefore encounter new territory first. The evidence that filopodia are important for pathfinding derives from several sources. If selected filopodia are mechanically detached from the substratum, growth cone extension occurs in the direction of the remaining attached filopodia (Wessells & Nuttall, 1978). Furthermore, if all filopodia are removed from growth cones by *in vivo* application of cytochalasins (see Chapter 2), pathfinding is disrupted (Bentley & Toroian-Raymond, 1986; Chien *et al.*, 1993). At choice points, growth cones enlarge and become more filopodial, and their rate of growth may slow, as if they are searching for some clue as to the correct path (see above). As expected for structures that have a role in pathfinding, certain receptor molecules for guidance cues are concentrated in filopodia (Bozyczko & Horwitz, 1986; Letourneau & Shattuck, 1989). Filopodia contain the appropriate intracellular signalling mechanisms and these can respond independently of the rest of the growth cone (Davenport *et al.*, 1993; see Chapter 4). Finally, growth cone behaviour can change as a result of a single filopodium contacting another cell (Kapfhammer & Raper, 1987a,b; Bantlow, Zachleder & Schwab, 1990; Moorman & Hume, 1990; O'Connor, Duerr & Bentley 1990; Bastmeyer & Stuermer, 1992; Myers & Bastiani, 1993) or substratum (Gomez & Letourneau, 1994; Fan & Raper, 1995; Kuhn, Schmidt & Kater, 1995).

Guidance cues

Early ideas about how the nervous system is connected up during development supposed that axon outgrowth is initially random and that only those connections that are functionally appropriate are retained (Weiss, 1924, 1928). This idea was supplanted by a chemical (molecular) based hypothesis, the so-called 'chemoaffinity hypothesis', as a result of the landmark experiments of Roger Sperry (Sperry, 1943a,b, 1963; reviewed in Hunt & Cowan, 1990). Sperry took advantage of the fact that the optic nerves of adult amphibians regenerate when severed. The optic nerve contains the axons of retinal ganglion cells. In non-mammalian vertebrates, retinal ganglion cells project their axons to the tectum (superior colliculus in mammals) to form a map of the visual world on the tectum. The projection maintains the neighbour relationships of retinal ganglion cells in the retina with their target neurons in the tectum. Such a projection is said to be topographic. The tectal map is inverted with respect to the map on the retina because the dorsal retina projects to the ventral tectum and vice versa, and the nasal part of the retina projects to the posterior tectum while the temporal retina projects to the anterior tectum. Sperry severed the optic nerve of adult newts and rotated the eye by 180°, so that temporal became nasal and dorsal became ventral, or left it in its normal position. When the optic axons had regenerated and re-innervated the tectum, he found that newts with rotated eyes behaved as if their world was upside down and reversed left to right, whereas in newts in which the eye had not been rotated, behaviour was normal. Sperry concluded that the axons of the retinal ganglion cells had grown back to their original locations in the tectum and formed functional connections, and therefore had not compensated for the eye rotation. He suggested that individual retinal ganglion cell axons had 'identification tags' or unique molecular markers that matched markers on an appropriate cell in the tectum. It was envisaged that the molecular markers played a role either in pathfinding by retinal ganglion cell axons or in synapse formation by these axons in the tectum (Attardi & Sperry, 1963). Testing this hypothesis has occupied several generations of developmental biologists in increasingly ingenious interventions (for a review, see Purves & Lichtman, 1985).

A convincing demonstration of the correctness of the chemoaffinity hypothesis in its original form would require showing that each retinal ganglion cell has a unique marker and that modifying individual markers affects pathfinding of the growth cone bearing that particular marker. It is not yet technically feasible to undertake such a test in a vertebrate, although this level of precision is being approached in invertebrates (see below). However, the overwhelming weight of evidence accumulated since Sperry's work suggests that growth cones do pathfind by detecting extrinsic, or extracellular, molecular cues in their local environment, so-called 'guidance cues', rather than growing out randomly, as suggested by Weiss (1924, 1928), or proceeding through a pre-determined set of manoeuvres. Although strongly opposed to the idea of specificity in growth cone pathfinding, Weiss was among the first to suggest that selective axonal adhesion (fasciculation, see below) might contribute to pathfinding (Weiss, 1941). Guidance cues are thought to alter the motile behaviour of the growth cone, re-directing the growth cone on to a new pathway or reinforcing an existing one. In this scheme, pathway decisions are made locally, at the level of the growth cone, which has the ability to respond rapidly and independently of the parent cell body. In other words, the path of the growth cone is directed by external influences. Furthermore, while it is now thought unlikely that there are as many molecular markers in the retino-tectal system as there are axons in the pathway or target cells in the tectum, as Sperry proposed, the consensus view is that such guidance cues are molecular in nature, although other kinds of cues, such as electromagnetic fields, have been considered (see below). Also, neuronal activity is now known to play a role in shaping and refining synaptic connections during development (see Chapter 5), and this was not fully appreciated in Sperry's time.

A wide spectrum of different types of molecular guidance cues are thought to exist and, increasingly, are being characterised at the level of the gene (Tables 3.1–3.4). Most of these molecules are proteins and many are members of large families, the diversity of which may be generated by alternative splicing of the gene transcript coding for the protein and by post-translational modification. Guidance cues may originate and operate locally in the immediate environment of the growth cone (short range) or they may derive from a distant source and operate over a long range. They may either attract or repel growth cones, and the same molecule may have opposite effects at different locations or time of development. At any one time, growth cones are probably responding to more than one guidance cue. Although considerable progress has been made recently in identifying and characterising guidance molecules and the growth cone receptors, for which they are ligands, the intracellular events consequent upon receptor activation, are largely unknown (see Chapter 4). In some cases, growth cones capitalise on the pathfinding decisions of earlier growth cones, or 'pioneers', in the pathway. In these cases the guidance cues are being provided by earlier pathfinders.

Pioneers

Some long projection pathways in vertebrate nervous systems and inter-ganglionic nerves in invertebrates are pioneered by growth cones that are followed later by other growth cones that grow along and form bundles or fascicles with

Table 3.1. *Extracellular matrix molecules implicated in growth cone pathfinding*

Name	Distribution	Effect	Reference
Chondroitin sulphate proteoglycans (e.g. neurocan, phosphacan)	Nervous system	Inhibits neurite outgrowth	Brittis *et al.*, 1992; Rauch *et al.*, 1992; Maurel *et al.*, 1994
Collagens	Peripheral tissues	Promotes neurite outgrowth	Kleitman *et al.*, 1988; Carbonetto, Gruver & Turner 1983
Fibronectin	Central nervous system and peripheral tissues	Promotes neurite outgrowth	Letourneau *et al.*, 1994; Akers, Mosher & Lilien, 1981
F-spondin	Spinal cord floor-plate	Promotes neurite outgrowth	Klar, Baldassare & Jessell, 1992
Heparan sulphate proteoglycans	Some axonal pathways	Promotes neurite outgrowth	Hantaz-Ambroise *et al.*, 1987
Keratan sulphate proteoglycans	Some axonal pathways	Inhibits neurite outgrowth	Cole & McCabe, 1991
Classical laminin (laminin-1)	All basal laminae	Promotes neurite outgrowth	Thompson & Pelto, 1982
Merosin (laminin-2)	Restricted basal laminae, astrocytes of CNS, Schwann cells of PNS	Promotes neurite outgrowth	Engvall *et al.*, 1992; Ehrig *et al.*, 1990; Morissette & Carbonetto, 1995
s-Laminin (laminin-3)	Neuromuscular junction, spinal cord floor-plate, cerebral cortex sub-plate	Inhibits neurite outgrowth	Hunter *et al.*, 1989a,b, 1992
Tenascin R[a]	CNS white matter	Inhibits neurite outgrowth	Pesheva, Spiess & Schachner, 1989; Norenberg *et al.*, 1992
Tenascin C[b]	CNS and PNS	Promotes and inhibits neurite outgrowth	Kruse *et al.*, 1985; Jones *et al.*, 1989
Thrombospondin	?	Promotes neurite outgrowth	Neugebauer *et al.*, 1991
Unc-6	*C. elegans* epidermal basal lamina	Promotes neurite outgrowth	Hedgecock *et al.*, 1990
Vitronectin	?	Promotes neurite outgrowth	Neugebauer *et al.*, 1991

[a] Also known as: janusin, J1-160/180, restrictin.
[b] Also known as: cytotactin, J1 glycoproteins, hexabrachion, myotendinous antigen, and glioma mesenchymal ECM antigen.
Abbreviations: CNS, central nervous system; PNS, peripheral nervous system.

Table 3.2. *Immunoglobulin superfamily glycoproteins involved in pathfinding*

Name (homologues)	Plasma membrane attachment	Binding	Distribution	Effect	Reference
Vertebrates					
DM-GRASP (SC1, BEN)	Transmembrane	Homophilic	Subsets of axons	Axon outgrowth	Burns *et al.*, 1991; Laessing *et al.*, 1994
F3/F11 (contactin)	GPI-linked	Heterophilic	Subsets of axons	Axon fasciculation	Chang *et al.*, 1987; Gennarini *et al.*, 1991
L1 (NILE, NgCAM, 8D9, G4)	Transmembrane	Homo- and heterophilic	Axons, immature astrocytes, Schwann cells	Axon fasciculation and outgrowth	Moos *et al.*, 1988
NCAM	Transmembrane or GPI-linked	Homo- and heterophilic	CNS, PNS, muscle, glia	Axon fasciculation and outgrowth	Cunningham *et al.*, 1987
Nr-CAM (Bravo)	Transmembrane	?	Retinal portion of optic axons		Grumet *et al.*, 1991
TAG-1 (axonin-1)	GPI-linked	Heterophilic	Spinal cord interneurons, motoneurons		Dodd *et al.*, 1988
Invertebrates					
Fasciclin I	GPI-linked	Homophilic	Subsets of axons		Bastiani *et al.*, 1987
Fasciclin II (NCAM homologue)	Transmembrane and GPI-linked	Homophilic	Subsets of axons		Harrelson & Goodman, 1988
Fasciclin III	Transmembrane	Homophilic	Subsets of axons		Patel *et al.*, 1987
Fasciclin IV (renamed Sema I)	Transmembrane	Heterophilic	Subsets of axons and epithelial cells		Kolodkin *et al.*, 1992
Neuroglian (L1 homologue)	Transmembrane	Homophilic	Subsets of axons		Bieber *et al.*, 1989
unc-5	Transmembrane	Not known	Pioneer neurons		Hedgecock *et al.*, 1990

Abbreviations: CNS, Central nervous system; GPI, glycosyl-phosphatidylinositol; NCAM, neural cell adhesion molecule; Ng-CAM, neuron–glial cell adhesion molecule; NILE, NGF-inducible large external glycoprotein; PNS, peripheral nervous system; unc, uncoordinated.

Table 3.3. *Cadherins involved in pathfinding*

Name	Distribution	Effect/function	Reference
DN-cadherin	*Drosophila* central nervous system	Axon fasciculation	Iwai *et al.*, 1997
E-cadherin[a]	Epithelia. Trigeminal ganglion, dorsal root ganglion	Blastula, embryonic endoderm cell adhesion	Johnson *et al.*, 1986
N-cadherin	Embryonic mesoderm, neural ectoderm, retina, brain, peripheral ganglia, cardiac and skeletal muscle	Axon fasciculation	Drazba & Lemmon, 1990; Rathjen, 1991; Redies *et al.*, 1992, 1993; Honig & Rutishauser, 1996
P-cadherin	Placenta	Cell adhesion	Nose & Takeichi, 1986
R-cadherin	Retina, visceral motor axons	Axon fasciculation	Redies *et al.*, 1993; Arndt & Redies, 1996
T-cadherin	Brain, retina, spinal cord, posterior somitic sclerotome	Inhibits motoneuron axon outgrowth	Vestal & Ranscht, 1992; Fredette *et al.*, 1996

[a]Also known as uromorulin, cell CAM120/80.

the axons of the pioneers, a process known as fasciculation. Axon tracts in the central nervous system and nerves in the peripheral nervous system are formed in this way. In some cases, once the pathway has formed, the pioneers die. The word 'pioneer' was first proposed by Ross Harrison based on observations that he made in the frog (Harrison, 1910). Pioneer axons occur in invertebrates (LoPresti *et al.*, 1973; Bate, 1976; Bentley & Keshishian, 1982; reviewed in Bastiani *et al.*, 1985) and in vertebrates, for instance, in the mammalian cerebral cortex (McConnell, Ghosh & Shatz, 1989; Kim *et al.*, 1991) and in the zebrafish forebrain (Wilson & Easter, 1991). Pioneer growth cones tend to be larger and more complex than follower growth cones (Kim *et al.*, 1991). In some cases, the presence of pioneer axons is essential for later following axons to find their targets (Klose & Bentley, 1989; Bastiani *et al.*, 1985). During the development of the mammalian cerebral cortex, a transient population of neurons form a laminar sheet of cells called the sub-plate, superficial to the ventricular zone. These sub-plate neurons pioneer a pathway between the cortex and the thalamus. At later times, neurons in the thalamus send their axons to the cortex along the route pioneered by axons of the sub-plate neurons. Although it is not yet clear whether the later growing thalamic axons fasciculate on the pioneering sub-plate axons, it is apparent that their presence is necessary for correct pathfinding by the thalamic efferents (Molnár & Blakemore, 1995). When sub-plate neurons are destroyed using the neurotoxin

Table 3.4. *Properties of chemotropic molecules*

Name	Expression domain	Effect	Type	Homologues and relatives	Reference
Netrin-1	Vertebrate embryonic spinal cord floor-plate cells	Chemoattractant for spinal cord commissural axons Chemorepellant for trochlear motor axons	Secreted protein, 78 kD	Unc-6	Serafini *et al.*, 1994; Kennedy *et al.*, 1994; Colamarino & Tessier-Lavigne, 1995
Netrin-2	Vertebrate embryonic ventral spinal cord excluding floor-plate	Chemoattractant for spinal cord commissural axons	Secreted protein, 75 kD		Serafini *et al.*, 1994; Kennedy *et al.* 1994
Collapsin-I	Dermamyotome	Collapses or repulses growth cones of dorsal root ganglion	Secreted glycoprotein, 100 kD	Sema III, Sema D, Sema I (fasciclin IV)	Kolodkin *et al.*, 1993; Luo *et al.*, 1993

kainic acid, the axons of thalamic neurons cannot find their cortical targets (Ghosh *et al.*, 1990). In the fish central nervous system, some pioneer pathways are essential for later arriving growth cones to pathfind correctly (Kuwada, 1986; Chitnis & Kuwada, 1991). In the spinal cord of the fish, Rohon-Beard neurons pioneer the dorsolateral fascicle by growing along the end-feet of the neuroepithelial cells (Kuwada, 1986). Laser ablation of Rohon-Beard neurons compromises pathfinding by later arriving growth cones that normally fasciculate with the Rohon-Beard pioneer axons (Chitnis & Kuwada, 1991). Similarly, in the embryonic nervous system of the grasshopper, ablation of some pioneer neurons causes later, following, growth cones to misroute (Raper *et al.*, 1983a,b, 1984; Bastiani, Raper & Goodman, 1984a,b). However, this is not always so and, in some cases, although destroying the pioneers causes initial confusion among later projecting axons, they can recover and make normal connections (Lin, Auld & Goodman, 1995). There are also clear-cut examples where pioneer axons are not necessary for later growing axons to pathfind correctly in vertebrates. In the zebrafish embryo, ablation of a pioneer motoneuron in the spinal cord does not prevent later arriving motoneurons, which normally fasciculate upon the pioneer axon, from finding the correct pathway (Eisen, Pike & Debu, 1989; Pike & Eisen, 1990). This suggests that in some cases there are 'fail-safe' mechanisms to ensure that, if fasciculation is disrupted, pathfinding is maintained.

How can guidance cues be identified?

Although great progress has been made recently in identifying and characterising guidance cues and their receptors at the molecular level (see below), there are strong arguments that many still remain to be discovered. Furthermore, many of those guidance cues that have been identified probably have other functions in neural development that have not yet been discovered. A particularly important goal is to identify all of the guidance cues for a specific axonal pathway so that we can understand how different cues are orchestrated in space and time, their hierarchical organisation and the level of redundancy of function. It is highly likely that the first pathway that will be completely understood in molecular terms will be in an invertebrate, perhaps the Ti1 pioneer sensory neurons in the grasshopper limb bud (Bentley & Keshishian, 1982; Bentley & O'Connor, 1991). Before a candidate molecule can be accepted as a growth cone guidance cue, a number of criteria have to be met. The molecule has to be expressed at the right time and in the right place. The growth cone must respond to the molecule in an appropriate way, and interventions that block the action of the molecule or remove it must affect pathfinding.

A wide variety of assays and experimental approaches have been used to try to identify guidance mechanisms and molecules. Prominent amongst these are: transplantation experiments, in which tissue is either re-orientated or grafted into ectopic positions; tissue culture bioassays for candidate molecules; and abrogation studies, in which antibodies or genetic mutations, are used to compromise the action of the candidate molecule. This can be done either with cells *in vitro* or, with the advent of transgenic animals, in the whole organism. There are two approaches to identifying molecules involved in pathfinding using genetics: so-called 'reverse' genetics, in which mutations are made in the gene of a known protein, and 'forward' genetics, in which random mutations are made in a genome and a mutant phenotype is screened for using a suitable assay. In the case of reverse genetics, the gene can be modified to produce a gain or loss-of-function mutation or deleted, so-called 'gene knockouts'. The application of these techniques has uncovered a number of mechanisms and molecules that appear to be important in growth cone pathfinding. However, it is highly unlikely that a single mechanism can explain pathfinding for any particular axon and that several of the proposed mechanisms, and probably others yet to be discovered, are operating over a particular route.

Channels and tunnels

One mechanism that might play a part in guiding growth cones to their targets is a preformed extracellular tunnel or channel. Such a tunnel may be formed from the spaces between neuroepithelial cells or glial cells and would act as a passive, physical guide. This form of growth cone guidance is known as stereotropism and is amongst the simplest mechanisms that have been proposed to explain pathfinding.

Stereotropism was first proposed by Wilhelm His in 1886 (see Jacobson, 1991, p. 211 and references therein). Weiss showed that growth cones can grow preferentially along scratches in the surface of a culture dish and thus can respond to physical features in their environment (Weiss, 1934). Growth cones will also grow along collagen fibres in an oriented collagen gel (Ebendal, 1976) and this effect has been used as the basis for strategies to repair severed peripheral nerves. Intercellular spaces in advance of migrating growth cones have been seen by electron microscopy in the developing *Xenopus* and newt spinal cord, in the regenerating spinal cord of partially amputated tails in adult lizards and newts (Egar, Simpson & Singer, 1970; Singer, Nordlander & Edgar, 1979; Nordlander & Singer, 1978, 1982a,b), and in the developing optic nerve head of the mouse (Silver & Sidman, 1980) and chick (Krayanek & Goldberg, 1981). These spaces were transiently formed between adjacent neuroepithelial cells at the peripheral margin of the neural tube and were longitudinally oriented, the same orientation as the early axon tracts formed by the ingrowing axons. These observations led Singer to propose the 'blueprint hypothesis' which essentially states that neuroepithelial cells form intercellular spaces or tunnels that are appropriately located in time and space to orientate growth cones. Although there has been no direct experimental test of this hypothesis, such as removal or re-orientation of the tunnels, in a mutant mouse that lacks neuroepithelial spaces, axon growth is compromised (Silver & Robb, 1979; Silver & Sidman, 1980).

However, there are a number of problems associated with stereotropism as an explanation for growth cone pathfinding. There is no orientational information in a simple tunnel and therefore a growth cone finding its way into such a tunnel somewhere along its length is equally likely to proceed in either direction along the tunnel. What directs neuroepithelial cells to form appropriately oriented tunnels in the first place? Furthermore, tunnels have only been seen in a few locations and growth cones have not been seen directly to grow along tunnels *in vivo*. Clearly, if tunnels do play a role, there must also be other mechanisms guiding growth cones.

The blueprint hypothesis is concerned with those early forming axons that pioneer a pathway in territory previously unencountered by axons. Such pioneers are followed by later arriving axons and these often grow along, or fasciculate with, the axons of the pioneers (see 'Cell surface molecules' below). In this way axon tracts in the central nervous system and nerves in the peripheral nervous system are formed. Many of the questions relating to growth cones growing along tunnels are also applicable to axon fasciculation. The interactions between growth cones and the axons with which they fasciculate appear to be mediated by cell-adhesion molecules and are discussed in detail in the section entitled 'Cell surface molecules'.

In addition to pre-formed tunnels, it has also been suggested that growth cones can create tunnels, which they subsequently grow along. This might be achieved by secreting metalloproteases or activators of proteases, such as plasminogen activator, that digest the extracellular matrix between cells (Krystosek & Seeds, 1981a,b, 1984; Pittman, 1985; Pittman & Williams, 1989; Pittman, Ivins & Buettner, 1989; reviewed in: Pittman *et al.*, 1989; Ivins & Pittman, 1992). Cultured sympathetic neurons contain a metalloprotease that can degrade extracellular matrix proteins, such as type I collagen and fibronectin, which they are likely to encounter during pathfinding (see below), and plasminogen activators that convert plasminogen into the trypsin-like protease, plasmin (Pittman, 1985; Pittman & Williams, 1989; Pittman *et al.*, 1989a,b). Both of these are probably constitutively released from the growth cones of these neurons. Surprisingly, inhibition of these proteases often increases the rate of axon growth of neurons in culture (Pittman *et al.*, 1989a,b). Whether proteases are released from growth cones *in vivo* and what their role might be is not known.

Electric fields

Cellular movement in response to an external electric field is known as galvanotaxis. Although there is no direct evidence *in vivo* for electric fields contributing to growth cone pathfinding, there is clear-cut evidence that electric fields can orientate growth cones *in vitro* and, since such fields exist *in vivo*, the possibility remains that electrical fields contribute to pathfinding (reviewed in: Patel *et al.*, 1985; McCaig, 1988; McCaig *et al.*, 1994; McCaig & Erskine, 1996; McCaig & Zhao, 1997). The first report of orientation of growing neurites in an applied electric field was made by Ingvar (1920) working in Ross G.

Harrison's laboratory. Since then there have been many reports showing that, in culture, growth cones from a wide variety of neuronal types orientate and increase their rate of growth when exposed to external (extracellular) direct current (DC) electric fields (Marsh & Beams, 1946a,b; Jaffe & Poo, 1979; Hinkle, McCaig & Robinson, 1981; Patel & Poo, 1982; McCaig, 1986, 1987; Davenport & McCaig, 1993). On most substrata *in vitro*, growth cones turn towards the cathodal (negative) end of an applied DC field and extend longer filopodia on this side of the growth cone. Neurite growth and the degree of branching is also increased on the cathodal side. DC electric fields can also enhance the rate of growth of regenerating axons (Borgens, Roederer & Cohen, 1981).

DC electric fields exist in embryos and, furthermore, the magnitudes of the electric fields necessary to produce reorientation of growth cones are similar to those measured directly in embryos (Jaffe & Nuccitelli, 1977). For instance, in chick embryos, there is current leakage from the hindgut owing to a low-resistance pathway formed as a consequence of apoptosis of cells in the cloacal region (Hotary & Robinson, 1992). This current gives rise to a voltage drop of 33 mV/mm, more than ample to drive growth cone orientation *in vitro*, which has a threshold of about 7 mV/mm (Nuccitelli & Smart, 1989; Hinkle *et al.*, 1981).

The mechanism underlying growth cone turning in an electric field has not been elucidated, although electric fields of similar strengths to those that can re-orientate growth cones are capable of electrophoresing molecules in the plane of the plasma membrane (Jaffe & Poo, 1979; Poo, 1981; Patel & Poo, 1982). It is conceivable that a differential accumulation of molecules in the growth cone plasma membrane may be associated with re-organisation of the underlying cytoskeleton and that this may explain re-orientation. Alternatively, electric fields may influence the behaviour of voltage-sensitive ion channels. There is evidence for both of these effects in cathodal re-orientation of embryonic *Xenopus* spinal cord motoneuron growth cones *in vitro* (Patel & Poo, 1982; Erskine & McCaig, 1995; Stewart, Erskine & McCaig, 1995). Galvanotropic turning of *Xenopus* motoneuron growth cones is completely blocked by the nicotinic acetylcholine receptor antagonist D-tubocurarine and partially inhibited by blockade of voltage-gated calcium channels (Patel & Poo, 1982; Erskine & McCaig, 1995; Stewart *et al.*, 1995). Since these growth cones are known to release acetylcholine spontaneously *in vitro* (Young & Poo, 1983) and electric fields are capable of inducing an asymmetry in the distribution of acetylcholine receptors in neuronal cell membranes (Patel & Poo, 1992), and therefore, presumably, in growth cone membranes, it has been proposed that activation of asymmetrically distributed acetylcholine receptors by an autocrine action of growth-cone-released acetylcholine produces a spatially regulated influx of calcium (McCaig & Zhao, 1997). Calcium influx could occur directly through the acetylcholine receptor complex or indirectly, via membrane depolarisation, through voltage-gated calcium channels (Stewart *et al.*, 1995). In support of this idea is the observation that growth cones can re-orientate under the influence of an externally applied gradient of neurotransmitter and that this depends on the establishment of an intracellular gradient of calcium ions (Zheng et al., 1994; Zheng, Wan & Poo, 1996).

Electric fields are also able to induce growth cone formation in the axon shaft in vitro, a process known as 'back-branching' or collateral sprouting, when it occurs in vivo (see below) (McCaig, 1990; Erskine, Stewart & McCaig, 1995; Erskine & McCaig, 1995; Williams et al., 1995). Steady ionic currents, probably carried by calcium ions, have been detected in cultured goldfish retinal ganglion cell growth cones (Freeman et al., 1985). The magnitude of the current was very small (10–100 nA/cm^2). The currents entered the tips of filopodia and emerged from their bases, suggesting that there is a concentration of calcium channels in the filopodium tip (see Chapter 4).

Although the in vitro evidence for a role for electric fields in pathfinding is compelling, there have been no in vivo tests of the idea and it is difficult to see how specificity can be incorporated into such a mechanism. How does a uniform electric field explain the diverse behaviours and routes taken by growth cones navigating through a choice point?

Extracellular matrix molecules

Since the pioneering experiments of Ross Harrison, described in Chapter 1, it has been apparent that growth cones cannot advance through a liquid medium: they require a substratum on which to extend (Harrison, 1914). In the embryo, the substrate for growth cone advance is either the surfaces of other cells or the extracellular matrix. Haptotaxis is the name given to cellular movement along an adhesive surface of substrate-bound molecules (Carter, 1967). Originally, it was thought that adhesivity of the substratum correlated with effectiveness of growth cone advance (Letourneau, 1975a,b, 1979; Bray, 1982). In this model, it was assumed that filopodia adhere to the substratum and exert tension on the growth cone and, therefore, the higher the adhesivity, the greater the tension generated and the greater the rate of growth cone advance (see Chapter 2). This idea was developed mainly from experiments in vitro in which a range of substrates were tested for their ability to promote neurite outgrowth (Letourneau, 1975a,b; reviewed in: Lockerbie, 1987; Sanes, 1989). This work suggested that the more adhesive the substrate, the better growth cones were able to advance upon it. However, many of the test substrates were synthetic molecules. For example, the vast majority of embryonic neurons that have been tested readily extend neurites on highly charged, synthetic polyamino acids such as poly-lysine and poly-ornithine, molecules that do not occur naturally (Letourneau, 1975a,b). These synthetic molecules are preferred over tissue-culture plastic and other non-adhesive surfaces, further correlating neurite growth with adhesivity (Letourneau, 1975a,b). However, adhesivity was rarely measured quantitatively in these experiments, for example, Hammarback et al. (1985). It was only when growth cones were given a choice between naturally occurring substrates and when the strength of attachment of growth cones was determined by trying to dislodge growth cones mechanically from the substratum that our views changed (Gundersen, 1987; Calof & Lander, 1991; Lemmon *et al.*, 1992). Gundersen (1987) showed that, whereas growth cones of chick dorsal root ganglia grow more rapidly on laminin than poly-lysine, they are more difficult to physically dislodge from the latter substratum than the former. Similarly, Lemmon *et al.* (1992) found that chick retinal ganglion cell growth cones grew more rapidly on laminin compared to the immunoglobulin superfamily member L1 (see below) but were more easily dislodged from laminin than L1 by a stream of culture medium. Also, there is no correlation between the closeness of contact between the growth cone and the substratum and preference of growth (Gomez & Letourneau, 1994). So, rather than differential adhesion between the substratum and the growth cone plasma membrane explaining growth cone guidance, it is now thought that haptotactic guidance molecules, and in fact all forms of guidance molecules, exert their effects through signal transduction mechanisms in growth cones, which ultimately lead to changes in the growth cone cytoskeleton and hence attachment (see Chapter 4).

The extracellular matrix is a complex association of glycoproteins and proteoglycans found in the interstices between cells (Fig. 3.1; for a compilation of components, see Kreis & Vale, 1993). Although many of the components of the extracellular matrix have been identified and characterised, our understanding of their function and how they interact with each other and with cells is poorly developed. The extracellular matrix is known to provide attachment for cells, both stationary and migrating, and to be involved in cellular differentiation and development. There is a considerable body of evidence that implicates the extracellular matrix in growth cone migration and pathfinding (reviewed in: Reichardt & Tomaselli, 1991; Letourneau, Condic & Snow, 1994; McKerracher, Chamoux & Arregui, 1996). Much of this evidence comes from *in vitro* experiments in which candidate molecules are tested for their ability to promote or sustain neurite outgrowth and elongation from cultured, dissociated neurons or explants. The test molecules are usually allowed to attach to the culture substratum, although the amount bound is rarely determined experimentally, and then neurite length is measured after plating out the neurons on to the substrate. Both stimulatory and inhibitory effects on neurite outgrowth have been observed. Although these assays are invaluable in identifying factors that promote neurite outgrowth and substratum attachment, they do not evaluate a molecule for a pathfinding role, in which the growth cone is directed on to the correct path. Furthermore, since some guidance molecules have an inhibitory or repellent action on growth cones (see 'Cell surface molecules' below), a neurite outgrowth assay does not identify all classes of guidance molecules. A guidance molecule may have neither inhibitory nor promotional effects on neurite outgrowth, but may cause neurites growing on a permissive substratum to turn when they encounter the molecule. Such an assay, in which the growth cone is offered two or more candidate substrates, is known as a 'choice assay' and a variety of configurations have been introduced including gradients and sharp-substrate borders (Baier &

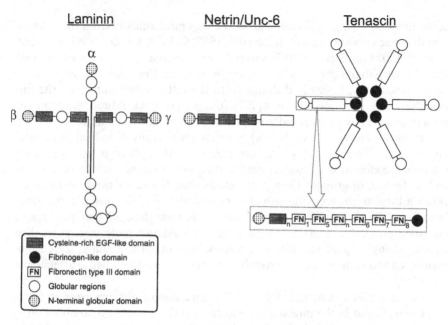

Figure 3.1 Diagrams showing the domain structure of representative members of extracellular matrix protein families implicated in growth cone pathfinding. The three laminin chains (α, β, γ) are linked together by disulphide bridges in the central region of the molecule, where the polypeptide chains are coiled into a triple helix. Members of the laminin family have a similar overall structure, a cruciform shape in rotary-shadowing images, but differ by having at least one genetically distinct chain. Netrins/Unc-6 have considerable sequence homology to the laminin $\gamma 1$ chain but their carboxy terminus is different. Tenascin molecules are linked together at their amino-terminal ends (black circles) by disulphide bridges into multimers. The hexabrachion of tenascin-C is shown. Alternative splicing of tenascin produces a variable number of epidermal growth factor (EGF)-like and fibronectin (FN)-like domains.

Bonhoeffer, 1992; Snow & Letourneau, 1992; Taylor, Pesheva & Schachner, 1993; Baier & Klostermann, 1994; Gomez & Letourneau, 1994). One variant of this class of assay which has contributed much to our understanding of guidance molecules is the 'stripe choice' assay (Walter, Henke-Fahle & Bonhoeffer, 1987a; Walter *et al.*, 1987b; see also Simon & O'Leary, 1992; Roskies & O'Leary, 1994). In the stripe choice assay, alternating stripes of test molecules or tissue extracts are absorbed on to a tissue-culture support on to which dissociated neurons or explants, at right-angles (Walter *et al.*, 1987a,b) or parallel (Roskies & O'Leary, 1994) to the stripes, are subsequently placed. Axons then grow out across the stripes and growth cones and axons are then scored for which stripes they prefer to grow along (see below).

The claim for a role for extracellular matrix components in pathfinding derives from several independent lines of evidence. First, several of the molecules of the extracellular matrix are spatially and temporally regulated in developing neural tissue in such a way as to suggest a role in pathfinding. The central nervous system has a rich and varied extracellular matrix during development, whereas in the adult it is less complex and mainly restricted to the larger blood vessels and the pia. The peripheral nervous system is different in this respect and peripheral nerves and ganglia, for instance, have considerable amounts of extracellular matrix in the adult, particularly collagens and basal lamina proteins such as laminin (see below). During development, the extracellular matrix is mainly synthesised and secreted by embryonic glia, upon whose surface membrane it appears (Letourneau *et al.*, 1994). Secondly, it has long been established that neurite elongation by neurons in culture is enhanced by extracellular matrix substrata. This is such a powerful effect in the case of one member of a family of extracellular matrix proteins, classical laminin (laminin-1), that it is routinely used as a substrate for dissociated neuronal cultures. Finally, perturbation experiments with antibodies to extracellular matrix molecules, and other interventions, affect pathfinding. In principle, extracellular matrix substrata may be either permissive or instructive for neurite elongation. Permissive substrata provide a suitable surface for attachment and elongation. The adhesivity of the surface is neither too high as to preclude elongation nor too low for attachment. An instructive substratum, on the other hand, contains directional information, for instance, it may be present in a concentration gradient to which the growth cone can respond. Both roles for extracellular matrix in pathfinding have been suggested.

Several molecules that are expressed in the extracellular matrix during the development of the nervous system have been shown in *in vitro* assays to promote or inhibit neurite outgrowth. Prominent among these are laminins, particularly for central neurons, tenascins, fibronectins and proteoglycans (Table 3.1).

Laminins

Laminins are large (> 850 kD), multifunctional, extracellular matrix glycoproteins (reviewed in Timpl & Brown, 1994; Fig. 3.1). They are composed of three polypeptide chains (α, β and γ) joined covalently by disulphide bridges in a triple-coiled region of the molecule (Fig. 3.1). Vertebrates have at least five α, three β, and two γ chains, which, when assembled in different combinations, produce at least twelve different laminins. Rotary shadowing experiments in the electron microscope show that laminins have a cruciform shape. Laminins are multidomain proteins with binding sites for other extracellular matrix proteins, including glycosaminoglycans, and for a family of membrane receptors called integrins that are widely expressed on neuronal and non-neuronal cell surfaces (reviewed in Mercurio, 1995). Laminins are major components of basal laminae, where they contribute to the integrin-mediated attachment of epithelial, muscle and other cell types to the extracellular matrix. They also promote neuronal cell adhesion and neurite outgrowth. Laminins are numbered in the order of their discovery and they differ by having at least one genetically distinct polypeptide chain (Fig. 3.1). The laminin family of proteins includes classical laminin (laminin-1, $\alpha1\beta1\gamma1$), merosin (laminin-2, $\alpha2\beta1\gamma1$, Ehrig *et al.*, 1990) and s-laminin (laminin-3, $\alpha1\beta2\gamma1$, Hunter *et al.*, 1989a,b) (Table 3.1). Laminin-1, the most comprehensively understood family member, is found in all basal laminae, whereas the distribution of the other family members is restricted. Laminin-2 is produced by certain astrocytes in the developing central nervous system (Morissette & Carbonetto, 1995) and by Schwann cells in the peripheral nervous system, where it is a component of the basal lamina of peripheral nerves and is probably an important factor in peripheral nerve regeneration (Ehrig *et al.*, 1990). In the retina, optic nerve and track of birds and mammals, laminin-2 is expressed at an appropriate place and time to interact with retinal ganglion cell growth cones (Cohen *et al.*, 1987; Morissette & Carbonetto, 1995). Cell culture experiments show that retinal ganglion cell axon growth is promoted by both laminin-1 and laminin-2, and it may be that the biologically relevant form in the retina and optic nerve is laminin-2 (Cohen & Johnson, 1991). Laminin-3 (s-laminin) is present in the basal lamina at the neuromuscular junction (Hunter *et al.*, 1989a,b) and that part of the neural tube basal lamina that abuts the floor-plate (Hunter *et al.*, 1992). In the developing and adult vertebrate central nervous system, basal laminae, with associated laminins, are restricted to the pial surface at the glial limitans, and to large blood vessels. However, laminins are also transiently expressed in a restricted fashion in the nervous system during development in a form not associated with the basal lamina, and in locations that suggest a role either in cell migration or axon outgrowth (reviewed in: Reichardt & Tomaselli, 1991; McKerracher *et al.*, 1996; Luckenbill-Edds, 1997). Examples include, in the peripheral nervous system, developing muscle (Chiu & Sanes, 1984), the peripheral pathway of the trigeminal system (Riggott & Moody, 1987) and the spinal nerve roots (Rogers *et al.*, 1986; Westerfield, 1987) and, in the central nervous system, the optic nerve (Cohen *et al.*, 1987; McLoon *et al.*, 1988; Morissette & Carbonetto, 1995), the ventral longitudinal tract of the spinal cord (Letourneau *et al.*, 1988), the cerebral cortex (Liesi, 1985; Hunter *et al.*, 1992) and the hippocampus (Gordon-Weeks *et al.*, 1989b; Zhou, 1990). In some regions of the nervous system in lower vertebrates, laminin expression persists into adulthood and this expression correlates with robust axonal regeneration following injury (e.g., the goldfish optic nerve, Hopkins *et al.*, 1985). It is plausible that growth cones would encounter laminin at some of these sites, given the timing of its expression and location, although this has not been demonstrated directly. On the other hand, contact between growth cones and the basal lamina of the glial limitans, and therefore presumably laminin, has been documented in a number of locations in both vertebrates (e.g. Wilson & Easter, 1991) and invertebrates (e.g. Hedgecock, Culotti & Hall, 1990; Condic & Bentley, 1989a,b).

Laminin-1 is a potent promoter of neurite outgrowth *in vitro*, particularly from central nervous system neurons, although virtually all neuronal types respond to it (Baron van Evercooren *et al.*, 1982; Thompson & Pelto, 1982; Lander *et al.*, 1983; Manthorpe *et al.*, 1983; Rogers *et al.*, 1983; Edgar, Timpl & Thoenen, 1984; Faivre-Bauman *et al.*, 1984; Liesi, Dahl & Vaheri, 1984; Adler,

Jerden & Hewitt, 1985; Engvall *et al.*, 1986; reviewed in: McKerracher, Chamoux & Arregui, 1996; Luckenbill-Edds, 1997). In general, comparisons of the neurite outgrowth promoting potency of laminin and other extracellular matrix molecules such as fibronectin, collagen or proteoglycans, show that laminin is the best substrate for most classes of neuron (Rogers *et al.*, 1983; Smalheiser, Crain & Reid, 1984; Gundersen, 1987; Hall, Neugebauer & Reichardt, 1987; Tomaselli, Damsky & Reichardt, 1987; Tomaselli & Reichardt, 1988). If the laminin substratum is patterned rather than being uniformly distributed, growth cones will remain on the laminin and hence their axons will trace out the laminin pattern (Letourneau, 1975a,b; Hammarback *et al.*, 1985; Hammarback & Letourneau, 1986; Gundersen, 1987). In assays where there is a choice between laminin and another molecule, growth cones usually prefer laminin; however, there are exceptions. For instance, in chick dorsal root ganglia there is a small population of neurons whose growth cones prefer fibronectin to laminin (Gomez & Letourneau, 1994). Also, retinal ganglion cell growth cones show a slight preference for the cell adhesion molecules L1 or N-cadherin over laminin (Burden-Gulley, Payne & Lemmon, 1995). In these experiments, growth cones were observed by time-lapse microscopy as they encountered substrate areas coated with laminin while growing on L1 or N-cadherin. Filopodial contact with the laminin caused a rapid (within 1 minute) change in the morphology of the growth cone. Rapid growth cone responses to laminin have also been seen when soluble laminin is applied directly to the growth cones of sympathetic neurons in culture by micropipette (Rivas, Burmeister & Goldberg, 1992). Growth cone behaviour can also be altered following a single filopodial contact with laminin-coated polystyrene beads (Kuhn *et al.*, 1995). In these experiments, laser tweezers were used to position laminin-coated beads near to chick dorsal root ganglion cell growth cones growing on fibronectin and the behaviour of the growth cones was observed. Initially, growth cone filopodia repeatedly made contact with beads and then retracted. This was followed by a sustained contact during which the filopodium enlarged and the growth cone turned toward the bead. Eventually, the growth cone advanced beyond the bead. Growth cones contacting laminin-coated beads not only changed their direction of growth but also their speed. Changes in growth cone speed were maintained long after contact with the bead had been lost. Increases in growth cone speed have also been seen when soluble laminin is added to growth cones (Rivas *et al.*, 1992), and filopodial enlargement is a feature of the interaction between Ti1 pioneer growth cones in the grasshopper limb bud and guide-post cells (O'Connor *et al.*, 1990; see Chapter 2). These experiments demonstrate that laminin has the potential to act as an intermediate target during pathfinding. The long-lasting changes may relate to intracellular signalling events associated with the activation of laminin receptors (see the section entitled 'Laminin receptors: integrins').

Laminin-2 (merosin) also stimulates neurite outgrowth (Cohen & Johnson, 1991; Engvall *et al.*, 1992), whereas laminin-3 (s-laminin), although strongly adhesive, apparently does not (Hunter *et al.*, 1989a,b, 1991) and may in fact be used as a stop-signal for motoneuron growth cones during synaptogenesis at the neuromuscular junction (see Chapter 5). A tripeptide motif, leucine-arginine-glutamate (LRE in the single letter code), within the $\beta 2$ polypeptide chain of laminin-3 is important for neuronal adhesion (reviewed in Sanes, 1993). When presented to growth cones as a substrate, this tripeptide prevents neurite outgrowth (Hunter et al., 1991). Consistent with the idea that laminin-2 is involved in the development of the neuromuscular junction is the finding that transgenic mice lacking the laminin-2 gene show structural and functional defects at their neuromuscular junctions (Noakes et al., 1995; see Chapter 5).

Fruit flies have at least one type of laminin (Montell & Goodman, 1989) and genetic studies in Drosophila suggest that laminin may have an axon growth-promoting role (García-Alonso, Fetter & Goodman, 1996). In the Drosophila larva, there are small sheets of epithelial cells, called imaginal discs, that give rise to a variety of adult structures, such as legs, at metamorphosis. In the eye-antenna imaginal disc, there is a group of sensory neurons that innervate the three simple eyes (ocelli) by pioneering a pathway to the brain. The growth cones of these neurons pass over the extracellular matrix of the disc epithelium, which is rich in laminin, before turning ventrally toward the brain. The pioneering ocellar neurons in flies with a mutation in the laminin A gene show striking pathfinding defects (García-Alonso et al., 1996). Instead of extending over the extracellular matrix, they grow along the epithelium and sometimes fasciculate with the axons of bristle neurons. However, although the initial direction of axon extension is appropriate, they seldom reach the brain. These findings suggest that the disc epithelium and bristle cell axons can provide adequate support for axon extension of the ocellar pioneers, but they cannot compensate for a loss of guidance cues that ensure that the brain is reached. Whether or not laminin is providing such cues is not clear from these experiments. It is assumed that ocellar pioneer growth cones are interacting directly with the laminin in the disc extracellular matrix, although this will only become clear if functional laminin receptors can be demonstrated on these growth cones and their function perturbed.

Laminin receptors: integrins

If laminins are involved in growth cone pathfinding, then we might expect to find growth cones expressing receptors for laminin in their plasma membranes. The best-characterised membrane receptors for laminins are the integrins (reviewed in: Hynes, 1992; Juliano & Haskill, 1993; Clark & Brugge, 1995; Mercurio, 1995; Fig. 3.2). They are a large family of heterodimeric transmembrane receptors comprised of a non-covalently associated α- and β-subunit, both of which contribute to the binding of the extracellular matrix protein. Integrins are widely expressed on the surfaces of neuronal and non-neuronal cells in vertebrates and invertebrates (reviewed in: Rathjen, 1991;

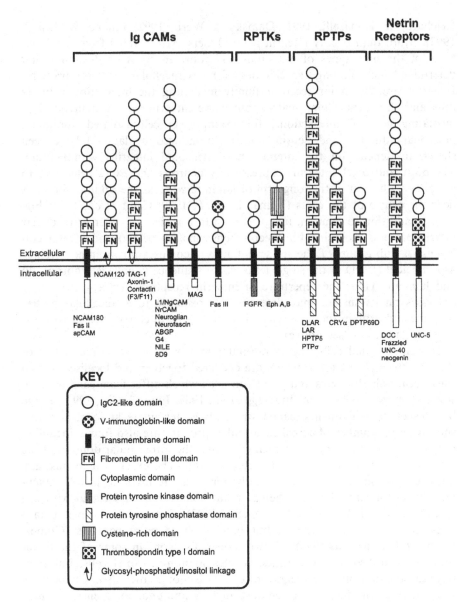

Figure 3.2 Diagrams showing the domain structure of representative members of the immunoglobulin superfamily of membrane proteins implicated in growth cone pathfinding. There are four main subfamilies of the immunoglobulin superfamily that have been implicated in axon growth and growth cone pathfinding, either as receptors or ligands (or both): immunoglobulin cell adhesion molecules (IgCAMs), receptor protein tyrosine kinases (RPTKs), receptor protein tyrosine phosphatases (RPTPS) and netrin receptors. The names of some examples of related molecules and homologues are given, but the lists are not exhaustive. Abbreviations: CAM, cell adhesion molecule; DCC, deleted in colon cancer; Fas, Fasciclin; FGFR; fibroblast growth factor receptor; Ig, immunoglobulin; MAG, myelin-associated glycoprotein; PTP, protein tyrosine phosphatase; UNC, uncoordinated.

Reichardt & Tomaselli, 1991; Damsky & Werb, 1992; Juliano & Haskill, 1993). More than twenty integrin heterodimers exist, formed from combinations of the nine types of α-subunit and fourteen types of β-subunit that determine the binding specificities for extracellular matrix proteins. Integrins are the major receptor family mediating the interaction between cells and the extracellular matrix, but some integrins are also involved in mediating cell–cell interactions, for example, in cell-mediated immunity. Neuronal adhesion and neurite outgrowth induced by laminin have been shown to depend on the interaction of neuronal integrins with laminin. The $\alpha3\beta1$ and $\alpha6\beta1$ integrins expressed by neurons and PC12 cells bind to a region at the end of the long arm of laminin, whereas the $\alpha\beta$ integrin binds near the centre of the cross (Tomaselli et al., 1990; Calof et al., 1994; Fig. 3.1). In addition to laminin, integrins bind to a variety of other extracellular matrix proteins including fibronectin, tenascin (see below) and several collagens. A sub-group of the integrin family recognises an arginine–glycine–aspartic acid (RGD in the single letter code) sequence in collagen, fibronectin and laminin. Thus, the repertoire of integrin receptors that a growth cone expresses on its surface may determine its response to extracellular matrix proteins and explain the preference that some growth cones show in vitro for different substrata (see above).

In non-neuronal cells, the extracellular matrix is linked to the cell cytoskeleton through integrins, which are clustered together and localised at two sites: focal adhesion sites and point contacts, which differ from each other in size and shape (reviewed in: Burridge et al., 1988; Luna & Hitt, 1992). Large bundles of actin filaments, called stress fibres, terminate at focal adhesion sites, where a number of cytoskeletal linker proteins are also found, including α-actinin, talin, and vinculin, which cross-link the intracellular domain of the β-subunit of integrins to the actin cytoskeleton. The exact protein linkages, however, are not yet clear. Focal adhesion sites are involved in cell attachment in non-neuronal cells, whereas point contacts are associated with cell spreading (Tawil, Wilson & Carbonetto, 1993). Growth cones do not have stress fibres or focal contacts, but they do have point contacts (Gomez, Roche & Letourneau, 1996). Growth cones also contain some of the linker proteins identified at focal adhesion sites in non-neuronal cells, but their physical relationship to integrin receptors is not known (see Chapter 2). Integrin receptors have been demonstrated by immunofluorescence on chick dorsal root ganglion cell growth cones, where they are distributed throughout the growth cone but concentrated in the filopodia (Bozyczko & Horwitz, 1986; Letourneau & Shattuck, 1989), in the growth cones of chick sympathetic ganglia (Wu et al., 1996; Grabham & Goldberg, 1997), and biochemically by immunoblotting, in a subcellular fraction from chick brain enriched in growth cones (Cypher & Letourneau, 1991). In the neuroblastoma cell lines SY5Y and NG108-15, and in PC12 cells, integrins are distributed in clusters in the growth cone (Arregui, Carbonetto & McKerracher, 1994; McKerracher et al., 1996).

Functional studies *in vitro* have indicated that integrins mediate neurite outgrowth on extracellular matrix proteins. Neurite outgrowth on laminin, collagen, tenascin and fibronectin *in vitro* is attenuated by antibodies to integrins (Bozyczko & Horwitz, 1986; Cohen *et al.*, 1987; Wehrle-Haller & Chiquet, 1993; Tomaselli, Reichardt & Bixby, 1986; Tomaselli *et al.*, 1987; Kuhn *et al.*, 1995).

The neuronal expression of integrins during development correlates well with the period of axon outgrowth (Cohen *et al.*, 1986; Kawasaki, Horie & Takenaka, 1986; Hall *et al.*, 1987; Tomaselli & Reichardt, 1988). Chick retinal ganglion cell neurons express integrins on their axons while they are growing towards the tectum, but these receptors lose their ability to bind laminin, or to respond appropriately, between embryonic days 6 and 12, at a time when their axons have arrived at the tectum (Cohen *et al.*, 1989). Ablation of the optic tectum reverses the down-regulation of laminin receptors on retinal ganglion cells, suggesting that a target derived factor regulates laminin receptors on afferent growth cones (Cohen *et al.*, 1989). Recent *in vitro* experiments show that laminin receptor (integrin) levels in growth cones can be regulated by laminin itself (Condic & Letourneau, 1997).

The uncoordinated (*unc*) mutants in the nematode *C. elegans* were first identified by Sydney Brenner (1974) and some of them define genes that play important roles in growth cone pathfinding (reviewed by Culotti, 1994). These mutant genes are now being cloned and are providing powerful insights into molecular aspects of pathfinding. A gene called *unc-6* codes for an extracellular matrix protein, Unc-6, that is a component of the epidermal basal lamina over which pioneering dorsoventral motoneuron axons pathfind (Hedgecock *et al.*, 1990). Unc-6 is a secreted protein, which has sequence homology with the $\gamma 1$ polypeptide chain of laminin (Fig. 3.1) and mutations in the gene cause errors in pathfinding (Ishii *et al.*, 1992). The receptor for Unc-6 may well be Unc-5, a transmembrane glycoprotein with immunoglobulin and thrombospondin type 1 extracellular domains coded for by the *unc-5* gene (see Fig. 3.2 and Table 3.2). The pioneer motoneurons that are affected by mutations in the *unc-5* gene send their axons dorsally around the body wall in *C. elegans* (Leung-Hagesteijn *et al.*, 1992). Ventral and longitudinal growth cone migrations are not affected by the *unc-5* mutation. Ectopic expression of Unc-5 in touch receptor neurons, whose axons normally grow ventrally or longitudinally, causes their axons to take a dorsal path, similar to that of the motoneurons endogenously expressing Unc-5 (Hamelin *et al.*, 1993). Strikingly, in the absence of Unc-6 expression, ectopic expression of Unc-5 does not lead to abnormal pathfinding. Although there is no direct evidence that these two proteins interact with each other, these results provide compelling indirect evidence that Unc-6 is the ligand for Unc-5 and, as such, forcefully demonstrate a crucial role for cell surface receptors in growth cone pathfinding, at least in the nematode. Remarkably, Unc-6 has sequence homology to a recently discovered secreted protein involved in pathfinding in vertebrates called netrin-1 (see the section entitled 'Netrins and their receptors').

Tenascins

Tenascins are large, multifunctional glycoproteins found in the extracellular matrix (Fig. 3.1 and Table 3.1). There are four members of the tenascin gene family: tenascin X and tenascin Y, which are not expressed in the central nervous system: tenascin R, which is mainly expressed postnatally in higher vertebrates: and tenascin C, which is expressed in the developing nervous system in vertebrates and down-regulated after axon tract formation (reviewed in Erickson, 1993; Faissner, 1997). Tenascin C is transiently expressed in the developing central and peripheral nervous system (Kruse et al., 1985; Grumet et al., 1985; Crossin et al., 1986; Hoffman, Crossin & Edelman, 1988; Steindler et al., 1989; Wehrle & Chiquet, 1990; Wehrle-Haller et al., 1991; Gonzalez, Malemud & Silver, 1993; Gonzalez & Silver, 1994; reviewed in: Martini, 1994; Faissner, 1997). In the central nervous system, tenascin C is mainly produced by astrocytes. The expression is spatially regulated and the pattern, which is often remarkably boundary-like in form, suggests that tenascin C may play a role in segregating axonal pathways. In the developing mouse somatosensory cortex, when axons are growing into the cortex, tenascin expression delineates the barrel fields, the terminal sites for whisker afferents (Steindler et al., 1989; Hoffman et al., 1988). Similarly, in the chick retinotectal system, tenascin is expressed in areas bordering the optic nerve and tract and the tectum (Perez & Halfter, 1993). However, there are examples where tenascins are expressed within regions where axons are growing. For example, tenascins are expressed in the molecular layer of the postnatal cerebellum when the parallel fibres of the granule cells are elongating and in the mouse optic nerve (Bartsch et al., 1992). Tenascin R expression is less widespread than tenascin C and is confined to the central nervous system, where it is produced by oligodendrocytes (Pesheva, Spiess & Schachner, 1989; Nörenberg et al., 1992; Bartsch et al., 1994). The expression data suggest that tenascins may be involved in both defining axon tracts and in axon outgrowth.

The effects of tenascin on neurite outgrowth *in vitro* depends critically on how growth cones encounter tenascin. Both stimulation and inhibition of neurite growth have been reported (reviewed in: Chiquet-Ehrismann, 1995; Faissner, 1997). Neurite outgrowth from a wide range of neuronal types, both from the central and peripheral nervous system, is generally stimulated when explants or dissociated cells are placed directly on to a tenascin substratum (Crossin et al., 1990; Wehrle & Chiquet, 1990; Lochter et al., 1991; Husmann, Faissner & Schachner, 1992; Taylor et al., 1993; Wehrle-Haller & Chiquet, 1993). However, not all embryonic neural explants extend axons when plated onto tenascin (Taylor et al., 1993; Perez & Halfter, 1993). In contrast, in choice assays where neurons extend neurites on permissive substrates, such as poly-ornithine or laminin, and encounter substrate areas of tenascin, growth cones avoid growing on the tenascin and remain on the permissive substratum (Crossin et al., 1990; Faissner & Kruse, 1990; Perez & Halfter, 1993; Taylor et al., 1993; Götz et al., 1996; Williamson et al., 1996). At the border between the permissive substrate and tenascin, growth cones extend filopodia on to the tenascin region. However, these filopodia are not stabilised, and therefore retract, and the growth cone turns to grow along the side of the border coated with the permissive substrate. It is important to emphasise that the behaviour of growth cones at tenascin borders is different from the response of growth cones to collapsing factors, at least when they are applied globally to the growth cone (see below). It seems, therefore, that growth cones will extend on tenascins if given no other choice but choose not to if an alternative, permissive substrate is offered. Interestingly, a small population of growth cones from dorsal root ganglion explants will cross on to tenascin at borders between permissive molecules and tenascin, but once growing on the tenascin their morphology and behaviour changes (Taylor *et al.*, 1993). Growth cone deflections have also been seen in co-cultures of neurons and glia cells producing tenascin-C (Grierson *et al.*, 1990; Meiners, Powell & Geller, 1995).

There are a number of possible explanations for these, apparently contradictory, effects of tenascins on neurite outgrowth. Tenascins are multifunctional molecules that exist in alternatively spliced forms with different domains that have opposite effects on neurite outgrowth (Götz *et al.*, 1996; reviewed in Faissner, 1997; Fig. 3.1). Antibody perturbation studies and experiments with recombinant fragments of tenascin have shown that several of the fibronectin type III repeat domains and some of the epidermal growth factor-like domains are associated with supporting neurite outgrowth, while other fibronectin-type repeats and epidermal growth factor-like repeats in the molecule inhibit outgrowth (Fig. 3.1; Lochter *et al.*, 1991; Phillips, Edelman & Crossin, 1995; Dorries *et al.*, 1996; Götz *et al.*, 1996). There have been two, independently produced mutant mice lacking the tenascin gene by homologous recombination and neither 'knockout' shows any gross phenotype (Forsberg *et al.*, 1996).

There are several candidate receptors for tenascin (Chiquet-Ehrismann, 1995; Faissner, 1997). Of these, perhaps the most interesting from the standpoint of growth cone pathfinding is contactin (F3/F11), a cell adhesion molecule that belongs to the immunoglobulin superfamily and is expressed on some growing axons and their growth cones (Gennarini *et al.*, 1989, 1991; Brümmendorf *et al.*, 1989, 1993; Zisch *et al.*, 1992; see Fig. 3.2 and Table 3.2). Contactin may bind to fibronectin-like repeats 5 and 6 (see Fig. 3.1) in tenascin C (Vaughan *et al.*, 1994). These repeats flank an alternatively spliced region of the molecule and binding of contactin to the spliced variants is weak. However, although a detailed study of the developmental expression of the spliced and non-spliced variants of tenascin C in relation to contactin expression by growth cones has not been done, the spliced variants predominate during axonogensis over the non-spliced forms (Crossin *et al.*, 1989; Faissner & Steindler, 1995). Antibodies to contactin abolish the growth cone turning response of cerebellar neurons at tenascin-R borders *in vitro* (Gennarini *et al.*, 1989, 1991; Pesheva *et al.*, 1993).

Embryonic chick motoneurons and dorsal root ganglion neurons may bind to tenascin C via their $\alpha 8\beta 1$ integrin receptors (Varnum-Finney *et al.*, 1995). The tenascin binding site of the $\alpha 8\beta 1$ integrin receptor has been mapped to a region within the fibronectin type III repeats and antibodies to $\beta 1$ integrins block binding of neurons to the same tenascin region (Phillips et al., 1995).

Proteoglycans

Proteoglycans are a diverse group of extracellular matrix molecules with a small, protein core and covalently attached glycosaminoglycans (Table 3.1). They are highly charged and bind cell adhesion molecules, other extracellular matrix proteins and growth factors. Proteoglycans can have either promotional or inhibitory effects on neurite outgrowth in vitro depending on the neuron type and the circumstances under which the growth cone encounters the molecule (reviewed in: Lander, 1993; Letourneau et al., 1994; Margolis et al., 1996; Margolis & Margolis, 1997). Heparan sulphate proteoglycan is transiently expressed in some developing axonal pathways (Halfter, 1993) and promotes neurite outgrowth in vitro (Hantaz-Ambroise, Vigny & Koenig, 1987). Keratan sulphate proteoglycan is also developmentally regulated and expressed in some axonal pathways but inhibits neurite outgrowth in culture (Cole & McCabe, 1991). Several members of the aggrecan family of hyaluronan-binding chondroitin sulphate proteoglycan family, including neurocan (Table 3.1), are implicated in axon growth and growth cone pathfinding (reviewed in Margolis & Margolis, 1997). In the retina, the spatio-temporal expression of chondroitin sulphate proteoglycan correlates with the differentiation of retinal ganglion cells, and may contribute to the mechanism that ensures that the axons of these cells grow towards the optic nerve head and not towards the margin of the retina (Snow et al., 1991; Brittis, Canning & Silver, 1992). This same proteoglycan inhibits neurite outgrowth from PC12 cells (Oohira, Matsui & Katoh-Semba, 1991). Proteoglycan expression in the roof-plate of the developing spinal cord may act as a barrier to axon growth across the dorsal midline (Snow, Steindler & Silver, 1990).

Do extracellular matrix molecules have permissive or instructive roles in pathfinding?

An important issue for those molecules of the extracellular matrix that are involved in pathfinding is whether or not they act instructively, for example, by causing growth cone turning, or merely provide a permissive, or non-permissive (inhibitory), substratum for axon growth. At present, the balance of experimental data points to the latter role but an instructive role has not been completely ruled out. One indication that extracellular matrix proteins might be able to act instructively would be their presence in a gradient, rather than being uniformly expressed. In the retina, chondroitin sulphate proteoglycans and laminin-2 are expressed as reciprocal gradients in the basal lamina at the end-feet of the radial glia, where retinal ganglion cell growth cones grow toward the optic nerve head (Snow et al., 1991; Brittis et al., 1992;

Morissette & Carbonetto, 1995). Retinal ganglion cell neurons are born in circumferentially arranged concentric rings beginning in the centre of the retina and radiating outwards. These neurons send their axons centrally toward the optic nerve head. The concentration of chondroitin sulphate proteoglycan is highest at the periphery of the retina and lowest at the optic nerve head, where retinal ganglion cell growth cones leave the retina to enter the optic nerve, whereas laminin-2 concentration is highest at the optic nerve head and declines toward the periphery of the retina. This reciprocal gradient expression has led to the suggestion that retinal ganglion cell growth cones in the retina are guided to the optic nerve by a combinatorial action of these two extracellular matrix proteins, one inhibitory (proteoglycan) and the other stimulatory (laminin-2). However, experiments *in vitro* with laminin gradients suggest that some vertebrate neurons do not respond to extracellular matrix gradients (McKenna & Raper, 1988). In embryonic grasshopper limb buds, the pioneer Ti1 neurons extend growth cones on the limb epithelium basal lamina (Bentley & O'Connor, 1991). Enzymatic removal of the basal lamina *in vivo* causes growth cone retraction, suggesting that the basal lamina provides an essential adhesive substrate (Condic & Bentley, 1989a,c). However, if the basal lamina is removed after the Ti1 growth cones have made contact with guide-post cells in the limb (see above), retraction does not occur, suggesting that the basal lamina is not providing essential guidance cues in this instance (Condic & Bentley, 1989a,b,c). Other attempts to block the action of laminin have also failed to affect axonal pathfinding. When antibodies to laminin were injected into avian embryos, no effect was observed on the innervation pattern of sympathetic preganglionic axons (Yip & Yip, 1992). In a similar vein, abrogation of integrin receptor function impairs axon extension but not growth cone pathfinding. In *Xenopus* embryos, retinal ganglion cell axons express the $\alpha 6 \beta 1$ integrin receptor, which normally binds laminin. When dominant negative forms of this integrin were expressed in retinal ganglion cells, many of these neurons failed to extend axons and those that did grew far more slowly than normal (Lilienbaum *et al.*, 1995). Despite the slower growth rate, these axons took the appropriate route to the tectum.

Thus, while there is overwhelming evidence that extracellular matrix proteins such as laminin and tenascin promote or inhibit axon growth during growth cone pathfinding, it remains unclear whether they have an instructive role.

Cell surface molecules

Growth cone behaviour can be modified by contact with the surface of another cell. The cell involved may be either another neuron or a non-neuronal cell, and the behaviour elicited may range from growth cone retraction and collapse to adhesion and fasciculation. A single filopodial contact with another cell can be sufficient to elicit a behavioural change in the growth cone (Kapfhammer & Raper, 1987a,b; O'Connor *et al.*, 1990; Bastmeyer & Stuermer, 1993). For instance, in the grasshopper limb bud during the formation of the peripheral nervous system, the growth cones of a pair of pioneering sensory neurons in the limb, the Ti1 neurons, pathfind a route toward the central nervous system partly by contacting a series of specific, individual epithelial cells, called guide-post cells, that act as intermediate targets for the migrating growth cones (Bate, 1976; Taghert *et al.*, 1982; Raper *et al.*, 1983a). Single filopodial contact of the Ti1 pioneer growth cones in the limb bud with guide-post cells is sufficient to produce re-orientation of the growth cone (reviewed in: Bentley & O'Connor, 1991; Palka, Whitlock & Murray, 1992). In many cases in invertebrates, guide-post cells are also neurons, or subsequently develop into neurons after they have functioned as guide-posts. Guide-post cells were originally called 'stepping stones' by Bate (1976), who discovered them in *Drosophila*, but they are not the sole mechanism used by the pioneering growth cones in these pathways. Although the filopodia of the Ti1 pioneers are extremely long – many tens of micrometres, these growth cones do not 'step' from one guide-post cell to the next; other intermediate cues, including bands of epithelial cells, are important. Growth cones also receive contact-mediated guidance instructions from non-neuronal guide-post cells, for instance, floor-plate cells in the neural tube and their midline equivalents in *Drosophila* (reviewed in: Klämbt, Jacobs and Goodman, 1991; Tear, Seeger & Goodman, 1993; Goodman & Tessier-Lavigne, 1997). In these examples, the guide-post cells are not present as a single, isolated cell, as in the grasshopper limb bud, but as groups of cells. In vertebrates there are no examples of single guide-post cells.

Growth cone cell-to-cell interactions are mediated by molecules expressed on the cell surface, either as membrane-bound or secreted components, which subsequently bind to the membrane surface. Many of the cell surface molecules that affect growth cones have been identified and their genes cloned (Table 3.2). A high proportion belong to the immunoglobulin superfamily and are highly conserved in evolution (reviewed in Sonderegger & Rathjen, 1992; Walsh & Doherty, 1993; Brümmendorf & Rathjen, 1996; Fig. 3.2).

Neural cell adhesion molecule and the immunoglobulin superfamily

The archetypal neuronal cell adhesion molecule found in the developing nervous system of vertebrates is the neural cell adhesion molecule, or NCAM, a member of the immunoglobulin superfamily of cell adhesion molecules (Cunningham *et al.*, 1987; Rutishauser, 1993). Another, well-characterised, vertebrate cell adhesion molecule of the immunoglobulin superfamily involved in neural development is L1 (Moos *et al.*, 1988; see Table 3.2 and Fig. 3.2). These vertebrate cell adhesion molecules have invertebrate homologues, for example, fasciclin II (see the section entitled 'Fasciclins') and neuroglian are insect homologues of NCAM and L1, respectively. There is considerable evidence that cell adhesion molecules have important roles in promoting axon extension and fasciculation, although direct evidence that NCAM, or any other cell adhesion molecule, has a growth cone pathfinding function is lacking (reviewed in: Edelman & Crossin, 1991; Rutishauser, 1991, 1993; Goodman & Shatz, 1993; Doherty & Walsh, 1994; Walsh & Doherty, 1997).

There are three major isoforms of NCAM produced by differential splicing of a single gene: NCAM-120, NCAM-140 and NCAM-180, with the numbers indicating the approximate molecular weights in kilodaltons. In addition, phosphorylation and glycosylation generates further isoforms. Early binding studies *in vitro* suggested that the receptor for NCAM was NCAM itself, that is to say, a homophilic interaction in which the NCAMs in the cell membranes of adjacent cells interact. More recently, however, it has emerged that NCAM may also interact with other molecules in the same plasma membrane, a so-called *cis*-interaction, such as the cell adhesion molecule L1 (Kadmon *et al.*, 1990a,b; Horstkorte *et al.*, 1993) and also with the fibroblast growth factor receptor (Williams *et al.*, 1994c, 1995a,b; Brittis *et al.*, 1996; Saffell *et al.*, 1997; reviewed in Doherty, Smith & Walsh, 1997). The interaction of NCAM with the fibroblast growth factor receptor opens up the possibility for NCAM to utilise the intracellular signalling pathways used by the receptor (see Chapter 4).

NCAM and L1 show widespread distribution on axon tracts throughout the central nervous system during development, after which their expression is generally down-regulated, except that L1 continues to be expressed on small, unmyelinated axons in the adult nervous system and NCAM-180 is present in postsynaptic densities at brain synapses (Silver & Rutishauser, 1984; Thanos, Bonhoeffer & Rutishauser, 1984; Stallcup, Beasley & Levine, 1985; Martini & Schachner, 1988). The 180 and 140 isoforms of NCAM are present on neurons, including their growth cones, whereas glial cells mainly express the 140 iso-form. L1 is expressed on many axons of developing neurons in the central nervous system (e.g. Persohn & Schachner, 1990) and on axons and growth cones of peripheral sensory neurons (Honig & Kueter, 1995). L1 is also expressed on Schwann cells. Other immunoglobulin superfamily cell adhesion molecules may show a more restricted expression, for instance, contactin (F3/ F11) is found on longitudinal axon tracts in the spinal cord (Rathjen *et al.*, 1987a,b). Commissural axons in the rat spinal cord express TAG-1, or its homologue axonin-1 in the chick, another member of the immunoglobulin superfamily of cell adhesion molecules (Table 3.2 and Fig. 3.2), while they are growing towards the floor-plate in the ventral midline, but down-regulate the molecule as they pass through the floor-plate so that the part of the axon that ascends toward the hindbrain on the contralateral side of the cord no longer expresses TAG-1 (Dodd *et al.*, 1988; Furley *et al.*, 1990). When they reach the contralateral side, commissural axons express L1.

Purified cell adhesion molecules, particularly L1, are very good substrates for supporting neurite growth *in vitro* (Lagenaur & Lemmon, 1987; Lemmon, Farr & Lagenaur, 1989; Grumet *et al.*, 1991; Kuhn *et al.*, 1991; Lemmon *et al.*, 1992; Felsenfeld *et al.*, 1994). Growth cones encountering a border of L1 or N-cadherin, while growing on laminin undergo rapid morphological changes, consistent with alterations in the cytoskeleton mediated by binding of the cell adhesion molecule to the growth cone (Burden-Gulley, Payne & Lemmon, 1995; Burden-Gulley & Lemmon, 1996).

Antibody perturbation experiments, both *in vivo* and *in vitro*, support the idea that cell adhesion molecules such as NCAM and L1 are important for axon fasciculation and axon growth (Rutishauser, Gall & Edelman, 1978; Silver & Rutishauser, 1984; Thanos *et al.*, 1984; Stallcup & Beasley, 1985; Chang, Rathjen & Raper, 1987; Bixby *et al.*, 1987; Harrelson & Goodman, 1988; Grenningloh, Rehm & Goodman, 1991). In cell culture experiments, antibodies to L1 and NCAM have been shown to inhibit neurite outgrowth on Schwann cells (Bixby, Lilien & Reichardt, 1988), astrocytes (Williams *et al.*, 1995b) and retinal glial (Müller) cells (Drazba & Lemmon, 1990). In general, *in vivo* studies have confirmed the *in vitro* findings that antibody perturbation of the immunoglobulin superfamily of cell adhesion molecules leads to axonal de-fasciculation. For example, injection of Fab fragments of antibodies to NCAM – which, being monovalent, are preferable to the whole antibody because they do not cause cross-linking effects – into the developing eye causes pathfinding errors of retinal ganglion cell growth cones. However, it is not clear whether this is due to interference with axon fasciculation or to the interaction between growth cones and the end-feet of radial glial fibres (Thanos *et al.*, 1984; Silver & Rutishauser, 1984). Similarly, Fab fragments of antibodies to L1 perturb axon growth of retinal ganglion cells in the rat retina (Brittis *et al.*, 1995; see also Bastmeyer *et al.*, 1995). Significantly, de-fasciculation of chick spinal cord commissural axons occurs after application of antibodies to NgCAM, the chick L1 homologue (Brümmendorf & Rathjen, 1996; Hortsch, 1996; see Fig. 3.2), but this was not associated with a failure of pathfinding across the floor-plate (Stoeckli & Landmesser, 1995; Stoeckli *et al.*, 1997).

NCAM is expressed during the early part of neural development in an iso-form which has large amounts of poly-sialic acid, a type of polysaccharide that is rare in eukaryotes (but not prokaryotes), covalently bound to the molecule (Hoffman *et al.*, 1982; Rutishauser *et al.*, 1988; Rutishauser, 1991; reviewed in Fryer & Hockfield, 1996). When hydrated, poly-sialic acid occupies a large volume and may interfere sterically with other cell adhesion molecules. Its high charge may also contribute to its function. Poly-sialation of NCAM reduces the strength of its homophilic binding. In the limb plexuses, moto-neuron axon growth cones fasciculate with each other as they exit the spinal cord but de-fasciculate in the plexus as they sort out into their appropriate branches (Tosney & Landmesser, 1985a,b,c). De-fasciculation is associated with an increase in the amounts of poly-sialic acid on motoneuron growth cones (Tang, Landmesser & Rutishauser, 1992). Enzymatic removal of poly-sialic acid from NCAM disrupts pathfinding of motoneuron axons as they negotiate the limb plexus (Tang *et al.*, 1992; see also: Doherty, Cohen & Walsh, 1990b; Boisseau *et al.*, 1991) and inhibits the branching of their axons in developing muscle (Landmesser *et al.*, 1990). These results suggest that post-translational modifications of NCAM can profoundly alter the prop-erties of the molecule (see also Doherty *et al.*, 1992a,b). Alternative splicing of the NCAM gene has also been shown to affect the function of the molecule dramatically. An alternatively spliced exon in NCAM-140, known as VASE (variable alternative spliced exon), has an additional ten amino acids in the fourth immunoglobulin-like domain of the extracellular region of NCAM (the immunoglobulin-like domain that is most distal to the plasma membrane is number 1, see Fig. 3.2; Doherty & Walsh, 1994). The effect of VASE is to reduce axon outgrowth in response to NCAM (Doherty et al., 1990a,b; Doherty et al., 1992a,b; Liu et al., 1993; Saffell, Walsh & Doherty, 1994).

Originally it was proposed that cell adhesion molecules promote axon elongation and growth by modulating growth cone and axon adhesion to cell surfaces, presumably by a similar mechanism proposed for extracellular matrix proteins (see above). Largely as a result of the work of Doherty and Walsh, this idea has now been supplanted by the notion that intracellular signalling events, presumably leading to modulation of the growth cone cytoskeleton, underlie the effects of cell adhesion molecules on neurite outgrowth. The experimental basis for this includes: the lack of a correlation between adhesion and neurite outgrowth for different isoforms of NCAM: the demonstration that cell adhesion molecules, such as NCAM, L1 and N-cadherin, can activate second-messenger pathways within PC12 cells and neurons in culture, and that this, rather than cell adhesion *per se*, correlates with their effects on neurite growth; and the finding that recombinant, soluble forms of cell adhesion molecules mimic the effects of membrane-bound forms (reviewed in: Doherty & Walsh, 1994; Walsh & Doherty, 1997). To investigate the mechanism by which NCAM and other cell adhesion molecules promote axon growth, Doherty and Walsh developed an *in vitro* assay in which PC12 cells and neurons were assessed for their ability to extend neurites on a monolayer of fibroblast-like cells (3T3) transfected with physiological levels of cell adhesion molecules (Doherty et al., 1990a,b, 1991b). The assay was quantified by measuring the length of the neurites extended on the 3T3 cells. There is a basal level of neurite outgrowth on untransfected 3T3 cells, probably mediated by integrin receptors, but this is augmented by a modest but significant increase (about two-fold) when neurites are extending on 3T3 cells transfected with cell adhesion molecules. This simple assay has provided evidence that NCAM, L1 and N-cadherin stimulate neurite outgrowth from PC12 cells and several types of neuron, including cerebellar granule cells and hippocampal neurons, by activating a common second-messenger pathway (reviewed in: Doherty & Walsh, 1994; Baldwin, Walsh & Doherty, 1996; see Chapter 4). These studies imply that cell adhesion molecules promote neurite growth by modulating intracellular events, presumably within growth cones, rather than changing the adhesion between the neurite and growth cone and cell surfaces.

Transgenic mice in which the gene for the highest molecular weight isoform of NCAM (NCAM-180) has been deleted by homologous recombination, a so-called knockout, show no gross abnormalities in their nervous system (Tomasiewicz et al., 1993). Similarly, when transgenic mice lacking all of the isoforms of NCAM are produced, they show no gross abnormal phenotype (Cremer et al., 1994). Although there are subtle differences between the nervous systems of these mutants and wild-type mice, they are far less dramatic or widespread than one might have predicted from the *in vitro* evidence for the function of these molecules. In the hippocampus of mice lacking all isoforms of NCAM, there is a reduction in the number of mossy fibres (Cremer *et al.*, 1996). Most interestingly, in the complete NCAM-deficient mouse, there is a profound loss of granule cells in the olfactory bulb and those that are present are disorganised (Ono *et al.*, 1994). This is due to a failure of migration of the granule cell precursors from the subependymal zone of the lateral ventricles. In both transgenic mice there is an almost entire lack of poly-sialic acid in the central nervous system, suggesting that NCAM bears most of this moiety.

The interpretation of the function of a protein in cases where the transgenic 'knockout' of the protein has no phenotype is problematic. Because developmental mechanisms may compensate for loss of the protein, more especially if the function of the protein is important in development, lack of a phenotype does not necessarily imply that the protein is unimportant. One way in which this may occur is by a hierarchy of guidance cues so that, if one level of cues fails, the next level can act as a 'fail-safe'. Although redundancy of this kind has not been directly demonstrated, and will not be demonstrated until we have a complete understanding of the guidance mechanisms operating over an identified pathway, there is some circumstantial evidence for it. For example, as pointed out by Goodman and Tessier-Lavigne (1997), the two *Drosophila* netrins appear to play functionally redundant roles in growth cone pathfinding at the midline of the central nervous system (see below). There may also be functional redundancy between the EphB2 and EphB3 receptors (see the section entitled 'Eph receptors and their ligands (ephrins)').

More often, transgenic animals in which the protein of interest is overexpressed, either as the normal protein or mutated so that it is constitutively active (so-called gain-of-function) or inactive (so-called 'dominant negative') or ectopically expressed, show phenotypes that are more revealing about the function of the protein.

Consistent with a role for L1 in axon outgrowth and fasciculation, mutations in the L1 gene in humans are associated with a wide range of neurological disorders, including mental retardation and limb spasticity (reviewed in: Wong *et al.*, 1995). Postmortem examination of some of these patients has revealed several neuroanatomical abnormalities, such as an absence or reduction of the corticospinal tracts and corpus callosum. However, the cause of such abnormalities is not known and may equally result from defects in neurogenesis and axonogensis as in pathfinding. Significantly, a phenotype similar in some respects to that in humans occurs in the L1 knockout mouse (Cohen *et al.*, 1997b; Dahme *et al.*, 1997). L1 is X-linked and so hemizygous males in the first generation lack L1. Although born with reduced weight, these animals survive to adulthood. They often had weak hind limb function. In adult mice lacking L1, abnormalities were found in the corticospinal tract, including a failure of many axons to cross the midline at the medullary pyramids in the caudal hindbrain and enter the contralateral dorsal columns. This was due to a pathfinding error because the failure occurred during the development of the pyramidal decussation (Cohen *et al.*, 1997b). During development, L1 is expressed on corticospinal tract axons. However, it is also expressed on many other axon tracts in the central nervous system, such as the corpus callosum and spinal commissural projection, but no abnormalities were seen in these tracts in mice lacking L1 (Cohen *et al.*, 1997b; Dahme *et al.*, 1997). Whether these pathfinding errors in mice lacking L1 are due to abnormalities in axon fasciculation or in other aspects of pathfinding is not clear. A more detailed, anatomical analysis of L1 mutant mice may help to distinguish between these possibilities.

The labelled pathways hypothesis

In invertebrates such as grasshopper and the fruit fly *Drosophila*, individual neurons and their growth cones can be identified by their morphology and location. This experimental advantage has made it possible to document the behaviour of individual, identified growth cones as they pathfind. Growth cone behaviour is remarkably stereotyped in these insects and this has provided an extremely fruitful basis for unravelling the mechanisms of growth cone pathfinding. In grasshopper and *Drosophila*, the central nervous system consists of a series of re-iterated, segmental ganglia that are bilaterally represented and connected to each other within a segment by commissural nerves, which cross the midline, and, between segments, by intersegmental nerves. Many of the neurons in the grasshopper nervous system have been characterised and named, and their presence is invariant. Remarkably, homologous neurons can be found in the *Drosophila* nervous system. The segmental network of axons is built up during embryogenesis by pioneer neurons forming an axonal framework or scaffold upon which later arriving axons fasciculate (Fig. 3.3). Growth cones growing along pioneer axons face decisions at junction-points of the scaffold. Striking examples of active selection of particular axons by growth cones have been documented in the grasshopper nervous system by Goodman and colleagues (Raper *et al.*, 1983a,b,c; Bastiani *et al.*, 1984a,b). For example, the growth cones of the G and C neurons migrate across the midline together, fasciculating on the axons of the posterior commissure. However, when they reach the anterior/posterior fascicle, which contains four identified axons, the C and G growth cones actively select one of these axons and their pathway then diverges: the G growth cone turns anteriorly while the C growth cone turns posteriorly (Raper *et al.*, 1983a). Selective removal of individual, identified neurons, and thus their axons, by laser ablation, impairs the pathfinding of those growth cones that normally fasciculate on the ablated axons (Goodman *et al.*, 1984; Grenningloh & Goodman, 1992). How do growth cones determine the axons with which to fasciculate? The behaviour of growth cones such as the C and G neurons led Goodman's group to propose that individual axons bear labels that can be recognised by particular growth cones, which respond to them by selective fasciculation – the so-called 'labelled pathways hypothesis' (reviewed in Goodman, Grenningloh & Bieber, 1991), a variant of Sperry's chemoaffinity hypothesis (see earlier text). Goodman's

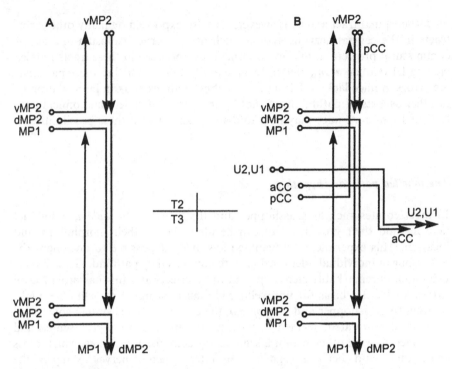

Figure 3.3 Formation of longitudinal fascicles in the grasshopper embryonic central nervous system. Schematic diagram of the formation of the first three longitudinal axon fascicles to form in the grasshopper embryo central nervous system and the seven neurons whose axons fasciculate to pioneer the three fascicles. Neurons from the right-hand half of two adjacent thoracic segments are shown and their cell bodies are represented as open circles. Many of these neurons are siblings, as indicated by their names, e.g the anterior corner cell (aCC) and the posterior corner cell (pCC). Lateral is to the right and anterior is up. The position of the second (T2) and third (T3) thoracic segment boundary and the midline is indicated in B. (A) The first two longitudinal pathways are pioneered by the growth cones of neurons MP1, dMP2 and vMP2. (B) The third longitudinal pathway is pioneered by the U1 and U2 growth cones. The aCC growth cone fasciculates with the U1/U2 axons while its sibling, the pCC growth cone, grows anteriorly and fasciculates with the dMP2 and MP1 axons. (Modified from Bastiani, Pearson & Goodman, 1984a, with permission of the Company of Biologists Ltd.)

group then set out to identify such hypothetical molecules, first by producing monoclonal antibodies against them and later by genetic screens.

Molecules mediating growth cone–cell surface interactions in invertebrates

To identify axonal surface markers that might be involved in growth cone pathfinding in insects, Goodman's group made monoclonal antibodies against embryonic grasshopper central nervous system, and screened the antibodies for labelling of single or subsets of axons in the embryo (Bastiani *et al.*, 1987; Patel, Snow & Goodman, 1987; Bieber *et al.*, 1989). This screen has led to the identification of a number of cell surface glycoproteins that are differentially expressed on axons and growth cones in insect embryos, and that have subsequently been shown to be involved in axonogenesis and pathfinding.

Fasciclins

Four members of a family of transmembrane glycoproteins, called fasciclins I–IV, have now been identified from the insect screens and their genes cloned (Table 3.2). They are all members of the immunoglobulin superfamily of cell adhesion molecules and have sequence homology with cell adhesion glycoproteins found in vertebrates (Fig. 3.2).

Antibody perturbation studies and analysis of mutant fruit flies have provided insights into the function of fasciclins in axon fasciculation and growth cone guidance in invertebrate neural development. Fasciclin II, the insect NCAM homologue, is expressed on a subset of fasciculating axons and growth cones in grasshopper and *Drosophila* central nervous system, and on all efferent axons, including motoneurons, in the peripheral nervous system of *Drosophila* (Bastiani *et al.*, 1987; Harrelson & Goodman, 1988; Grenningloh, Rehm & Goodman, 1991; Lin *et al.*, 1994b). Like NCAM, fasciclin II interactions are homophilic (Grenningloh *et al.*, 1991). When antibodies to fasciclin II are introduced into cultured grasshopper embryos, disruption of the recognition and subsequent fasciculation of the growth cones of the MP1 neuron, which expresses fasciclin II, with the dMP2 fascicle occurs (Harrelson & Goodman, 1988; see Fig. 3.3). Null mutants for fasciclin II in *Drosophila* have no apparent phenotype in the peripheral nervous system but do show axonal fasciculation errors in the central nervous system: the MP1/dMP2 fascicle fails to develop (Grenningloh *et al.*, 1991; Lin *et al.*, 1994b). When fasciclin II is overexpressed in all neurons, Lin and Goodman (1994) found that there was a dose-dependent increase in axon fasciculation. This caused a failure of motoneuron axons to de-fasciculate near their targets and they grew to inappropriate body wall regions. In later stages of embryogenesis, the fasciclin II levels declined and then some motoneuron axons attempted to grow towards and innervate their appropriate targets from inappropriate directions. These findings strongly support the suggestion from *in vitro* studies that fasciclin II functions to fasciculate axons, and that de-fasciculation prior to target innervation is brought about by the subtle interplay between attraction to other axons and to the target.

Fasciclin III is expressed on subsets of motoneurons and muscle fibres in *Drosophila* (Halpern *et al.*, 1991; Jacobs & Goodman, 1989), including the RP3 motoneuron and its targets, the ventral longitudinal muscle fibres 6 and 7 of the body wall. Since fasciclin III binds homophilically in *in vitro* binding assays (Snow, Bieber & Goodman, 1989), it may play a role in the interaction between the RP3 motoneuron and its muscle targets. Unexpectedly, however, mutations that cause loss of function in the fasciclin III gene in *Drosophila* have no phenotype in the peripheral nervous system (Chiba *et al.*, 1995). When fasciclin III is ectopically expressed on all body wall muscle fibres, the RP3 motoneuron axon establishes novel synapses with a large number of body wall muscles (Chiba *et al.*, 1995). These experiments show that, although fasciclin III can induce novel synapses, in its absence the RP3 motoneuron can still find its appropriate muscles. This suggests that there are other guidance cues that ensure correct pathfinding and target recognition by the RP3 growth cone. These may include fasciclin I and II, which are also expressed by RP3. However, embryos carrying null mutations in either fasciclin I or fasciclin II show no abnormalities in RP3 motoneuron pathfinding (Grenningloh *et al.*, 1991; Keshishian *et al.*, 1993; Lin *et al.*, 1994b).

A null mutant for neuroglian, the insect homologue of L1, in *Drosophila*, shows only subtle abnormalities in neural development (Bieber *et al.*, 1989). However, a more recent analysis of other mutations in the gene encoding neuroglian revealed errors in motoneuron axon pathfinding (Hall & Bieber, 1997).

A major drawback of antibody perturbation and transgenic experiments is that there is very little control over the precise location and timing of the perturbation. A recently introduced technique goes some considerable way to overcome this problem. In micro-chromophore-assisted laser inactivation (micro-CALI), a laser beam, localised under visual control, is used to inactivate a protein by highly localised production of short-lived free radicals generated by laser irradiation of a dye (malachite green) covalently coupled to an antibody bound to the protein. Photochemical damage produced by micro-CALI has a half-maximal radius of about 1.5 nm, which makes it possible to inactivate discrete functional domains in a protein. This technique has been used to inactivate fasciclin I in grasshopper embryos, which resulted in disruption of pathfinding by pioneer axons (Jay & Keshishian, 1990), and to inactivate calcineurin (Chang *et al.*, 1995) and myosin V (Wang *et al.*, 1996a) in growth cones in culture (reviewed in Jay, 1996; see Chapter 2).

Cadherins

Not all cell adhesion molecules involved in growth cone pathfinding are members of the immunoglobulin superfamily. The cadherins are a multifunctional family of plasma membrane glycoproteins that promote cell adhesion in a calcium-dependent manner (reviewed in: Takeichi, 1988; Ranscht, 1991; Rathjen, 1991; Geiger & Ayalon, 1992; Grunwald, 1993; Brümmendorf & Rathjen, 1996; Redies, 1997; Table 3.3). Three cadherins were originally described and named after their tissue localisation: N-cadherin (neuronal), E-cadherin (epithelial) and P-cadherin (placental), but the family has since been extended and at least twelve cadherins are expressed in brain (Redies, 1997; Table 3.3). Most of the cadherins are single-pass transmembrane proteins. In the extracellular domain of vertebrate cadherins there are five repeat regions ('cadherin repeats'), each of about 110 amino acids, which contain the histidine–alanine–valine (HAV in the single letter code) motif which is required for cell adhesion. A conserved region in the intracellular domain associates with a family of cytoplasmic proteins called catenins (α, β and γ) that function to cross-link cadherin to the actin cytoskeleton and as nuclear signalling molecules (Rathjen, 1991; Aberle, Schwartz & Kemler, 1996). This interaction is essential for adhesion. T-Cadherin is unusual among cadherins in that it is anchored to the plasma membrane by a glycosyl-phosphatidylinositol linkage and therefore does not have an intracellular domain (Vestal & Ranscht, 1992). T-Cadherin may have a role in repelling sensory and motoneuron growth cones from the posterior sclerotome (Fredette, Miller & Ranscht, 1996; see section entitled 'Collapsing factors'). Cell aggregation experiments suggest that N- and E-cadherin interactions are predominantly homophillic, but it is not clear if this is also true of other family members. Cadherins are generally important for histogenesis in a wide variety of developing tissues, mediating strong interactions involved in cell sorting and tissue boundaries (Geiger & Ayalon, 1992).

Cadherins appear to have a role in axon fasciculation during neural development. Consistent with this view is their expression patterns in developing nervous systems: specific cadherins are transiently expressed in restricted axon tracts. In the developing spinal cord, N-cadherin is expressed on subsets of somatic sensory axons, and on sensory axons in the cranial nerves, including the trigeminal and the glossopharyngeal (Redies, Inuzuka & Takeichi, 1992). In the developing retina, N-cadherin is expressed on retinal ganglion cell axons and R-cadherin is expressed on the radial glial fibres on which they migrate, suggesting that cadherins may be involved in mediating this cellular interaction (Redies, Engelhart & Takeichi, 1993; Redies & Takeichi, 1993). Cadherins are present on growth cone filopodia of dorsal root ganglion cells in culture (Letourneau et al., 1990) and, when presented to growth cones as a substrate, they affect growth cone morphology (Payne et al., 1992). Purified N-cadherin (Bixby & Zhang, 1990; see also Paradies & Grunwald, 1993) and R-cadherin (Redies & Takeichi, 1993) are good substrates for promoting neurite outgrowth and elongation from a wide variety of neuronal types in vitro. Antibody perturbation experiments in vitro suggest that N-cadherin is important in axon fasciculation, and in the adhesion between axons and astrocytes (Matsunaga et al., 1988; Tomaselli et al., 1988; Drazba & Lemmon, 1990; Inouye & Sanes, 1997). Antibody experiments in vivo also support a role for cadherins in axonal fasciculation, and possibly in pathfinding. Honig and Rutishauser (1996) found that blocking antibodies to N-cadherin induced axonal de-fasciculation and defects in axon projections in the hind limb nerve plexus of the chick. A dominant negative form of N-cadherin lacking a large portion of the extracellular domain disrupts axonal and dendritic outgrowth from retinal ganglion cells in Xenopus embryos (Riehl et al., 1996). A cadherin family member (DN-cadherin) has been identified in Drosophila and genetic analysis suggests that it is involved in axon fasciculation in a subset of developing neurons in the central nervous system (Iwai et al., 1997). However, although there were defects in axon fasciculation in DN-cadherin mutants, growth cone pathfinding was mostly normal.

Growth cone collapse and retraction

Cell surface interactions between growth cones and axons in culture may lead, in some circumstances, to filopodial and lamellipodial retraction and to a collapse of the growth cone (reviewed in: Walter et al., 1990; Fawcett, 1993; Luo & Raper, 1994). Growth cone collapse is often followed by neurite retraction. Hughes (1953) was the first to observe growth cone retraction on contact with another neurite (see Chapter 1). Since that time, growth cone collapse and neurite retraction have been observed in a number of situations in culture (Bray et al., 1980; Kapfhammer et al., 1986, Kapfhammer & Raper, 1987a,b; Schwab & Caroni, 1988; Fawcett, Rokos & Bakst, 1989; Cox et al., 1990; Davies et al., 1990; Moorman & Hume, 1990; Raper & Grunewald, 1990; Johnston & Gooday, 1991; Bantlow et al., 1993; Honig & Burden, 1993; Kobayashi, Watanabe & Murakami, 1995; Igarashi et al., 1996; Li et al., 1996; reviewed in: Patterson, 1988; Walter et al., 1990), and, more recently, in vivo (Halloran & Kalil, 1994; reviewed in Stirling & Dunlop, 1995). A form of growth cone collapse occurs when regenerating axons of adult dorsal root ganglion cells growing in the dorsal root toward the central nervous system encounter astrocytes at the dorsal root entry zone (Liuzzi & Lasek, 1987). In general, growth cones collapse following filopodial contact with the surfaces of certain cell types, for instance, when retinal ganglion cell growth cones encounter sympathetic neurons in culture (Fig. 3.4; Raper, Chang & Raible, 1992), or with extracellular matrix molecules. Collapse is frequently preceded by an inhibition of motility and followed by neurite retraction. Often, growth cones will retain a single filopodial contact with the cell, recover and make additional attempts to cross the inhibitory surface, a behaviour which is probably mediated by specific, cell surface receptor interactions (Fig. 3.4, see below). Growth cone collapse is specific in the sense that, for any particular neuron, only certain other neurons or non-neuronal cells will elicit collapse. For example, dorsal root ganglion cell growth cones collapse on encountering oligodendrocytes, whereas retinal ganglion cell growth cones do not (Kapfhammer & Raper, 1987a,b; Kobayashi et al., 1995). Growth cone collapse can occur following contact by a single filopodium (Kapfhammer & Raper, 1987a,b; Bantlow et al., 1990), which implies that an intracellular signalling mechanism is involved.

In addition to cell-mediated growth cone collapse, a wide variety of exogenously applied substances, defined and undefined, can cause reversible growth cone collapse in culture, including the phospholipid lysophosphatidic acid, the major mitotic agent in serum (Jalink et al., 1993), thrombin (Jalink & Moolenaar, 1992; Suidan et al., 1992), and the free-radical gas nitric oxide (Hess et al., 1993; Renteria & Constatine-Paton, 1996).

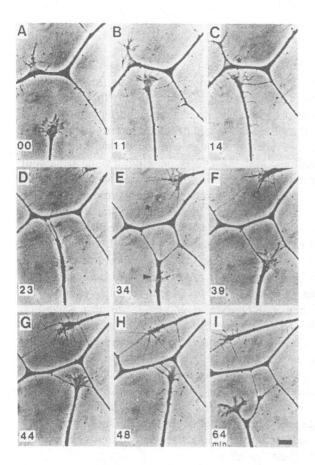

Figure 3.4 Growth cone collapse. Phase-contrast photomicrographs showing the response of a retinal ganglion cell growth cone to a sympathetic axon in culture. When the growth cone comes into filopodial contact with the axon, it collapses and retracts but retains filopodial contact (B–E). Within minutes the growth cone recovers from the collapse while retaining filopodial contact with the sympathetic axon (F), suggesting that there is a desensitisation to the collapsing signal. After recovery from collapse (F), the growth cone makes another attempt to cross the axon but collapses a second time (G–I). Notice that the growth cone first entering the field at D also undergoes a collapse on contacting the sympathetic axon. Numbers in the lower left-hand corner indicate elapsed time in minutes. Bar in I = 10 μm. (From Raper *et al.*, 1992, with permission.)

The phenomenon of growth cone collapse *in vitro* has been exploited in a number of cell culture assays as a basis for the identification of collapsing factors. The involvement of cell-surface contact in growth cone collapse suggested that the collapsing agent was associated with the plasma membrane of cells. It was therefore of some surprise when the first collapsing factor to be identified turned out to be a secreted, soluble glycoprotein (Luo *et al.*, 1993; see below). Most other collapse-inducing factors are, however, membrane proteins.

Collapsing factors

Collapsins/semaphorins

Raper's group have developed an *in vitro* assay as a basis for purifying a collapsing factor that affects dorsal root ganglion cell growth cones in the chick (Raper & Kapfhammer, 1990). In their assay, neuronal membranes from chick brain, presumed to contain the collapsing factor, are incorporated into lipid micelles and these are pipetted on to growth cones in culture or added directly to the culture medium, and the effects on growth cone behaviour assessed. Various subfractions of the neuronal membranes were then tested in the assay until the collapsing activity correlated with a prominent protein band seen on a polyacrylamide gel. Raper's group then obtained amino acid sequence data from the protein and this allowed them to clone the gene encoding the protein, which they called, not surprisingly, collapsin (Luo *et al.*, 1993). When the amino acid sequence of collapsin was determined, it turned out to be closely related to fasciclin IV (Table 3.2), a protein that had been previously implicated in sensory axon guidance in the peripheral nervous system of the grasshopper embryo (see above, Kolodkin *et al.*, 1992; see also Kolodkin, Matthes & Goodman, 1993). However, unlike fasciclin IV, which is a transmembrane glycoprotein, collapsin is a secreted glycoprotein. In support of the identity of collapsin as the chick brain collapsing factor was the demonstration that recombinant collapsin and collapsin produced by transfected non-neuronal cells in culture can induce growth-cone collapse of dorsal root ganglion cells in the collapse assay (Luo *et al.*, 1993). Interestingly, collapsin did not cause collapse of retinal ganglion cell growth cones, as did chick brain membranes, suggesting that other collapsing factors might exist (see the section entitled 'Eph receptors and their ligands').

This prediction has now been confirmed and further work has identified other members of this very large family of proteins, now known as the semaphorins (Kolodkin *et al.*, 1993; Luo *et al.*, 1995; reviewed in Kolodkin, 1996; Culotti & Kolodkin, 1996). The semaphorins number at least thirty members, include both transmembrane and secreted proteins, and have in common a large, extracellular domain, of about 500 amino acids in length, known as the semaphorin domain. The semaphorin/collapsin family is highly conserved from insects to humans and chicken collapsin, now renamed collapsin-I, is homologous to human Sema III and rodent Sema D, while fasciclin IV has been renamed Sema I (see Table 3.4). Some semaphorins have a single immunoglobulin-like domain in their extracellular region (collapsin-I, Sema III, II and A-E), and one subtype has six thrombospondin type 1 domains (Messersmith *et al.*, 1995; Püschel *et al.*, 1995; Luo *et al.*, 1995). Semaphorins are expressed widely and with complex patterns within the developing nervous system. Their effects on growing axons *in vitro* are inhibitory (Luo *et al.*, 1993; Messersmith *et al.*, 1995; Püschel, Adams & Betz, 1995, 1996), i.e. they are chemorepellents and, since there are both secreted and membrane-bound forms, they presumably act both locally and over a distance.

Genetic and antibody perturbation experiments also suggest, in the main, that the semaphorins function *in vivo* as chemorepellents. In the grasshopper limb bud, Sema I (fasciclin IV) is expressed on transverse stripes of epithelial cells. The Ti1 pioneer sensory neuron growth cones stall and then make a sharp, ventral turn at the distal border of one of these stripes (Kolodkin *et al.*, 1992). When blocking antibodies to Sema I are applied to grasshopper limb bud explants in culture, the growth cones of the Ti1 pioneers fail to make a sharp turn at the distal border of Sema I expressing epithelial cells and instead de-fasciculate and extend on the cells, their axons becoming highly branched. Interestingly, the growth cones still make a ventral turn, although far less sharply. These findings suggest that Sema I functions to stall Ti1 pioneer growth cones and prevent their de-fasciculation and axonal branching. Since growth cones still make a ventral turn, some other guidance cue must help them to distinguish ventral from other directions. In *Drosophila*, Sema II also seems to be involved in confining axonal branching, in this case of motoneur-ons to specific muscles (Matthes *et al.*, 1995). These experiments suggest that Sema I and II are growth cone chemorepellents. However, there are indications that semaphorins may also act as permissive guidance cues. At a later time in grasshopper limb development, a second pair of sensory axons, from the sub-genual organ cell body neurons, which lie more distal to the Ti1 pioneers, encounter a second stripe of Sema I across which they migrate before fascicu-lating on the Ti1 pioneer axons (Wong, Yu & O'Connor, 1997). Significantly, function-blocking antibodies to Sema I prevent the extension of these growth cones across the Sema I stripe, suggesting that, in this case, Sema I may be acting as a permissive/attractive guidance cue.

There is some evidence that Sema III in vertebrates may be involved in patterning the projection of sensory afferents in the developing spinal cord. When sensory afferents from the dorsal root ganglion first enter the dorsal spinal cord, there is high expression of Sema III in a region of the dorsal spinal cord that flanks the incoming axons (Messersmith *et al.*, 1995; Shepherd *et al.*, 1996). Since this expression of Sema III occurs at a time when both nerve growth factor-dependent nociceptive afferents and neurotrophin 3-dependent afferents can respond to the chemorepellent activity of Sema III (Shepherd *et al.*, 1997), it may confine these axons to the dorsal spinal cord. At later times, the neurotrophin 3-dependent axons, but not the nerve growth factor-depen-dent ones, lose their sensitivity to Sema III, and this is correlated with their growth towards the ventral part of the neural tube, where their motoneuron targets are (Messersmith *et al.*, 1995). In support of the idea that Sema III transiently confines the neurotrophin 3-dependent axons to the dorsal spinal cord is the finding that ventral neural tube explants release a factor which repels sensory axons *in vitro* (Fitzgerald *et al.*, 1993) and, importantly, this factor can be inhibited with antibodies to Sema III (Shepherd *et al.*, 1997).

Two transgenic mouse lines lacking Sema III have been produced (Behar *et al.*, 1996; Taniguchi *et al.*, 1997). However, these two animals have different phenotypes. In one knockout (Behar *et al.*, 1996), nerve growth factor-depen-dent primary sensory axons project abnormally into ventral spinal cord, as would be predicted from the Sema III experiments discussed above. The other animal (Taniguchi *et al.*, 1997) has no abnormalities of sensory axon projections in the spinal cord but does show severe abnormalities throughout the peripheral nervous system, including abnormal nerve branches from the dorsal root ganglia and abnormal nerve branches in the branchial arches and limb buds. The reason for this discrepancy is not understood.

Whether the collapsins/semaphorins cause frank growth cone collapse *in vivo* or elicit more subtle changes in growth cone behaviour is not yet clear. In the *in vitro* growth cone collapse assay these chemorepellents are applied globally to the growth cone and this probably does not mimic the way in which growth cones encounter such factors *in vivo*. When chick dorsal root ganglion cell growth cones contact collapsin-I coated beads *in vitro*, they do not collapse but instead steer away from the beads (Fan & Raper, 1995). Thus, growth cone behaviour towards chemorepellents depends critically on how they are encountered.

The growth cone collapse assay has also been used to examine the posterior half-somite collapsing factor in vertebrates (Davies *et al.*, 1990) and collapsing factors from the tectum in the midbrain (Cox, Müller & Bonhoeffer, 1990). Spinal nerves are segmentally organised, in the sense that they are paired structures found only in the anterior half of the sclerotome, despite their rootlets exiting all along the spinal cord (Keynes & Stern, 1984). This arrangement is not brought about by an intrinsic segmentation mechanism in the spinal cord, but by segmental differences in the paraxial mesoderm and the somites that develop from it. This results in sensory and motoneuron growth cones only growing through the anterior half of the somite (Keynes & Stern, 1988; Jacobson, 1991). The migration pathways of neural crest cells are also segmented; they also only migrate through the anterior half-somite. It has been proposed that a factor, expressed by posterior half-somite cells during development, is inhibitory to axon growth (Keynes & Stern, 1984). Consistent with this idea is the finding that membrane fractions derived from the posterior half-somite, but not those from the anterior half-somite, cause dorsal root ganglion cell growth cones to collapse in the Raper collapsing assay (Davies *et al.*, 1990). Also, when motoneuron growth cones encounter posterior sclerotome cells in culture, growth cone advance is inhibited and they turn away or branch, whereas interactions with anterior sclerotome cells enhance growth cone advance (Oakley & Tosney, 1993). The identity of the molecule responsible has not been unequivocally established, although there are several candidate molecules that are differentially expressed in the posterior half-somite at appropriate times during development (reviewed in Tannahill, Cook & Keynes, 1997).

In vertebrates, near the neural tube and notocord, the somites develop into sclerotome, which gives rise to the vertebral column and, more laterally, dermatome and myotome (together = dermamyotome), which develop into dermis and skeletal muscle respectively. Although collapsin-I is expressed in the dermamyotome during the time when growth cones are migrating through the anterior sclerotome, it is only restricted to the posterior sclerotome after axonal migrations (Wright *et al.*, 1995; Adams, Betz & Püschel, 1996; Shepherd *et al.*, 1996). Furthermore, there are no abnormalities in the segmentation of spinal nerves either in Sema III/Sema D knockout mice (Behar *et al.*, 1996; Taniguchi *et al.*, 1997) or in animals in which the dermamyotome has been surgically removed (Tosney, 1987).

The calcium-dependent, glycosyl-phosphatidylinositol-linked cell adhesion molecule T-cadherin (see above and Table 3.3) is expressed in the posterior half-sclerotome, but not the anterior half-sclerotome, and on the surface of motoneuron growth cones (Ranscht & Bronner-Fraser, 1991; Fredette & Ranscht, 1994). Significantly, T-cadherin inhibits motoneuron axon growth cones *in vitro* when presented as a substratum at a stage when they would be growing through the sclerotome (Fredette, Miller & Ranscht, 1996).

Some members of the Eph receptor tyrosine kinase family and their ligands are also expressed in the posterior sclerotome and are the best candidates for the posterior half-somite collapsing factor (see below).

Collapsin/semaphorin receptors
A receptor family for some members of the semaphorin family, called neuropilins, has been discovered (Chen *et al.*, 1997; Feiner *et al.*, 1997; He & Tessier-Lavigne, 1997; Kolodkin *et al.*, 1997; reviewed in Kolodkin & Ginty, 1997). Neuropilin-1 was originally identified with a monoclonal antibody that recognised a membrane protein in the developing nervous system of vertebrates and stained specific subsets of axons, including those of spinal and cranial motor and sensory neurons, the visual system and the olfactory system (Takagi *et al.*, 1987, 1991, 1995; Kawakami *et al.*, 1996). Neuropilins are single-pass transmembrane proteins with large and complex extracellular domains containing motifs that suggest protein–protein interactions and a small intracellular domain of about forty amino acids with no known homologies. Overexpression of neuropilin-1 in chimeric mice caused ectopic sprouting and de-fasciculation of axons in the peripheral nervous system (Kitsukawa *et al.*, 1995). These observations suggest that neuropilins may be involved in axon growth or guidance. This idea has been confirmed by the finding that Sema III binds to neuropilin-1 with high affinity and that antibodies to the extracellular domain of neuropilin-1 can prevent the collapse of dorsal root ganglion cell growth cones produced by Sema III in culture (He & Tessier-Lavigne, 1997; Kolodkin *et al.*, 1997). Neuropilin-2 binds with high affinity to Sema E and IV but not Sema III (Chen *et al.*, 1997). Neuropilin-1 is expressed on Sema III responsive axons, including those of peripheral sensory and sympathetic neurons and spinal motoneurons (He & Tessier-Lavigne, 1997; Kolodkin *et al.*, 1997). The neuropilin knockout mouse (Kitsukawa *et al.*, 1997) has a similar phenotype to the Sema III knockout mouse, described by Taniguchi *et al.* (1997), in which sensory axon projections in the spinal cord are normal whereas peripheral projections are disrupted. These observations provide strong evidence for a role for Sema III in peripheral pathfinding by sensory axon growth cones.

Now that a family of semaphorin receptors has been identified, we can expect rapid progress to be made in identifying the intracellular signalling pathways that lead to semaphorin-mediated growth cone collapse. Before it was realised that the neuropilins are semaphorin receptors, Goshima *et al.* (1995) used a *Xenopus* oocyte expression screen to identify a 62 kD protein called CRMP-62, for *c*ollapsin *r*esponse *m*ediator *p*rotein, that is a potential intracellular signalling component in the collapsin/semaphorin pathway in growth cones. CRMP-62 is highly expressed in the chick developing nervous system and, when antibodies against CRMP-62 are introduced into chick dorsal root ganglion cells *in vitro*, they prevent growth cone collapse induced by collapsin-I. Interestingly, CRMP-62 has sequence homology to the product of the *unc-33* gene of *C. elegans* (Li, Herman & Shaw, 1992), a gene involved in axonal pathfinding in many neuronal classes in these animals (McIntire *et al.*, 1992).

The growth cone collapse induced *in vitro* by collapsin-I in chick dorsal root ganglion cells is associated with a selective loss of actin filaments in the peripheral domain of the growth cone (Fan *et al.*, 1993). Collapsin-I also causes axon retraction in these neurons and this is probably associated with changes in the microtubule cytoskeleton (see Chapter 2), although this has not been examined directly.

Eph receptors and their ligands (ephrins)
In birds and amphibians, retinal ganglion cell axons form connections with the tectum in the midbrain in such a way as to produce a topographic map of the retina on the tectum; nasal axons of retinal ganglion cells project to posterior tectum, and temporal, dorsal and ventral retinae are connected to anterior, ventral and dorsal tectum, respectively. To study the mechanisms underlying the specificity of the retino-tectal projections, Bonhoeffer's group have developed an *in vitro* assay called the 'stripe choice' assay, in which explants of embryonic chick retina are allowed to extend ganglion cell axons over a substratum consisting of alternating stripes of test material (Walter *et al*, 1987a,b). In the most common arrangement, the stripes consist of membrane fractions derived from either the posterior or the anterior part of the tectum. Using such an assay system, Bonhoeffer's group have shown that retinal ganglion cell axons derived from the temporal part of the retina only grow on stripes containing membranes derived from the anterior part of the tectum. In contrast, nasal axons grow equally well on membranes from either anterior or posterior tectum. The main mechanism confining temporal growth cones to anterior stripes seems to be that posterior stripes cause them to collapse. The collapsing factor is unlikely to be collapsin-I because it does not cause the collapse of retinal ganglion cell growth cones *in vitro* (Luo *et al.*, 1993; see earlier text). Furthermore, the tectal factor is a glycosyl-phosphatidylinositol-linked membrane glycoprotein, whereas collapsin-I is a secreted protein (Stahl *et al.*, 1990). Bonhoeffer's group have identified a protein factor from posterior tectum, which is responsible for causing the collapse of temporal retinal ganglion cell growth cones (Drescher *et al.*, 1995). Biochemical evidence indicated that the

These observations are consistent with a role for Eph receptors and ephrins in confining or restricting axonal projections by chemorepulsion.

Chemotropic (diffusible) factors

Soluble factors (molecules) that act as guidance cues and originate from a source at some distance from the growth cone, and are delivered to the growth cone by diffusion are known as chemotropic molecules. Such molecules should be distinguished from chemotrophic factors, which are molecules that nourish and sustain cells but do not produce directed or vectorial growth. It is assumed that chemotropic molecules are continuously released into the extracellular space by target cells and that molecular diffusion creates a gradient of concentration in the embryo that growth cones can detect and to which they can respond. As originally conceived, chemotropic factors were assumed to exert a positive effect on growth cone motility, and thus to induce growth towards the source of the chemoattractant – a movement known as chemotaxis. Recent experiments, however, show that some chemotropic factors may be inhibitory and therefore induce growth away from the source of the chemotropic factor, in other words a chemorepellent. In either case, chemotropism is distinct from other mechanisms of growth cone guidance in that the factor operates over relatively long distances. Although the existence of chemotropic factors has been suspected for a long time (Ramón y Cajal, 1892; Langley, 1895, 1897), it is only in recent years that their action has been directly demonstrated, mainly by the use of *in vitro* assays (Coughlin, 1975; Lumsden & Davies, 1983, 1986; Tessier-Lavigne *et al.*, 1988; Placzek *et al.*, 1990a,b; Pini, 1993; Fitzgerald *et al.*, 1993; reviewed in: Tessier-Lavigne, 1992, 1994). Chemotropic factors are now being characterised at the level of the gene (Kennedy *et al.*, 1994; Serafini *et al.*, 1994).

Chemotropism as a mechanism for growth cone pathfinding during development was first proposed by Ramón y Cajal on the basis of a wealth of circumstantial evidence (Ramón y Cajal, 1892). The evidence included the observation that inappropriately located or ectopic neurons and disoriented neurons still managed to direct their axons during development to appropriate targets despite having to take abnormal routes to do so (Ramón y Cajal, 1960). It has also been known since the end of the last century that the distal stump of a severed adult nerve releases chemotropic factors that attract the growth cones of regenerating axons (reviewed in Ramón y Cajal, 1928). Forssman coined the word *neurotropism* to describe the chemoattractant emanating from the distal stump of injured nerves (Forssman, 1898, 1900). Ectopically located neurons occur naturally in normal animals and also in some neurological mouse mutants (see Jacobson, 1991). In the reeler mouse, for instance, neurons in the cerebral cortex migrate abnormally and thus become inappropriately located in the adult. Despite this, and the fact that their axons take abnormal routes, correct connections between these neurons and their appropriate synaptic partners in the thalamus are established (Caviness, 1976; Caviness & Yorke,

1976; Dräger, 1981; Simmons *et al.*, 1982). Furthermore, if embryonic neurons
are ectopically transplanted they can still, in some cases, find their correct
targets despite taking abnormal routes to do so (Lance-Jones & Landmesser,
1980, 1981b; Harris, 1986). However, these experiments do not always distinguish between guidance cues diffusing over long distances from local, more
stable influences, since an apparent 'homing behaviour' may also be produced
by displaced growth cones serendipitously encountering axon tracts and other
structures along which they can grow (Harris, 1989).

Nerve growth factor

Nerve growth factor (NGF) is a protein growth factor secreted by target cells
of sympathetic and sensory axons in the peripheral nervous system (reviewed
in: Levi-Montalcini & Angeletti, 1968; Barde & Thoenen, 1980; Barde, 1989).
It is a member of a growing family of related proteins, the neurotrophins, that
prevent genetically programmed, or apoptotic, cell death of distinct classes of
embryonic neurons and influence innervation patterns in target tissues. The
discovery of NGF in the 1950s led to the idea that it may be a chemotropic
factor (reviewed in: Davies, 1987; Ernsberger & Rohrer, 1994; Tessier-Lavigne,
1994). This was supported by the finding that, when injected into the brain
stem of neonatal mice and rats, it caused sympathetic axons from the para-
vertebral sympathetic ganglia in the peripheral nervous system to grow abnormally through the dorsal roots into the spinal cord and through the dorsal
funiculus toward the injection site (Menesini-Chen, Chen & Levi-Montalcini,
1978; see also Hassankhani *et al.*, 1995). Further support came from the direct
demonstration by Gundersen and Barrett that growth cones can re-orientate
rapidly (within minutes) toward a pipette releasing NGF *in vitro* (Fig. 3.5;
Gundersen & Barrett, 1979, 1980; see also: Campenot, 1977; Ebendal &
Jacobson, 1977; Letourneau, 1978; Gallo, Lefcort & Letourneau, 1997) or
adsorbed gradients (Gundersen, 1985). The growth cones of PC12 cells and
sympathetic neurons can also respond rapidly to applied NGF (Seeley &
Greene, 1983; Connolly, Seeley & Greene, 1985; Aletta & Greene, 1988;
Grabham & Goldberg, 1997). This local growth cone response is due, at
least in part, to a response to substrate-adsorbed NGF (Gundersen, 1985).

Although these experiments show that growth cones can respond to chemical gradients with turning responses, they were performed using far higher
concentrations of NGF than growth cones are likely to encounter *in vivo*
(reviewed in Tessier-Lavigne & Placzek, 1991). NGF is released from the target
tissues of sensory and sympathetic neurons and, when the source of NGF is
removed by X-irradiation of somitic mesoderm, much of the pattern of nerve
branching is normal (Lewis *et al.*, 1981). Only at the level of the innervated
tissue is the innervation pattern abnormal or absent (Lewis *et al.*, 1981; Martin,
Khan & Lewis, 1989). Finally, in the trigeminal system at least, neurons do not
respond to NGF and NGF is not produced by the target tissue until after the
innervating axons have arrived (Davies *et al.*, 1987).

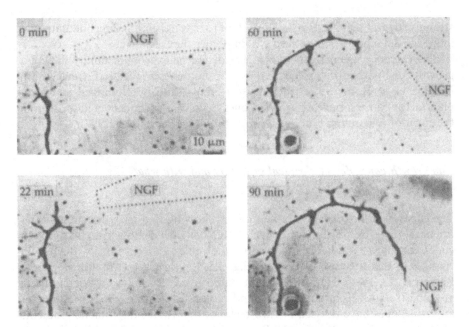

Figure 3.5 Chemotropic effects of nerve growth factor. Photomicrographs showing the response of a dorsal root ganglion cell growth cone to nerve growth factor diffusing from a micropipette tip (dotted line). Numbers in the upper left-hand corner indicate elapsed time. The growth cone grows towards the micropipette tip under the influence of nerve growth factor (approximately 2.5 nM). Ninety minutes after the start of the experiment, the axon has grown by 108 μm and turned an angle of almost 160° from its original direction of growth. The turning response of the growth cone to NGF is rapid, of the order of minutes, suggesting that it is a direct, tropic effect and not mediated by a transcription event in the nucleus, i.e. a trophic action. (Reproduced with permission from Gundersen & Barrett, 1979. Copyright 1979 American Association for the Advancement of Science.)

While these observations preclude a role for NGF in pathfinding, they do not rule out a local role in determining innervation patterns, such as axonal density and branching, within the target tissue. Consistent with this view are experiments using transgenic mice in which NGF was ectopically expressed in sympathetic neurons, which are normally exposed to NGF only after reaching their target tissues. Axonal pathfinding by these neurons in the transgenic mice was normal but the extent of terminal branching within target tissues was dramatically reduced (Hoyle *et al.*, 1993). The simplest interpretation of this finding is that NGF, secreted by the sympathetic axons as they invade their normal targets, compromised the target-derived NGF gradient and that it is the latter which is responsible for determining the extent of axonal branching within the target. To test this idea, Hoyle *et al.* (1993), in an elegant experiment, artificially increased the amount of NGF produced by a target tissue and found that this restored the normal innervation pattern by sympathetic axons that were ectopically expressing NGF, presumably by re-establishing the gradient. Collectively, these results strongly suggest that NGF is not important for axonal pathfinding but that it can influence the innervation pattern within tissues that secrete it.

The evidence against other members of the neurotrophin family having a growth cone pathfinding role is far less substantial than for NGF (reviewed in McFarlane & Holt, 1997). Both neurotrophin-3 and brain-derived neurotrophin factor can re-orientate growth cones in culture when delivered to them in the form of a standing gradient *in vitro* (Song, Ming & Poo, 1997).

Can growth cones detect chemical gradients and with what steepness?

The experiments of Gundersen and others showing that growth cones will re-orientate towards a diffusing source of NGF *in vitro* demonstrate that growth cones can detect gradients of diffusing molecules and respond to them with changes in the direction of growth (Fig. 3.5; Letourneau, 1978; Gundersen & Barrett, 1979, 1980). Haptotaxis is the name given to cellular movement along an adhesive surface of substrate-bound molecules (Carter, 1967). Bonhoeffer and colleagues have devised *in vitro* techniques to make haptotactic gradients of molecules that influence pathfinding (Baier & Bonhoeffer, 1992; Baier & Klostermann, 1994). This has made it possible to determine the steepness of the gradient to which growth cones will respond. For these experiments they grew chick retinal ganglion cells from the temporal side of the retina on a substratum made of membranes from the posterior tectum – a region of the tectum that contains a factor which repels growth cones of temporal retinal ganglion cells (see above). The membranes were deposited in a concentration gradient along which the retinal growth cones were allowed to grow. At particular locations along the gradient the growth rate of the retinal axons slowed, presumably because they were responding to the repelling factor. By determining the steepness of the gradient where growth cones responded, Baier and Bonhoeffer (1992) calculated that a concentration difference of just over 1% over 25 μm could be detected by the growth cones. Growth cones responded to the slope of the gradient and not its absolute value. This degree of gradient is remarkably shallow and implies that growth cones contain an internal amplification system to enhance the steepness of the gradient effectively (Walter *et al.*, 1990). Presumably a spatially regulated gradient of some factor must be established transiently within growth cones that reflects the gradient of the extrinsic guidance cue.

Demonstration of chemotropic factors in vitro

Tissue culture studies have contributed greatly in demonstrating and characterising chemotropic factors. The general approach is to place in culture small pieces (explants) of neural tissue in close proximity to explants of target tissue or other tissue to be assayed, and to monitor the behaviour of neurites emerging from the neural explant (Fig. 3.6). In many cases, explants of developing neural tissue do not extend axons *in vitro* unless an appropriate target tissue is also placed in the culture dish within about 1 mm of the neural explant. For instance, Coughlin (1975) found that foetal mouse submandibular ganglion explants only extended axons when co-cultured with submandibular gland epithelium – their natural target – separated by a distance of not more than 0.5 mm. Other salivary glands could substitute for the submandibular, with varying efficiency, but not a range of other embryonic tissue. The chemotropic factor passed through a 0.1 μm filter, suggesting that it was a molecule and not a cell process. Furthermore, the factor was not NGF, as indicated by a lack of effect of either NGF itself or immunoprecipitating antibodies to NGF. From similar experiments using a range of neural tissues, it is now clear that developing neurons from both the peripheral nervous system (Chamley, Goller & Burnstock, 1973; Ebendal, 1976; Ebendal & Jacobson, 1977; Lumsden & Davies 1983, 1986) and the central nervous system (Gähwiler & Hefti, 1984; Gähwiler & Brown, 1985; Tessier-Lavigne *et al.*, 1988; Bolz *et al.*, 1990; Heffner, Lumsden & O'Leary, 1990; Placzek *et al.*, 1990a,b; Molnár & Blakemore, 1991) can respond *in vitro* to chemoattractants derived from appropriate target tissues.

In these *in vitro* assays, it is necessary to distinguish between a chemotropic and a chemotrophic effect on the neurons in the explant. The key feature is growth cone turning. If the factor released into the culture medium by the target tissue is a chemotrophic molecule, then it will enhance survival of the neurons in the explant and they will, therefore, extend axons from the explant – but these axons will not orientate towards the target explant, instead they will extend radially from the explant (Fig. 3.6). Those neurons in the explant that are nearest to the target explant will be exposed to a higher concentration of the chemotrophic factor than those further away, and therefore their response may be more vigorous or occur earlier, so that they extend neurites more rapidly or sooner and therefore have longer neurites.

Growth cone turning towards the target, and hence axon re-orientation, is only seen when chemotropic factors are acting (Fig. 3.6). This distinction was first made in the experiments of Lumsden and Davies (1983, 1986). In their assay, explants of the chick trigeminal ganglion were co-cultured with maxillary epithelium, the natural target for these neurons, and limb bud, an unnatural target. The explants were cultured in a collagen gel, a three-dimensional extracellular matrix, which has been shown to support diffusion gradients of chemoattractants (Ebendal & Jacobson, 1977). Axons emerging from the trigeminal explants turned towards the maxillary epithelium explants and not towards the limb bud explants. The chemoattractant released by maxillary epithelium, which is sometimes known as Max Factor, has not been identified. That it might be NGF has been ruled out (Lumsden & Davies, 1986; Davies *et al.*, 1987).

Since the pioneering observations of Lumsden and Davies (1983, 1986), there have been many demonstrations of chemotropic effects *in vitro* from other parts of the nervous system (Tessier-Lavigne *et al.*, 1988; Heffner,

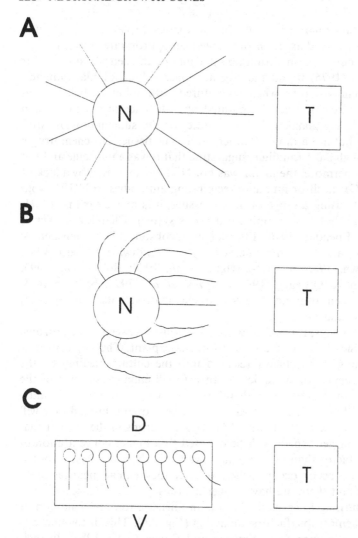

Figure 3.6 Diagrams illustrating *in vitro* assays for chemotropic factors. (A) Absence of a chemotropic effect. In this case neurite outgrowth (straight lines) from the neural tissue (N) is radially symmetrical, indicating that there is no chemo*tropic* factor being released from the target tissue (T). However, the length of neurite outgrowth from the target side of the neural explant is greater than elsewhere, implying that the target tissue may be releasing a chemo*trophic* factor. (B) Neurite outgrowth from the neural tissue (N) is directed towards the target tissue explant (N). This behaviour is consistent with the release of a chemotropic factor from the target. (C) In this example, an explant of spinal cord has been placed near to the target explant (T), oriented so that dorsal (D) is up and ventral (V) is down. Commissural interneurons (circles) within the neural explant normally send axons ventrally. In the presence of a target tissue releasing chemotropic factors for these neurons, axon outgrowth is re-oriented toward the target. The magnitude of the chemotropic effect is directly proportional to the distance between the individual neuron and the target. The chemotropic effect can be seen *in vitro* over several hundred micrometres (Tessier-Lavigne *et al.*, 1988).

Lumsden & O'Leary, 1990; Placzek *et al.*, 1990a,b; Yamamoto *et al.*, 1992; Fitzgerald *et al.*, 1993; Pini, 1993; Colamarino & Tessier-Lavigne, 1995; Guthrie & Pini, 1995; Messersmith *et al.*, 1995; Shirasaki *et al.*, 1995; Tamada, Shirasaki & Murukami, 1995; Sugisaki *et al.*, 1996).

In vertebrate cerebral cortex, cortico-spinal neurons from layer V send an axon into the spinal cord and then, at some later stage, a collateral or 'back' branch develops from this axon in the hindbrain, which then invades the basilar pons (O'Leary & Terashima, 1988). Explant culture experiments in collagen gels suggest that the pons secretes a factor that can induce collateral branching in cortico-spinal axons and also attract the collateral branch (Heffner, Lumsden & O'Leary, 1990; O'Leary *et al.*, 1990; Sato *et al.*, 1994). By developing an explant culture system consisting of a slice of brain tissue from neonatal mice containing the entire pathway of cortico-spinal axons, Bastmeyer and O'Leary (1996) were able to image collateral branching of cortico-spinal axons to the basilar pons using time-lapse video microscopy. The first signs of collateral branching from axon shafts is an alternating thickening and thinning of the shaft, and the appearance of varicosities and filopodia; these events are followed by axon outgrowth toward the pons. In this example, a clear-cut growth cone at the tip of the collateral branch does not appear to form.

Early in the development of the spinal cord of higher vertebrates a population of commissural interneurons in the dorsal and lateral mantle zone extend axons ventrally along the neuroepithelial end-feet of the mantle zone to the mid-dorsoventral level, and then turn obliquely, deep to the motoneuron pool, and grow towards the midline (see Fig.1.3 and Chapter 1). They cross the ventral midline through the floor-plate, a region of the neural tube where a specialised group of neuroepithelial cells are found. On the ipsilateral side of the floor-plate, they turn and ascend to the brain. The behaviour of the commissural growth cones suggests that they are attracted to the floor-plate by a signal emanating from it (Ramón y Cajal, 1909). The first experimental demonstration that the floor-plate might attract commissural axons by chemotropism was achieved by mechanically disrupting the neural tube so that the commissural neurons were displaced (Weber, 1938). When this was done, it was found that commissural axons could still find the floor-plate, despite, in some cases, having to leave the neural tube and then enter it at another site.

To test the idea that a chemotropic factor for commissural axons is produced by the floor-plate, Tessier-Lavigne and colleagues, in landmark experiments, co-cultured rat neural tube floor-plate explants in collagen gels with explants from the region of the neural tube containing the commissural neurons (Tessier-Lavigne *et al.*, 1988; Placzek *et al.*, 1990a,b; Fig. 3.6C). They found that commissural growth cones were attracted to the floor-plate explants and that this attraction operated over a distance of about 250 μm (Fig. 3.6C). This is well within the distance a chemotropic factor might be expected to act in the spinal cord and in other locations during neural development. For example, in *Xenopus* embryos, pioneer axons in the spinal cord elongate at rates of 41–75 μm/hr and make target contact from between 1 and 3.5 hours after the initial outgrowth from the neuronal cell body (Jacobson & Huang, 1985); thus they would have travelled between 41 and 262.5 μm. Within the explant containing the commissural neurons, axons were seen to re-orientate toward the floor-plate explant to a degree that was directly proportional to their distance from the floor-plate explant (see Fig. 3.6). The attraction of the floor-plate factor was specific, since only floor-plate of appropriate age could attract commissural growth cones. Conditioned medium from floor-plate cultures had the same effect. A protein factor produced by floor-plate cells and which has many of the properties expected of the chemoattractant has been purified and its gene cloned (see below).

Not all long-range chemotropic factors appear to be attractants. In the mammalian olfactory system, olfactory bulb mitral and tufted neurons project their axons out of the olfactory bulb to form the lateral olfactory tract. Within the bulb these axons grow away from the septum at the midline. To test whether chemorepellent factors might originate from the septum to direct the axons away from the midline, Pini (1993) co-cultured olfactory bulb explants with septal explants in collagen gels. He found that olfactory bulb axons grew away from the septal explants, both within the explant and the collagen gel, consistent with the idea that a chemorepellent factor produced by the midline septum directs axons within the bulb away from the midline. Evidence for chemorepellent factors has also been obtained from other regions of the developing nervous system. For example, an inhibitory, chemorepellent factor for a population of sensory, dorsal root ganglion axons is released from ventral, but not dorsal, embryonic rat spinal cord in collagen gels (Fitzgerald *et al.*, 1993). The production of this factor is developmentally regulated. Since some classes of primary sensory axons terminate deep within the ventral spinal cord whereas other classes remain within superficial regions of the dorsal cord, these results suggest that the ventral cord chemorepellent plays a role in patterning of primary sensory afferents in the dorso-ventral axis of the spinal cord. The ventral cord chemorepellent factor is probably Sema III (see the section entitled 'Collapsing factors').

Experiments with tissue explants in collagen gels suggest that the peripheral projection pathways of primary sensory axons may also be influenced by chemorepellents (Keynes *et al.*, 1997).

Molecular characterisation of chemotropic factors

Netrins and their receptors

The first report of the isolation and characterisation by gene cloning of a diffusible chemotropic factor produced by developing vertebrate neural tissue was made possible by the use of an *in vitro* assay. This revealed that the molecules belonged to a family with representatives in invertebrates (Kennedy *et al.*, 1994; Serafini *et al.*, 1994; Ackerman *et al.*, 1997). Two molecules have been discovered, called netrin-1 and netrin-2 ('netr', from the Sanskrit meaning 'one who guides'), which were isolated because of their ability to induce and re-orientate the growth of spinal cord commissural axons *in vitro*. They therefore mimic the properties of the spinal cord chemotropic factor described above which derives from the floor-plate and attracts the axons of commissural interneurons from their origin in the dorsal spinal cord to the ventral midline in chicken and rat embryos (Tessier-Lavigne *et al.*, 1988; Placzek *et al.*, 1990a,b).

The netrins (Table 3.4) were isolated by classical biochemical means, not from the floor-plate – because of the small amount of material available – but from embryonic chick brain. Subcellular fractions were selected on the basis of their ability to promote axon outgrowth from embryonic spinal cord explants containing commissural neurons. Two proteins, of molecular weights 78 kD and 75 kD (netrin-1 and netrin-2, respectively; see Table 3.4), were purified to homogeneity from 10,000 embryonic chick brains using an army of students to dissect out the tissue. Microsequencing of the proteins led to the identification of the genes, which revealed that both proteins were probably secreted. This prediction was confirmed when COS cells, transfected with cDNA coding for netrins, released factors capable of inducing axon growth from dorsal spinal cord explants and of re-orienting these axons in collagen explants (Serafini *et al.*, 1994). Furthermore, recombinant netrins can also induce axon outgrowth from spinal cord explants in collagen gels (Kennedy *et al.*, 1994; Serafini *et al.*, 1994). These experiments show that netrins have direct chemoattractive effects on axons and stimulate axon outgrowth. Evidence consistent with the idea that netrin-1 is involved in attracting commissural axons to the floor-plate comes from expression studies of mRNA using *in situ* hybridisation (Kennedy *et al.*, 1994) and immunohistochemistry with netrin-1 antibodies (MacLennan *et al.*, 1997). Netrin-1 message is expressed in floor-plate cells precisely during the period when commissural axons are attracted to them (Kennedy *et al.*, 1994). However, in contrast, immunoreactivity for netrin-1 is not seen in the floor-plate until *after* the first commissural axons have crossed the midline (MacLennan *et al.*, 1997; see later text).

Netrin expression in the developing vertebrate nervous system is not restricted to the spinal cord floor-plate, but is found in the floor-plate at all axial levels and elsewhere in the central nervous system (Kennedy *et al.*, 1994; Shirasaki *et al.*, 1995; Deiner *et al.*, 1997; Livesey & Hunt, 1997; Métin *et al.*, 1997; Strähle, Fischer & Blader, 1997).

In transgenic mice in which the netrin-1 gene has been deleted by targeted mutagenesis there are severe abnormalities in pathfinding by spinal cord commissural axons, very few of which reach the floor-plate (Serafini *et al.*, 1996). Commissural axons appear to be produced normally and the initial part of their pathway is followed faithfully; however, in the ventral spinal cord, about 150 μm from the floor-plate, their behaviour becomes abnormal: many simply stall while others grow in inappropriate directions. This is within the distance that the floor-plate chemoattractant has been shown to act in collagen gels *in vitro* (see above). However, despite the absence of netrin-1 in these animals, a few commissural axons project normally and reach the floor-plate. There are also defects in axon projections in the brains of mice lacking netrin-1. Many, but not all, commissures are reduced or absent.

Floor-plate-derived netrin does not always attract growing axons. A group of vertebrate motoneurons, whose axons form the trochlear nerve (cranial nerve IV), uniquely project their axons dorsally and circumferentially from cell bodies located near the floor-plate at the hindbrain/midbrain junction. They exit the neural tube at the roof-plate near the dorsal midline to form the nerve to one of the extrinsic eye muscles. Colamarino and Tessier-Lavigne (1995) cultured explants of ventral regions of embryonic neural tube from the rat hindbrain/midbrain junction in collagen gels and found that the trochlear neurons extended axons dorsally in the explant and emerged from the dorsal surface of the explant to grow in the gel as a thick bundle of axons. The majority of these axons originated from the dorsal surface of their respective cell bodies, i.e. on the farthest side from the floor-plate. Occasionally, however, they emerge from the side of the cell body facing the floor-plate and initially grow toward the floor-plate. However, these axons invariably turned through 180° and then grew dorsally. These observations suggested that some factor, originating from the floor-plate, was repelling the axons. Further experiments showed that the factor is probably netrin-1 (Colamarino and Tessier-Lavigne, 1995). Most significantly, floor-plate cells and COS cells secreting netrin-1 suppress the formation of a dorsal axon bundle in ventral explants containing trochlear neurons in the collagen gel cultures.

When the gene for the vertebrate netrin-1 was cloned, it became apparent that it had homology to the laminin $\gamma 1$ chain (Fig. 3.1) and to a previously described gene in *C. elegans*, the *unc-6* gene, which had been implicated in growth cone pathfinding (see the section entitled 'Laminin receptors: integrins'; Hedgecock *et al.*, 1990; Ishii *et al.*, 1992). Genetic experiments suggest that Unc-6 is required for the appropriate migration of growth cones towards midline cells expressing Unc-6 (Hedgecock *et al.*, 1990; McIntire *et al.*, 1992; Wadsworth, Bhatt & Hedgecock, 1996). In *C. elegans*, mutations in the *unc-6* gene cause many of the axons that normally project circumferentially and ventrally, toward the ventral midline, to grow longitudinally (Hedgecock *et al.*, 1990). Fruit flies also have netrin homologues (netrin-A and netrin-B), which are normally expressed by midline cells (Harris, Sabatelli & Seeger, 1996). Loss of function of netrins in mutant flies is associated with pathfinding errors of axons that normally grow towards the midline (Harris *et al.*, 1996; Mitchell *et al.*, 1996). When both netrin genes are absent in *Drosophila*, there is a considerable reduction in the number of axons crossing the midline, resulting in thinner anterior and posterior commissures. Motor neuron pathfinding is also affected in this mutant. Importantly, the effects of the loss of netrins in *Drosophila* on axon projections can be rescued by re-introducing either netrin-A or netrin-B into midline cells, ruling out the possibility that the phenotype was due to some other molecule than netrin (Harris *et al.*, 1996; Mitchell *et al.*, 1996). Incidentally, this case may provide the first example of functional redundancy in pathfinding molecules. Mutant flies in which either netrin is ectopically expressed in neurons also results in pathfinding errors across the midline (Harris *et al.*, 1996; Mitchell *et al.*, 1996). In the zebrafish mutant *floating head*, netrin-1a, the zebrafish netrin, is absent in patches along the caudal spinal cord. Within these patches, spinal cord commissural neurons extend their axons aberrantly towards an adjacent patch of the spinal cord that does express netrin-1a (Lauderdale, Davis & Kuwada, 1997). Messenger RNAs encoding netrin-A and netrin-B are expressed in midline cells in *Drosophila* (Harris *et al.*, 1996; Mitchell *et al.*, 1996) but, as is the case for vertebrate netrins (see above), netrin protein expression by midline cells is not detectable until after the first axons have crossed (Klämbt, Jacobs & Goodman, 1991; Harris *et al.*, 1996). Perhaps the first few commissural pioneer axons are attracted to the floor-plate by a netrin-independent mechanism and then induce the expression of netrin in the floor-plate, which then attracts later arriving axons.

The discovery that vertebrate netrins are homologous to the *C. elegans* protein Unc-6 led to the suggestion that candidate Unc-6 receptors in *C. elegans*, Unc-5 and Unc-40, may be homologues of vertebrate netrin receptors. From this starting point, homology screening has led to the identification of a family of vertebrate netrin receptors (Keino-Masu *et al.*, 1996; Leonardo *et al.*, 1997). Like Unc-5 and Unc-40, they are all members of the immunoglobulin superfamily of transmembrane receptors (Fig. 3.2). The best characterised of the vertebrate netrin receptors is Dcc (*Deleted in colon carcinoma*), a homologue of the *C. elegans* gene Unc-40 (Chan *et al.*, 1996). Dcc was originally thought to be a tumour-suppressor gene partly because of the high frequency with which it is absent or mutated in human colorectal carcinomas (Fazeli *et al.*, 1997). The evidence for Dcc as a netrin receptor includes the observation that it is expressed on spinal cord commissural axons while they navigate toward the floor-plate and that it binds netrin-1 with high affinity *in vitro* (Fazeli *et al.*, 1997). Functional studies also implicate Dcc in netrin-mediated pathfinding. Monoclonal antibodies that recognise the extracellular domain of Dcc block the axon outgrowth of commissural neurons from spinal cord explants in collagen gels produced by netrin-1 (Keino-Masu *et al.*, 1996). However, these antibodies do not block the netrin-1-mediated re-orientation of commissural axons that occurs within the explant. This may be due to the inability of the antibodies to penetrate into the explant effectively.

In transgenic mice in which the Dcc receptor is absent through homologous recombination, there is no evidence of an increase in tumour predisposition, but there are striking abnormalities in spinal cord and brain development (Fazeli *et al.*, 1997). In the spinal cord, there is a reduction in the number of commissural axons. Of those axons that are present, many project normally into the ventral cord but then become misrouted, either laterally or medially, rather than projecting toward the floor-plate. Some axons reach the floor-plate normally. The projections of primary sensory neurons and motoneurons appear normal. There are also abnormalities in brain development. The corpus callosum and hippocampal commissure are completely absent, the axons that normally form these commissures remaining ipsilateral. However, as in the netrin-1 knockout mouse (Serafini *et al.*, 1996), not all commissures are affected, suggesting that netrins are not universally involved in commissure formation. The striking similarity in the neural defects seen in the Dcc gene knockout mouse to those seen in the netrin-1 gene knockout mouse (Serafini *et al.*, 1996) provide very strong evidence that Dcc is a physiological receptor for netrin function in axonal pathfinding.

Considerable evidence has accumulated to support a role for netrins in the developing retina. Netrins are expressed at the optic nerve head (disc) at a time when retinal ganglion cell axons leave the retina at the disc to enter the optic nerve, which also expresses netrins (Kennedy *et al.*, 1994; Serafini *et al.*, 1996; de la Torre *et al.*, 1997; Deiner *et al.*, 1997; Lauderdale *et al.*, 1997; Livesey & Hunt, 1997; Strähle *et al.*, 1997). Furthermore, Dcc is expressed on developing retinal ganglion cell axons (Pierceall *et al.*, 1994; Keino-Masu *et al.*, 1996; Deiner *et al.*, 1997; de la Torre *et al.*, 1997; Leonardo *et al.*, 1997; Livesey & Hunt, 1997). In the netrin-1 and Dcc knockout mice, many retinal ganglion cell axons do not exit through the optic disc correctly (Deiner *et al.*, 1997). The *noisthmus* (*noi:* the zebrafish ortholog of Pax2) mutant in zebrafish lacks netrin-1a expression in the optic disc and nerve, and also shows abnormal projections of retinal ganglion cell axons (MacDonald *et al.*, 1997). These findings suggest that retinal ganglion cell axons might be guided to the optic disc by netrins, an idea supported by the observation that netrin-1 can induce axon outgrowth from retinal ganglion cell explants *in vitro* (Wang *et al.*, 1996b; Deiner *et al.*, 1997). Furthermore, *Xenopus* retinal ganglion cell growth cones turn towards an *in vitro* gradient of netrin-1 and this is attenuated by antibodies against Dcc (de la Torre *et al.*, 1997). In these experiments, retinal ganglion cell growth cones, when uniformly exposed to netrin-1, become more complex with increases in the number and length of filopodia, and their speed *in vitro* increases by 50%. These findings suggest that netrin-1 may contribute to the changes in growth cone morphology seen in the embryo when growth cones are migrating through the optic disc (Holt, 1989; Brittis & Silver, 1995; Mason & Wang, 1997). When growth cones were presented with a gradient of netrin-1 diffusing from a point source (micropipette) directed at 45° to the direction of neurite growth, the growth cones turned rapidly (within 15 minutes) toward the netrin-1 source (de la Torre *et al.*, 1997). The netrin-1-mediated turning response was abolished by antibodies to Dcc.

In *Drosophila* there is a Dcc homologue called *frazzled*, which is expressed by embryonic axons in the central nervous system and by motoneuron axons (Kolodziej *et al.*, 1996). The null-mutant has the same phenotype as the netrin knockout in flies (see above) and the *frazzled* knockout can be rescued by ectopically expressing *frazzled* in all neurons (Kolodziej *et al.*, 1996).

Collectively, the available data on netrins show that these molecules function as long-range chemoattractants in guiding certain groups of axons in developing nervous systems and that they are phylogenetically conserved, as is their function. However, in their role in attracting growth cones to the midline, it is highly likely that they are not the sole mediators of this event, as indicated by the persistent success of some axons in crossing the midline in loss or gain-of-function mutants. It is not yet clear whether netrins act in a freely diffusing gradient or as a fixed (haptotactic) gradient, and very little is known about the intracellular signalling events associated with netrin action. Intriguingly, the chemoattractive response of *Xenopus* spinal cord neurons to gradients of netrin-1 *in vitro* can be converted to a repulsive response by reducing the intracellular levels of cAMP (Ming *et al.*, 1997; see Chapter 4). Both of these responses are mediated by the Dcc receptor, since they can be blocked by antibodies to the receptor, suggesting that Dcc underlies the chemoattractant and chemorepellent actions of netrin-1.

The best-characterised role of netrin-1 is in the attraction of commissural axons to the ventral midline in vertebrates and invertebrates. In vertebrates, the ventral midline is occupied by a palisade of floor-plate glial cells. Once they have reached the floor-plate, commissural axons cross the floor-plate and make a sharp rostral turn at the far border. The molecular mechanisms that direct commissural growth cones across the floor-plate must be distinct from those that attract these growth cones to the midline. Some of the molecules involved have been identified recently. As they approach the floor-plate, commissural growth cones in the chick express the cell adhesion molecule axonin-1, a homologue of TAG-1 in the rat (Dodd *et al.*, 1988; Furley *et al.*, 1990; see Fig. 3.2 and the section entitled 'Neural cell adhesion molecule and the immunoglobulin superfamily'), while floor-plate glia express neuronal cell adhesion molecule (NrCAM). When function-blocking antibodies against either axonin-1 or NrCAM or soluble axonin-1, which compete with the binding of growth cone axonin-1 to its receptors in the floor-plate, were injected into chick embryos *in ovo* during the migration of the commissural growth cones through the floor-plate, about 50% of the axons failed to enter the floor-plate. Instead, they grew along the ipsilateral border of the floor-plate, in either the rostral or, abnormally, the caudal direction (Stoeckli & Landmesser, 1995). Time-lapse video microscopy of commissural growth cones growing toward floor-plate explants in culture showed that blocking antibodies to axonin-1 caused growth cones to collapse when they came into contact with the floor-plate (Stoeckli *et al.*, 1997). These experiments suggest that axonin-1 expressed on commissural growth cones interacts, heterophilically, with NrCAM on floor-plate glia and that this interaction overcomes an intrinsic inhibitory factor, also expressed by floor-plate cells. This factor has not been identified but obvious candidates include members of the semaphorin/collapsin family and the Eph/ephrin family (see earlier text). The function of the inhibitory factor revealed by these experiments may be to prevent axons that have crossed the floor-plate once from re-crossing. If this were the case, then we might predict that growth cones would down-regulate their ability to respond to NrCAM, which implies a loss of TAG-1 or axonin-1. In the rat, TAG-1 is down-regulated on axons that have crossed the midline but in the chick embryo crossed axons still retain axonin-1. Alternatively, growth cones might express a receptor for the floor-plate inhibitory factor after they have crossed the midline. Evidence to support this mechanism and further demonstration of the growth cone's ability to integrate opposing signals has been obtained from studies of commissure formation in *Drosophila*. In a near-saturating, large-scale genetic screen in which chemically mutagenised embryos were examined for abnormalities in commissure formation in the central nervous system, Goodman's group identified two classes of mutation (Seeger *et al.*, 1993). In *roundabout* (*robo*) mutants, many axons cross the midline that do not normally do so. Furthermore, axons that do normally cross the midline re-cross it repeatedly. The protein encoded by *robo* is a novel transmembrane protein and a member of the immunoglobulin superfamily (see Goodman & Tessier-Lavigne, 1997).

The second class of mutants, known as *commissureless* (*comm*), are distinguished by having no axons crossing the midline and therefore no commissures (Seeger *et al.*, 1993). In these mutants, midline cells differentiate normally but no longer express the *comm* protein, as they do in wild-type embryos (Tear *et al.*, 1996). In *comm* mutants, commissural neurons project their axons toward the midline normally but do not cross it. Instead they project ipsilaterally. No vertebrate homologue of *comm* has been identified.

The protein encoded by *comm* is probably a ligand for a receptor on the growth cones of commissural axons that stimulates or facilitates midline crossing. In contrast, the *robo* protein is likely to be involved in inhibiting the midline crossing of axons. This suggests that *comm* is necessary to overcome the inhibitory effects of *robo*. The function of these proteins may become clearer when their expression patterns are determined.

Summary

All four possible categories of growth cone guidance molecule have been discovered: short- and long-range attraction and short- and long-range repulsion. Many of the molecules involved belong to phylogenetically conserved families. Growth cones use a number of cues simultaneously. Some guidance cues are used more than once and may have opposite effects: collapse vs adhesion.

4

Intracellular Signalling in Growth Cones

Introduction

Axonal guidance cues can influence growth cone behaviour by binding to specific, cell surface receptors on growth cones. Activation of such receptors produces intracellular signalling events, which leads to changes in growth cone behaviour. The principal target of intracellular signalling pathways in growth cones is the cytoskeleton – the 'final common path of action' of guidance cues because it underlies growth cone motility and neurite extension – although other growth cone functions, such as plasma membrane growth, may also be targeted. Our understanding of the intracellular signalling pathways in growth cones is still fairly rudimentary. There are three main reasons for this. Firstly, although several families of guidance molecules have been identified and characterised, the growth cone receptors for these molecules, and thus their intracellular signalling mechanisms, are only now being discovered (see Chapter 3). Secondly, most experiments on signalling mechanisms have been done *in vitro* and focus on assaying neurite outgrowth rather than the pathfinding behaviour of growth cones, such as growth cone turning, for instance, in 'choice' assays (see Chapter 3), which more closely resemble growth cone pathfinding *in vivo*. Thirdly, there is considerable reliance on the use of pharmacological reagents to inhibit or disrupt signalling pathways, an approach which is often compromised by lack of specificity or the unavailability of more than one reagent with a different mechanism of action.

Several methodological advances have had a significant impact on our understanding of the signalling events within neurites and growth cones. The first was the introduction of cell lines derived from neuroblastomas, such as PC12 cells (Augusti-Tocco & Sato, 1969; Schubert *et al.*, 1969; Greene &

Tischler, 1976). These cell lines are mitotic in culture but they can be induced to escape from the cell cycle and differentiate into a neuron-like phenotype by the action of specific agents, such as nerve growth factor (NGF) or dibutyryl cyclic-AMP. Differentiation is accompanied by the extension of neurites. Although a distinct axon and dendrite are not produced, the neurites most resemble axons. Since large numbers of cells can be obtained before differentiation, and differentiation is more or less synchronous, sufficient material can be obtained to investigate biochemical and molecular events associated with growth cone formation and neurite extension. Furthermore, cell lines and neuroblastomas, unlike neurons, are easily transfected with DNA.

The second advance was the development of subcellular fractionation techniques for the preparation of isolated growth cones (Pfenninger *et al.*, 1983; Gordon-Weeks & Lockerbie, 1984; Gordon-Weeks, 1987a,b; Cypher & Letourneau, 1991). In this technique, developing neural tissue is gently homogenised in isotonic buffers so that growth cones are sheared from the ends of their neurites. The broken plasma membranes of the isolated growth cones reseal and they can then be separated from other cellular elements by subsequent differential and gradient centrifugation (reviewed in Lockerbie, 1990). Growth cones have been isolated from embryonic (Pfenninger *et al.*, 1983) and early postnatal rat brain (Gordon-Weeks & Lockerbie, 1984), chick brain (Cypher & Letourneau, 1991) and cultured human SH-SY5Y neuroblastoma cells (Meyerson, Pfenninger & Påhlman, 1992; Meyerson *et al.*, 1994). The growth cones isolated from embryonic and early postnatal rat brain are axonal, rather than dendritic, in origin (Saito *et al.*, 1992; Lohse *et al.*, 1996) and the latter, but not the former, bear filopodia. The isolation of growth cones in a subcellular fraction has enabled the direct biochemical and pharmacological investigation of growth cones *in vitro* (reviewed in Lockerbie, 1987).

Finally, a powerful approach to understanding the role of intracellular signalling mechanisms in growth cone pathfinding is to use genetics (see also Chapter 3). Broadly speaking there are two approaches – so-called reverse genetics, in which mutations are made in the gene of a known protein, and forward genetics, in which random mutations are made in a genome and a mutant phenotype is screened for using a suitable assay. With the advent of transgenic animals, it is now possible to test the role of specific proteins, such as kinases, either in a null mutant or by loss- or gain-of-function mutations. A phenotype can then be looked for in the whole organism or in cells derived from the organism. Although this approach is still in its infancy, a number of studies where genetics has been used to address questions about intracellular signalling mechanisms in growth cones have appeared (see below).

A large number of intracellular signalling molecules and pathways have been identified in growth cones including: calcium and calmodulin, G proteins, inositol trisphosphate and diacyl glycerol, nitric oxide, cyclic-AMP, and various phosphoproteins and their associated kinases and phosphatases. Many of these molecules have been implicated in neurite growth and elongation, and various growth cone behaviours such as growth cone collapse and turning.

How these diverse second-messenger pathways regulate growth cone motility and neurite elongation is largely unknown.

Calcium

The most thoroughly studied second-messenger molecule in growth cones is calcium. Calcium has an essential signalling role in many intracellular processes as a second messenger and it is not surprising to find that it is a key signalling molecule in growth cones. Studies of intracellular calcium in growth cones have been greatly facilitated by the introduction of fluorescent calcium reporter dyes such as fura-2 (Grynkiewicz, Poenie & Tsien, 1985; Tsien, 1988). These dyes are sensitive to the minute quantities of free calcium found in cells. The dyes are available in a form (the acetoxymethyl ester) that is membrane permeant and can therefore be loaded into cells. Most cells contain endogenous esterases which rapidly hydrolyse the dye to the impermeant free acid, which is thus trapped within the cell. Such calcium indicators have allowed the direct determination of the spatial distribution and concentration of calcium within growth cones, and thereby its relation to growth cone behaviour.

Calcium is implicated in neurite growth

Of those behaviours thought to be regulated by calcium, that of neurite initiation and growth has been most extensively investigated and there is a considerable literature implicating intracellular calcium in these processes (Schubert *et al.*, 1978; Koike, 1983; Freeman *et al.*, 1985; Connor, 1986; Lipton, 1987; Mattson & Kater, 1987; Lipscombe *et al.*, 1988; Mattson, Guthrie & Kater, 1988b,c; Mattson, Taylor-Hunter & Kater, 1988e; Silver, Lamb & Bolsover, 1990; Bentley, Guthrie & Kater, 1991). The central piece of evidence is that changes in extracellular and/or intracellular calcium concentrations are associated with changes in the rate of neurite initiation and extension, although the direction of the change is inconsistent. Intracellular calcium concentrations in growth cones *in vitro* can be altered experimentally in a number of ways, for instance, by changing extracellular calcium, using calcium ionophores such as A23187, inhibiting calcium channels, or by electrical stimulation. Unfortunately, there are very few *in vivo* studies of calcium and neurite extension (Bentley *et al.*, 1991). Early experiments investigating the relationship between intracellular calcium concentration and neurite growth, before the introduction of calcium reporter dyes, produced contradictory results. Inhibition of calcium channels or removal of calcium from the culture medium, procedures expected to lower growth cone cytosolic calcium, either inhibited neurite growth (Nishi & Berg, 1981; Suarez-Isla *et al.*, 1984) or did not affect it, or even stimulated growth (Letourneau & Wessells, 1974; Grinvald & Farber, 1981; Anglister *et al.*, 1982; Bixby & Spitzer, 1984; Campenot & Dracker, 1989). Furthermore, procedures expected to raise growth cone cytosolic calcium, such as calcium ionophores, either inhibited outgrowth in some cases

(Haydon, McCobb & Kater, 1984; Suarez-Isla *et al.*, 1984; Hantaz-Ambroise & Trautmann, 1989; Robson & Burgoyne, 1989b) or stimulated it in others (Hinnen & Monard, 1980; Nishi & Berg, 1981; Anglister *et al.*, 1982).

These early experiments on the effects of calcium on neurite growth of neurons in culture were compromised by the assumption that changing extracellular calcium concentration, addition of calcium ionophores or membrane depolarisation, would produce long-lasting changes in intracellular calcium levels. With the availability of calcium reporter dyes, it became apparent that neurons differ in their ability to buffer changes in intracellular calcium. For instance, in the B19 neuron of the snail *Helisoma* and in rat hippocampal pyramidal neurons, calcium buffering is poor and imposed increases in intracellular calcium are maintained for long periods of time (Mattson, Guthrie & Kater, 1989; Mills & Kater, 1990). In contrast, in the B5 neurons of *Helisoma* (Mills & Kater, 1990) and in many neuroblastoma cell lines, calcium buffering is very efficient and imposed changes in intracellular calcium are transient (Bolsover & Spector, 1986; Ahmed & Connor, 1988). Therefore, there is a requirement to measure imposed changes in intracellular calcium concentrations in growth cones directly, if meaningful deductions are to be made about the role of calcium in growth cone behaviour.

Direct measurement of calcium levels in growth cones using calcium reporter dyes revealed that the intracellular calcium concentration in growth cones in culture changes continuously as the growth cone advances, pauses and retracts (reviewed in: Kater & Mills, 1991; Cypher & Letourneau, 1992; Davenport, 1996). However, as in the earlier studies (see above), although there is general agreement that growth cones that are actively moving forward in culture have different levels of cytosolic calcium from growth cones that are stationary or retracting, there is no agreement about the direction of the calcium change. In some experiments, higher levels of cytosolic calcium are found in motile growth cones (Cohan, Haydon & Kater, 1985; Connor, 1986), while in others lower levels are found (Silver, Lamb & Bolsover, 1989). To what extent this reflects differences in neuronal class, or some other variable, is not clear.

Contradictory results have also been seen in the effects of artificially changing calcium levels on neurite outgrowth. Inhibition of neurite elongation has been seen both by artificially increasing (Cohan, Connor & Kater, 1987; Connor, 1986; Lankford & Letourneau, 1989; Mattson & Kater, 1987; Mattson, Guthrie & Kater, 1988c; Mattson *et al.*, 1988d; McCobb *et al.*, 1988; Mills & Kater, 1990; Silver *et al.*, 1989) and decreasing (Anglister *et al.*, 1982; Suarez-Isla *et al.*, 1984; Connor, 1986; Mattson & Kater, 1987; Goldberg, 1988; Lankford & Letourneau, 1989) growth cone calcium levels *in vitro*. Conversely, increasing intracellular calcium concentrations in growth cones *in vitro* may produce no change or a stimulation of axon growth (Cohan *et al.*, 1985; Connor, 1986; Mattson, Guthrie & Kater, 1988b,c; Mattson *et al.* 1988e; Tolkovsky *et al.*, 1990; Garyantes & Regehr, 1992; Rehder & Kater, 1992). Inhibition of growth cone motility and axon growth at high levels of cytosolic calcium is a common finding *in vitro* and can be seen with a wide range of stimuli, including serotonin applied to specific *Helisoma* growth cones

(Haydon, McCobb & Kater, 1984; Cohan *et al.*, 1987) and glutamate to rat hippocampal pyramidal neuron growth cones (Mattson, Dou & Kater, 1988a; Mattson *et al.*, 1988d; Mattson & Kater, 1989), and membrane depolarisation (Neely, 1993).

The calcium 'set-point' hypothesis

To resolve the apparently contradictory results of the calcium experiments, Kater and colleagues have proposed that there is an optimal 'set-point' for calcium concentration in growth cones and axon growth, and that, when calcium levels fall below or above the set-point concentration, axon growth is inhibited (Kater *et al.*, 1988; Kater & Mills, 1991). There is experimental evidence to support the set-point model. Al-Mohanna, Cave and Bolsover (1992) experimentally altered the cytosolic calcium levels in the growth cones of rat dorsal root ganglion cells in culture using the membrane-permeable calcium chelator BAPTA. They found that the rate of axon growth and initiation reached a maximum at a certain calcium level, but declined either side of this value. Similarly, in chick dorsal root ganglion neurons in culture, Lankford and Letourneau (1991) found that growth cones continued to advance when exposed to the calcium ionophore A23187 when the extracellular calcium was 250 nM, but were inhibited when the extracellular calcium levels were either above or below 250 nM. Rehder and Kater (1992) found an optimal intracellular calcium concentration for filopodial extension in the large (up to 50 μm across) growth cones of neurons from the snail *Helisoma*. At values of calcium to either side of the optimum, between about 100 and 150 nM calcium, filopodial extension was reduced. These experiments lend support to the notion of a set-point for calcium concentration in growth cones (see also Fields *et al.*, 1993). One of the potential functional consequences of a set-point is that it may enhance the sensitivity of the growth cone to those external stimuli, such as guidance cues, that induce changes in growth cone calcium levels.

Calcium regulation in growth cones

How is calcium concentration regulated in growth cones so that the set-point is maintained? Cytosolic calcium concentrations in growth cones are probably maintained by the combined effect of several processes including a continual, inward leakage of calcium, the action of Ca^{2+}-ATPases in the plasma membrane and in the smooth endoplasmic reticulum that is abundant in growth cones (see Chapter 1). Mitochondria are also capable of accumulating calcium, by their Na^+/Ca^{2+} exchanger (McBurney & Neering, 1987; Blaustein, 1988) and mitochondria are invariably present in growth cones. However, although the calcium storage capacity of mitochondria is high, the affinity of the exchanger for calcium is low (mM) compared to that of smooth endoplasmic reticulum (μM) and therefore, under normal circumstances, calcium buffering by mitochondria in growth cones is probably not important. The growth cones of some neurons appear to be able to buffer changes in cytosolic calcium quite effectively (Bolsover & Spector, 1986; Ahmed & Connor 1988), while in other neurons, growth cone calcium buffering is poor (Mattson, Guthrie & Kater, 1989; Mills & Kater, 1990).

Calcium can enter cells through voltage-gated calcium channels and ligand-gated calcium channels, such as the glutamate-activated N-methyl D-aspartate (NMDA) receptor and the γ-amino butyric acid $(GABA)_A$ receptor (see Chapter 5). Developing neurons express voltage-gated calcium channels in their somal, axonal and growth cone membranes. There is no evidence that these channels differ in their properties in different regions of the developing neuron (Lipscombe et al., 1988) but there does seem to be a concentration of calcium channels in the growth cone, where they may be localised in clusters (Silver, Lamb & Bolsover, 1989, 1990; Zimprich & Bolsover, 1996). Indirect evidence from pharmacological experiments and direct measurement of membrane calcium flux using electrodes, suggests that growth cones possess L-type, N-type and T-type voltage-gated calcium channels (Anglister et al., 1982; Freeman et al., 1985; Goldberg, 1988; Lipscombe et al., 1988; Silver et al., 1989, 1990; Gottman & Lux, 1990; Doherty et al., 1991a; Reber & Reuter, 1991; Vigers & Pfenninger, 1991; Bedlack, Wei & Loew, 1992; Williams et al., 1992; Przywara et al., 1993). Calcium entry into growing neurons may be restricted or highly localised to growth cones. Silver et al. (1990) found a spatial correlation between local clustering of L-type voltage-gated calcium channels and membrane out-growth in the growth cones of N1E-115 neuroblastoma cells. In the vicinity of the clustered calcium channels the magnitude of the calcium concentration change could approach 1 μM. In growth cones from the same neuroblastoma cell line, Zimprich and Bolsover (1996) found a higher concentration of L-type calcium channels at the tip of the growth cone than in more proximal regions in neurites that were growing in a straight line, which explains the gradient of cytosolic calcium concentration seen in these growth cones. The mechanism of calcium-channel clustering and its regulation is not under-stood, but it could provide a means for initiating local morphological change in response to extracellular stimuli. For instance, the activation of calcium channels by cell-adhesion molecules (see later text).

There are gradients of calcium in growth cones. Lankford and Letourneau (1991) found calcium levels higher in the C-domain than in the P-domain. Others have observed higher concentrations of calcium in the P-domain (Connor, 1986), but this may be due to the presence of calcium-reporter dye in organelles, which have accumulated at the leading edge of the growth cone rather than a gradient of cytosolic calcium (Bolsover & Silver, 1991). Transient elevations in calcium levels, or 'spikes', have also been seen in growth cones and there is evidence that these depend upon calcium influx through non-voltage-gated calcium channels, since they are blocked by general calcium channel blockers but not by inhibitors of voltage-gated calcium channels (Gomez, Snow & Letourneau, 1995).

Do changes in growth cone calcium regulate pathfinding events?

In the majority of *in vitro* experiments in which calcium concentration in growth cones is changed experimentally, the imposed change usually occurs throughout the growth cone, more or less simultaneously, i.e. the changes are global. It is not clear to what extent this reflects the situation *in vivo* where guidance cues may induce changes in growth cone calcium concentrations. Since growth cones usually encounter guidance cues in a spatially restricted way (see Chapter 3), localised changes in growth cone calcium, and their behavioural consequences, may be more physiologically relevant. In a number of studies, localised changes in growth cone calcium levels have been produced by applying stimuli to local regions of the growth cone and this tends to be associated with less dramatic changes in growth cone behaviour, such as filopodial and lamellipodial extension (Goldberg, 1988; Silver *et al.*, 1990; Davenport & Kater, 1992; Rehder & Kater, 1992) or growth cone turning (Gundersen & Barrett, 1980; Bedlack *et al.*, 1992; Zheng *et al.*, 1994), rather than the growth cone collapse or retraction seen with large, global increases in growth cone calcium concentration.

There have been several reports of induced, localised changes in calcium concentration in growth cones and these have correlated closely with localised changes in growth cone morphology and behaviour. Chick dorsal root ganglion cell growth cones will re-orientate towards a micropipette releasing nerve growth factor in a culture dish (Gundersen & Barrett, 1979; see Chapter 3) and this behaviour depends on an influx of calcium into the growth cone (Gundersen & Barrett, 1980). Goldberg (1988) showed that focal application of calcium to *Aplysia* growth cones caused increased lamellipodial extension on that side of the growth cone where cytosolic calcium was presumed to have increased. Focal electric fields applied to growth cones in culture can cause growth cone re-orientation and increased filopodial extension in the direction of the turn, which can be inhibited by calcium channel blockers (McCaig, 1986, 1989; see Chapter 3). There have been several studies of the response of the growth cones of the neuroblastoma cell line N1E-115 to focally (Silver *et al.*, 1990) and globally applied electric fields (Bedlack *et al.*, 1992). Silver *et al.* (1990) found that calcium entry through locally clustered L-type voltage-dependent calcium channels in the growth cone was associated with lamellipodial extension, and Bedlack *et al.* (1992) extended this work by showing that N1E-155 growth cones re-orientate towards the cathode when an electric field is applied across the whole growth cone and that this behaviour was associated with the formation of a calcium concentration gradient within the growth cone. Both events, re-orientation and the calcium gradient, were abolished by the addition of calcium channel blockers to the culture medium, including L-type channel blockers. In growth cones from *Helisoma* neurons, fura-2 imaging of cytosolic calcium shows that focally applied electric fields can induce local rapid (< 1 second) increases in calcium concentration in the growth cone, which are associated with initiation and extension of filopodia in that region of the growth cone where the calcium increase occurs (Davenport & Kater,

1992). The local rise in calcium concentration can be as high as 1.5 μM and it eventually spreads across the growth cone. The electric field induced local rise in cytosolic calcium concentration and the subsequent increase in filopodia were both inhibited by removing extracellular calcium, suggesting that calcium influx into the growth cone drives both events (Davenport & Kater, 1992). Thus, local changes in growth cone morphology and growth cone re-orientation induced by focally applied electric fields may involve changes in cytosolic calcium (reviewed in McCaig & Erskine, 1996).

Physiological agents affecting growth cone calcium levels

Neurotransmitters, such as acetylcholine, applied as a localised gradient by micropipette to the growth cones of *Xenopus* spinal cord neurons in culture, cause growth cone turning up the acetylcholine gradient by a calcium-dependent mechanism (Zheng *et al.*, 1994). In other neurons, where application of neurotransmitters also causes increases in cytosolic calcium, inhibitory effects are observed. For instance, the growth cones of certain identified *Helisoma* neurons in culture respond to excitatory neurotransmitters by a large increase in cytosolic calcium and cessation of neurite elongation (Haydon *et al.*, 1984; Cohan *et al.*, 1987). Similarly, dendritic growth cones of cultured pyramidal neurons from the rat hippocampus show an increase in cytosolic calcium concentration and stop growing when the excitatory neurotransmitter glutamate is applied to them; both effects are abrogated by prior application of the inhibitory neurotransmitter γ-amino butyric acid (Mattson *et al.*, 1988a,d; Mattson & Kater, 1989. The isolated growth cone particles from early postnatal rat brain express GABA$_A$ receptors (Lockerbie & Gordon-Weeks, 1985; Fukura, Komiya & Igarashi, 1996), whose activation leads to a calcium influx through L-type voltage-sensitive calcium channels (Fukura *et al.*, 1996). Developing hypothalamic neurons in culture express GABA receptors in their growth cones and activation of these, by application of GABA to the culture medium, causes a rise in growth cone Ca^{2+} levels (Obrietan & van den Pol, 1995, 1996). Some members of the neurotrophin family of growth factors can re-orientate growth cones *in vitro* when applied as a gradient (Gundersen & Barrett, 1979, 1980; Song *et al.*, 1997). Growth cones from *Xenopus* spinal cord neurons are attracted toward a source of brain-derived neurotrophic factor *in vitro* and, since this response can be abolished by reducing the concentration of calcium ions in the culture medium, activation of the receptor for brain-derived neurotrophic factor, TrkB, on growth cones presumably results in an influx of calcium ions into the growth cone (Song *et al.*, 1997). Interestingly, not all growth cone turning responses elicited by neurotrophins depend on extracellular calcium because, although neurotrophin-3 can re-orientate growth cones *in vitro*, reducing the concentration of calcium ions in the culture medium has no effect on the turning response (Song *et al.*, 1997). The receptor for neurotrophin-3 is TrkC and, therefore, these results imply that different neurotrophin receptors utilise different intracellular signalling pathways in growth cones to elicit the same growth cone response: turning toward a source of neurotrophin.

These experiments demonstrate that external stimuli can induce discrete localised changes in growth cone calcium concentration that subsequently provoke changes in growth cone behaviour that are also localised. Whether growth cones encounter and respond to electric fields, neurotransmitters or neurotrophins during pathfinding *in vivo* is not clear (see Chapter 3) but at least these experiments demonstrate that, in principle, guidance cues could operate through calcium-dependent intracellular signalling pathways.

There have been a number of studies in which molecules suspected or known to have a growth cone guidance role have been investigated for their effects on intracellular calcium levels, and its relationship to neurite growth and growth cone behaviour. Cell adhesion molecules of the immunoglobulin superfamily, such as NCAM, and of the calcium-dependent cadherin family, such as N-cadherin (see Chapter 3), enhance neurite outgrowth *in vitro* by a calcium-dependent mechanism (reviewed in: Bixby & Harris, 1991; Doherty & Walsh, 1994; Baldwin, Walsh & Doherty, 1996; Kamiguchi & Lemmon, 1997). Walsh, Doherty and co-workers have developed an *in vitro* neurite outgrowth assay to investigate intracellular signalling mechanisms elicited by the immunoglobulin superfamily of cell adhesion molecules (Doherty *et al.*, 1990a,b, 1991a,b; see Chapter 3). They have produced a number of stably transfected fibroblast cell lines that express physiological levels of cell adhesion molecules on their surfaces (Doherty *et al.*, 1990a,b, 1991a,b). Neuroblastoma cells or dissociated primary neurons are grown on a monolayer of the transfected cells and neurite outgrowth is measured. Cell lines expressing different levels of cell adhesion molecules can be generated so that dose–response curves for concentration of cell adhesion molecule and neurite outgrowth can be constructed (Doherty *et al.*, 1990a). Using a predominantly pharmacological approach, Doherty and Walsh have exploited their transfected cell line system to identify a number of components of intracellular signalling pathways that are involved in neurite outgrowth on NCAM, L1 and N-cadherin (reviewed in: Doherty & Walsh, 1994; Baldwin *et al.*, 1996). Pharmacological antagonism of L- and N-type calcium channels inhibits outgrowth from primary neurons and PC12 cells growing on NCAM, N-cadherin and L1 expressing cell monolayers, suggesting that calcium influx is important for neurite outgrowth on these cell adhesion molecules (Doherty *et al.*, 1991a; Williams *et al.*, 1992).

A number of intermediate steps have been identified between cell adhesion molecule interaction and calcium influx including activation of the tyrosine kinase fibroblast growth factor receptor and generation of inositol phosphate second messengers (see later text and Fig. 4.1). These studies were important in being the first to demonstrate a link between cell adhesion molecules and an intracellular second messenger, calcium – one that had already been established as a regulator of neurite outgrowth (see above). Pharmacological inhibition of tyrosine kinases and phospholipase Cγ also inhibits neurite outgrowth on all three cell adhesion molecules as does inhibition of the enzymes diacylglycerol lipase, which catalyses the conversion of diacylglycerol into fatty acids such as

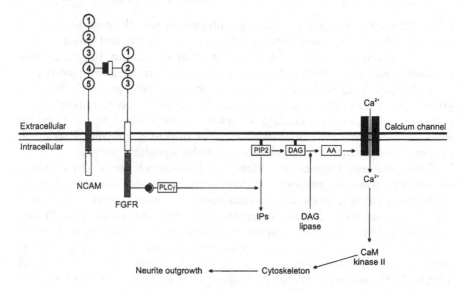

Figure 4.1 Diagram showing the scheme proposed by Doherty and Walsh to explain how cell adhesion molecules promote neurite outgrowth. An intracellular signalling pathway is activated as a result of a *cis*-interaction between cell adhesion molecules, such as NCAM, and the fibroblast growth factor receptor (FGFR). Binding takes place between the fourth extracellular immunoglobulin-like domain of NCAM and the second immunoglobulin-like domain of the FGFR, and leads to the tyrosine phosphorylation of the intracellular portion of the FGFR. Subsequent binding of phospholipase Cγ (PLCγ) to the tyrosine phosphorylated site of the FGFR initiates an intracellular signalling cascade, where PLCγ generates inositol phosphates (IPs) and diacylglycerol (DAG) from phosphatidylinositol-4,5-bisphosphate (PIP2). The DAG is then converted to arachidonic acid (AA) by the action of diacylglycerol lipase (DAG lipase). Arachidonic acid is then thought to open voltage-gated calcium channels, allowing an influx of calcium which, among other actions, activates, calcium/calmodulin-dependent protein kinase II (CaM kinase II). Although further steps in this pathway have not been identified, it seems likely that CaM kinase II activity ultimately affects the neurite cytoskeleton in order to enhace neurite outgrowth.

arachidonic acid, and calcium/calmodulin-dependent kinase (CaM kinase II; Williams, Walsh & Doherty, 1994a,b; Williams *et al.*, 1995a). Since NCAM, L1 and N-cadherin have no intrinsic kinase activity, these observations have led to the suggestion that the responses of the three cell adhesion molecules are mediated by the stimulation of a tyrosine kinase cascade that leads to activation of phospholipase C to generate diacylglycerol, which is hydrolysed to arachidonic acid by diacylglycerol lipase, and the subsequent increase in calcium-channel permeability induced by arachidonic acid leads to a calcium influx and an activation of CaM kinase II (Fig. 4.1; reviewed in: Doherty & Walsh, 1996; Walsh & Doherty, 1997). The substrates for CaM kinase II that mediate neurite outgrowth have not been identified and the identity of the tyrosine kinase is controversial. There is strong evidence for an involvement of both the tyrosine kinase fibroblast growth factor receptor (Williams *et al.*,

1995b; Saffell *et al.*, 1997) and non-receptor tyrosine kinases (see later text; Beggs *et al.*, 1994; Ignelzi *et al.*, 1994; Wong *et al.*, 1996). It is somewhat surprising that all three cell adhesion molecules operate through the same intracellular signalling pathway given that they have different effects on growth cone speed, morphology and cytoskeletal organisation *in vitro* (Payne *et al.*, 1992; Abosch & Lagenaur, 1993; Burden-Gulley *et al.*, 1995; Burden-Gulley & Lemmon, 1996; Drazba *et al.*, 1997; see Table 1.1).

A number of *in vitro* experiments strongly suggest that the cell-adhesion molecules NCAM, N-cadherin and L1 achieve their effects on neurite out-growth through a *cis*-interaction with the tyrosine kinase fibroblast growth factor receptor (reviewed in: Doherty *et al.*, 1997; Walsh & Doherty, 1997). The fibroblast growth factor receptor family is encoded by four separate genes and alternative splicing of these genes generates a large number of family members (Jaye, Schlessinger & Dionne, 1992; Green, Walsh & Doherty, 1996). The receptor is a single-pass membrane protein with three immunoglo-bulin-like domains in its extracellular region and an intracellular domain con-taining a tyrosine kinase (see Figs 3.2 and 4.1). Dimerisation of the receptor, produced by binding of fibroblast growth factor, leads to autophosphorylation of the receptor on its intracellular domain. This in turn creates a phospho-tyrosine site at which so-called Src-homology 2 domains within proteins can bind (reviewed in Schlessinger & Ullrich, 1992). The best established protein containing a Src-homology 2 domain that binds to the activated fibroblast growth factor receptor is phospholipase Cγ (Fig. 4.1), which has been impli-cated in cell adhesion molecule stimulated neurite outgrowth (see earlier text). There is now a considerable body of evidence from *in vitro* studies that impli-cate the fibroblast growth factor receptor in the stimulation of neurite out-growth produced by cell adhesion molecules. The extension of axons by cerebellar granule cells and retinal ganglion cells on fibroblast cell lines expres-sing NCAM, N-cadherin and L1 is blocked by antibodies that inhibit fibroblast growth factor receptor function (Williams *et al.*, 1994c; Brittis *et al.*, 1996). Furthermore, soluble recombinant NCAM and L1 chimeras consisting of the extracellular portion of the cell adhesion molecule fused to the Fc portion of an immunoglobulin are as effective at stimulating neurite outgrowth from these neurons as full-length NCAM expressed on the surface of transfected fibro-blasts, and this effect is blocked by antibodies to fibroblast growth factor receptor (Doherty, Williams & Walsh, 1995; Saffell *et al.*, 1997). Soluble forms of these cell adhesion molecules cause a rapid phosphorylation of the fibroblast growth factor receptor in PC12 cells, a hallmark of receptor activa-tion (Saffell *et al.*, 1997). Interestingly, not all soluble forms of cell adhesion molecules can reproduce the effects on neurite outgrowth exhibited by the membrane-bound forms (Volkmer *et al.*, 1996). Finally, PC12 cells and cere-bellar neurons expressing a dominant-negative form of the fibroblast growth factor receptor, in which the kinase domain is missing, no longer respond to NCAM, N-cadherin and L1 (Saffell *et al.*, 1997). These and other pharmaco-logical data support the notion that the cell adhesion molecules, NCAM, N-cadherin and L1 all mediate their *in vitro* effects on neurite outgrowth through a *cis*-interaction with the fibroblast growth factor receptor. How such an interaction could lead to dimerisation of the growth factor receptor or, indeed, if such dimerisation is necessary for intracellular signalling, is a topic for future study.

What evidence is there that an interaction between cell adhesion molecules and the fibroblast growth factor receptor also occurs *in vivo* during growth cone pathfinding? While there is no direct test of the role of the putative interaction between these cell adhesion molecules and the fibroblast growth factor receptor, there is clear-cut evidence that growth cone pathfinding by retinal ganglion cell axons in *Xenopus* and mammals partly involves the fibroblast growth factor receptor, although by a mechanism that is not understood (McFarlane, McNeill & Holt, 1995; Brittis *et al.*, 1996; McFarlane *et al.*, 1996). Expression of a dominant-negative form of the fibroblast growth factor receptor in *Xenopus* retinal ganglion cells during the period in which they are growing out of the retina to form the optic nerve impairs the ability of a significant portion of the axons from exiting the retina correctly (McFarlane *et al.*, 1995, 1996). Antibodies against L1 cause de-fasciculation and, in some cases, mistakes in pathfinding of rat retinal ganglion cell axons in retinal wholemount explants *in vitro* (Brittis *et al.*, 1995; see Chapter 3). A similar phenotype is seen when fibroblast growth factor receptor function is blocked, either pharmacologically or with antibodies, in rat retinal wholemounts (Brittis *et al.*, 1996). These observations are at least consistent with an interaction between cell adhesion molecules and the fibroblast growth factor receptor during growth cone pathfinding by retinal ganglion cells, and open up the possibility that growth factors, such as fibroblast growth factor, may influence cell adhesion molecule function in pathfinding.

The alteration in the growth cone cytoskeleton caused by cell adhesion molecules such as L1 and NCAM suggests that a major target of the intracellular signalling events activated by these cell adhesion molecules is the cytoskeleton. A clue to which part of the cytoskeleton has come from high-resolution studies of growth cones in *Aplysia*. The cell adhesion molecule apCAM in *Aplysia*, a homologue of NCAM, can direct the assembly of actin filaments in growth cones when it is clustered in the plasma membrane (Thompson, Lin & Forscher, 1996). apCAM clustering in growth cones was induced by adding apCAM antibody-coated beads to *Aplysia* growth cones in culture. Highly charged polycationic beads had previously been shown to cause actin-filament polymerisation localised to the region of the growth cone where the bead bound (Forscher *et al.*, 1992). This suggested that the charged beads could cluster membrane proteins, which were then somehow able to regulate submembranous actin filament polymerisation. At that time it was not known to which components of the growth cone membrane the charged beads were binding and, in the light of the more recent experiments, it now seems likely that apCAM was one of the molecules involved. There was a direct correlation between the concentration of apCAM antibodies bound to the bead and the time taken for actin filament assembly to occur.

Down-stream targets for calcium in growth cones

Down-stream targets for calcium in growth cones have not been identified with any certainty. Presumably some of the effects of calcium are mediated through calcium binding proteins such as calmodulin and through kinases, such as the calcium/phospholipid-dependent kinase, protein kinase C, CaM kinase II, and protein phosphatases, such as protein phosphatase 2B, also known as calcineurin, all of which are present in growth cones (see Table 4.1). There is pharmacological evidence for a role for calmodulin in calcium-dependent effects in growth cones. Pharmacological inhibition of calmodulin in *Helisoma* B19 neuron growth cones abrogated the calcium-dependent effects of serotonin on neurite growth and growth cone morphology (Polak *et al.*, 1991). More significantly, calmodulin, and hence presumably calcium, is involved in pathfinding as indicated by ingenious genetic experiments using dominant negative inhibitors of calmodulin function (VanBerkum & Goodman, 1995). In these experiments, transgenic *Drosophila* embryos were created in which a small subset of the developing neurons and their growth cones expressed two genetically engineered fusion proteins: the kinesin motor domain fused to a calmodulin antagonist peptide, which blocks calmodulin binding to proteins; or the kinesin motor domain fused with calmodulin itself, which acts as a calcium binding protein. Kinesin is a 'plus'-end directed micro-tubule motor protein and, when expressed in developing neurons, would be expected to accumulate in the growth cone by microtubule transportation. Specific genetic enhancer elements were used to drive the expression of the fusion proteins in a small subset of embryonic neurons whose growth cones pioneer the first two longitudinal axon pathways in the *Drosophila* embryo and whose behaviour has been extensively characterised (Lin *et al.*, 1994b; see Fig. 3.3, Chapter 3). The phenotype of both types of transgenic embryo was essentially similar, and consisted of inhibition of axon extension, caused by growth cone stalling and errors in pathfinding. The phenotype was seen only in those growth cones in which the fusion proteins were expressed. The initial formation of growth cones and their orientation in cells expressing fusion proteins appeared normal. Two main classes of errors in pathfinding were seen: a failure of growth cones to de-fasciculate from an axon bundle, and abnormal crossing of the midline. These findings strongly implicate calmodulin and cytosolic calcium in growth cone pathfinding, and thus in the intracellular response to guidance cues.

Calmodulin-binding proteins expressed in growth cones include growth-associated protein 43 (GAP-43, see later text), CaM kinase II and calcineurin (Ferreira, Kincaid & Kosik, 1993). Overexpression of CaM kinase II in neuro-blastoma cells in culture enhances neurite elongation (Goshima, Ohsako & Yamauchi, 1993) and pharmacological evidence implicates calcineurin in axon growth. The immunosuppressant drugs, cyclosporin A and FK506, have been shown to enhance neurite outgrowth from PC12 cells and embryonic dorsal root ganglion cells (Lyons *et al.*, 1994). The immunosuppressant action of these drugs is mediated by their effects on calcineurin in the T lymphocytes

Table 4.1. *Calcium/calmodulin-binding proteins in growth cones*

Protein	Function	Reference
Calcineurin (protein phosphatase 2B)	Phosphatase	Ferreira *et al.*, 1993; Chang *et al.*, 1995
Calcium/calmodulin-dependent protein (CaM) kinase II	Kinase	Ferreira *et al.*, 1993
Caldesmon	Actin binding	Kira *et al.*, 1995
Calmodulin	Calcium binding	VanBerkum & Goodman, 1995
Calspectrin (fodrin)	Actin binding Membrane linkage	Koenig *et al.*, 1985; Sobue & Kanda, 1989
GAP-43	Unknown	Alexander *et al.*, 1988
Protein kinase C	Kinase	Igarashi & Komiya, 1991

of the immune system (reviewed in Marks, 1996). Cyclosporin A and FK506 bind to specific binding proteins in these cells called immunophilins: FK506 binding protein and cyclophilin, respectively. The drug–immunophilin complex is then able to bind to calcineurin and inhibit its activity, which results in the accumulation of phosphorylated calcineurin substrates, including nuclear factor of activated T cells, that enter the nucleus to stimulate interleukin-2 expression. However, the immunophilins are far more abundant in the nervous system than in the immune system, although their function in the nervous system is not known (reviewed in Snyder & Sabatini, 1995). Another immunosuppressant, rapamycin, also enhances axon growth in PC12 cells, probably via an immunophilin. However, it does not inhibit calcineurin. Calcineurin may not be the relevant target for the action of immunosuppressant drugs since inactivation of growth cone calcineurin using micro-CALI causes filopodial retraction (Chang *et al.*, 1995; reviewed in Jay, 1996).

Growth cone calcium concentration and the cytoskeleton

Under circumstances in which growth cone calcium levels are raised, leading to growth cone retraction, it has been found that there is a selective loss of actin filaments, particularly in the peripheral domain of the growth cone (Lankford & Letourneau, 1989, 1991; Neely & Gesemann, 1994). In chick dorsal root ganglion cells growing in culture, raising or lowering growth cone cytosolic calcium levels above or below the set-point inhibits growth cone advance and concomitantly causes a selective loss of actin filaments from lamellipodia, but, surprisingly, not from filopodia (Lankford & Letourneau, 1991). Phalloidin, an actin filament stabilising agent (see Chapter 2), can antagonise the effect of elevated calcium, which substantiates the conclusion that raised intracellular calcium levels depolymerise actin filaments and that this inhibits growth cone advance (Lankford & Letourneau, 1989). Calcium is probably mediating these effects on actin filaments indirectly, for instance, by regulating calcium-sensitive actin-binding proteins such as gelsolin, which severs and caps actin filaments when calcium is bound to it. The expression of this class of actin-binding proteins in growth cones is poorly documented (see Chapter 3).

Calcium also appears to be involved in some (Bantlow et al., 1993; Moorman & Hume, 1993), but not all (Ivins, Raper & Pittman, 1991), forms of growth cone collapse.

Growth-associated protein 43

The growth-associated protein, GAP-43, is ubiquitously synthesised by developing neurons, transported along their growing axons by fast axonal transport and accumulates in the growth cone, where it is a major phosphoprotein present in concentrations of up to 100 μM (Skene et al., 1986; Meiri, Pfenninger & Willard, 1986; Meiri & Gordon-Weeks, 1990; Moss et al., 1990). The expression of the protein is down-regulated at the end of axonogenesis in many neurons. Furthermore, since it is re-expressed in those adult neurons that regenerate their axons after injury and therefore might be important in axon regeneration, it has received a considerable amount of attention (reviewed in: Benowitz & Routtenberg, 1987; Skene, 1989; Strittmatter & Fishman, 1991; Baetge et al., 1992).

GAP-43 was discovered independently in several different circumstances, and is referred to in the early literature by various acronyms and epithets including: F1 (Ehrlich & Routtenberg, 1974); B-50 (Zwiers et al., 1976; Zwiers, Schotman & Gispen, 1980); neuromodulin (Andreason et al., 1983; Cimler et al., 1987); GAP-43 (Skene & Willard, 1981a,b,c,d) and pp46 (Benowitz, Shashoua & Yoon, 1981; Benowitz & Lewis, 1983; Katz, Ellis & Pfenninger, 1985). One means by which GAP-43 was identified used a differential assay for proteins whose expression is up-regulated during axonal regeneration (Skene & Willard, 1981a,b,c,d; Benowitz & Lewis, 1983). In this assay, radiolabelled amino acids, usually ^{35}S methionine, are injected into the aqueous humour of the eye and, after a suitable delay, the optic nerve is removed, sliced transversely into short segments and the proteins isolated and separated by two-dimensional polyacrylamide gel electrophoresis. The injected amino acids become incorporated into newly synthesised proteins in the ganglion cell bodies of the retina and some of them are transported into the axons of these neurons, which form the optic nerve. Comparison of radiolabelled proteins in the optic nerve from adult, developing and regenerating axons can identify proteins whose synthesis is correlated with axon growth.

Widespread expression of GAP-43 in the developing nervous system

GAP-43 is expressed at very high levels in the developing nervous system, where its expression has been extensively mapped (Jacobson, Virág & Skene,

1986; Kalil & Skene, 1986; Oestreicher & Gispen, 1986; Gorgels et al., 1987, 1989; McGuire, Snipes & Norden, 1988; Moya et al., 1988; Fitzgerald, Reynolds & Benowitz, 1991; Reynolds, Fitzgerald & Benowitz, 1991). All vertebrate neurons examined express GAP-43 while extending neurites and it is particularly concentrated in axonal growth cones, where it is a major phosphoprotein (Meiri et al., 1986; Skene et al., 1986; Meiri & Gordon-Weeks, 1990; Moss et al., 1990). Immunogold electron microscopy of developing rat pyramidal tract axons and isolated growth cone particles from developing rat forebrain shows that GAP-43 is highly concentrated on the cytoplasmic surface of the plasma membrane of axons and growth cones (Gorgels et al., 1989; Van Lookeren Campagne et al., 1989). Dendritic growth cones do not contain appreciable amounts of GAP-43 (Goslin et al., 1988). Subcellular fractionation analysis has revealed that GAP-43 is highly concentrated in the cortical cytoskeleton within growth cones (Meiri & Gordon-Weeks, 1990; Moss et al., 1990), probably because of binding to other components of the membrane cytoskeleton such as actin or directly to the membrane (see below). Most neurons down-regulate their expression of GAP-43 once the growth cone has reached its target; however, there are well-documented examples where it persists in the adult nervous system (reviewed in Benowitz & Routtenberg, 1997).

Originally it was thought that GAP-43 was only expressed in neurons but it is now known to be synthesised in a variety of non-neuronal cell types in vitro including type-2 astrocytes, immature oligodendrocytes and their precursors (Vitkovic et al., 1988; da Cunha & Vitkovic, 1990; Deloulme et al., 1990; Curtis et al., 1991) and in Schwann cells in vivo following nerve injury (Curtis et al., 1992; Woolf et al., 1992). Thus there seems to be a correlation between GAP-43 expression and cell motility or process extension.

GAP-43 is a calmodulin and actin-binding phosphoprotein

The molecular properties of GAP-43 suggest that it is the target for a number of signalling pathways in growth cones and that it may act as a molecular switch. GAP-43 can bind to calmodulin via an isoleucine–glutamine (IQ in the single letter code) motif, a highly conserved calmodulin binding domain also found in type I and II myosins (Chenney & Mooseker, 1992), and the affinity is indirectly proportional to the Ca^{2+} concentration (Alexander et al., 1987, 1988; Chapman et al., 1991). GAP-43 is a phosphoprotein and can be phosphorylated in vivo by the neuron-specific form of protein kinase C (III) on serine-41 (Zwiers et al., 1976; Aloyo, Zwiers & Gispen, 1983; Akers & Routtenberg, 1985; Katz, Ellis & Pfenninger, 1985; Coggins & Zwiers, 1989; Apel et al., 1990; Spencer et al., 1992). In growth cones, GAP-43 is the major phosphorylation substrate for protein kinase C (Apel et al., 1990). Phosphorylation by protein kinase C on serine-41 blocks the binding of calmodulin (Alexander et al., 1987, 1988) and, conversely, calmodulin binding prevents protein kinase C phosphorylation on serine-41. GAP-43 also binds to actin filaments, at least *in vitro* (Moss *et al.*, 1990; Strittmatter, Vartanian & Fishman, 1992; Hens *et al.*, 1993; He, Dent & Meiri, 1997) and this property is also regulated by phosphorylation at the protein kinase C site (He *et al.*, 1997). GAP-43 does not appear to cross-link actin filaments *in vitro* (Hens *et al.*, 1993; He *et al.*, 1997), but the phosphorylated form can increase filament length whereas the non-phosphorylated form decreases it (He *et al.*, 1997). The interaction between GAP-43 and actin filaments in growth cones may be functional, as suggested by the observation that actin filaments are reduced in growth cones in which GAP-43 is down-regulated by antisense treatment of neurons in culture (Aigner & Caroni, 1995). Also, there is a co-localisation of the phosphorylated form with actin filaments in regions of the P-domain that are extending or stabilised, whereas the non-phosphorylated form is mainly found in regions that are retracting (Dent & Meiri, 1992; He *et al.*, 1997). These properties may enable GAP-43 to function as a molecular switch in the sense that protein kinase C phosphorylation prevents calmodulin binding and alters actin filament interaction. GAP-43 can be de-phosphorylated *in vitro* by the calcium and calmodulin-dependent phosphatase calcineurin (Liu & Storm, 1989; see also Meiri & Burdick, 1991).

A significant observation is that the protein kinase C phosphorylated form of GAP-43 is concentrated within growth cones as they approach their targets (Meiri, Bickerstaff & Schwob, 1991; Dent & Meiri, 1992). Using a monoclonal antibody specific for the protein kinase C phosphorylated form of GAP-43, Meiri and colleagues showed that, when axonal growth cones first emerge from the neuronal cell body, GAP-43 is mainly non-phosphorylated at the protein kinase C site but that, as the growth cone approaches its target, GAP-43 becomes phosphorylated in the growth cone (Meiri *et al.*, 1991; Dent & Meiri, 1992). The distribution of the protein kinase C phosphorylated form of GAP-43 in growth cones is not uniform and is associated with advancing, but not retracting, lamellipodia (Dent & Meiri, 1998). Thus protein kinase C phosphorylated forms of GAP-43 are restricted in their distribution to the distal axon and the growth cone. By inference, protein kinase C phosphorylation is also similarly restricted. These results imply that local extrinsic signals, perhaps derived from the target cells, can activate protein kinase C or inhibit a phosphatase in the growth cone, thus causing an increase in GAP-43 phosphorylation. One target-derived factor that may be involved is NGF (see Chapter 3) since GAP-43 phosphorylation in isolated growth cones is increased by NGF (Meiri & Burdick, 1991).

GAP-43 regulates phosphatidylinositol-4-phosphate kinase and G protein activity

There is evidence from *in vitro* studies that GAP-43 interacts with other signal transduction proteins and that these interactions are regulated by phosphorylation of GAP-43. Phosphorylated GAP-43 can inhibit phosphatidylinositol-4-phosphate kinase activity which results in a reduction in the concentration of the intracellular signalling molecule phosphatidylinositol-4,5-bisphosphate (Van Hooff et al., 1988). GAP-43 also binds to the hetero-trimeric GTP-binding protein G_o, a major component of the plasma membranes in isolated growth cones, and stimulates GTP binding to G_o (Strittmatter et al., 1990).

GAP-43 is dynamically palmitoylated in growth cones, on both of the two cysteines in the molecule, at position 3 and 4 of the N-terminus (Perrone-Bizzozero et al., 1989; Skene & Virag, 1989; Hess et al., 1993; Patterson & Skene, 1994). Palmitoylation involves the covalent attachment of long-chain fatty acids to cysteine residues. This post-translational modification is thought to target GAP-43 to functionally and compositionally distinct micro-domains in the plasma membrane (Zuber, Strittmatter & Fishman, 1989). The evidence for the existence of these microdomains rests on the finding that certain membrane proteins, including glycosylphosphatidyl inositol-anchored membrane proteins and lipid-linked non-receptor tyrosine kinases, glycosphingolipids and cholesterol, form detergent-insoluble complexes after extraction with the non-ionic detergent Triton X-100 (Simons & Ikonen, 1997). However, there is no direct evidence that such domains exist in cell membranes. Detergent-insoluble membrane fractions derived from isolated growth cone fractions and PC12 cells are highly enriched in GAP-43 (Meiri & Gordon-Weeks, 1990; Moss et al., 1990; Arni et al., 1998). Inhibition of the S-palmitoylation of GAP-43 in neurons in culture causes growth cone collapse and neurite retraction, suggesting that this modification of GAP-43, and probably other proteins within growth cones, is important for neurite advance (Hess et al., 1993; Patterson & Skene, 1994).

GAP-43 function

Despite intensive research, the function of GAP-43 remains somewhat enig-matic. Two broad functions for its role in axon growth have been proposed. There is evidence that it is required for axonal growth cone formation and axonal extension. For instance, over-expression of GAP-43 in PC12 cells enhances neurite formation (Yanker et al., 1990). In contrast, loading GAP-43 antibodies into neuroblastoma cells in culture inhibits neurite formation (Shea et al., 1991; Shea, 1994; Shea & Benowitz, 1995). Furthermore, antisense oligodeoxynucleotides to GAP-43 reduce neurite extension from PC12 cells (Jap Tjoen San et al., 1992) and axon extension from dorsal root ganglion cells in culture (Aigner & Caroni, 1993; 1995). GAP-43 depleted chick dorsal root ganglion cell growth cones spread and adhere poorly on a highly adhesive substratum, have highly dynamic but unstable lamellipodia, an altered actin filament distribution, and altered or failed responses to various receptor-mediated stimuli (Aigner & Caroni, 1993; 1995). However, PC12 cell lines with greatly reduced levels of GAP-43 are still able to extend neurites (Baetge & Hammang, 1991). When transfected into non-neuronal cells, GAP-43 can induce filopodia formation, an observation which has been interpreted to mean that GAP-43 contributes to this aspect of growth cone morphology (Zuber *et al.*, 1989; Widmer & Caroni, 1993; Strittmatter, Valenzuela & Fishman, 1994a). GAP-43 may exert this effect on filopodia formation directly by influencing the cytoskeleton or membrane dynamics. Alternatively, it could alter the response to filopodia-inducing stimuli from the local microenvironment of the growth cone (Strittmatter *et al.*, 1995).

An alternative proposal for the function of GAP-43 has been made by Strittmatter and Fishman on the basis of biochemical experiments with GAP-43 (Strittmatter *et al.*, 1990, 1991; Strittmatter, Igarashi & Fishman, 1994b). They suggested that GAP-43 is involved in the modulation of intracellular signalling events that are responding to extrinsic guidance cues. As discussed above, in support of this proposal is the finding that GAP-43 can interact with some of the components of intracellular signalling pathways, such as calmodulin (Alexander *et al.*, 1987), heterotrimeric G-proteins (Strittmatter *et al.*, 1990, 1991; Strittmatter, Igarashi & Fishman, 1994b) and protein kinase C (Zwiers *et al.*, 1976; Coggins & Zwiers, 1989; Apel *et al.*, 1990). It could be argued, of course, that such interactions might equally well be necessary for modulating growth cone formation and/or axon elongation. More significantly, when injected into *Xenopus* oocytes, GAP-43 considerably augments the action of G-protein coupled receptors (Strittmatter *et al.*, 1993). This proposal has gained additional force from recent important findings of abnormal pathfinding in a mouse line lacking a functional GAP-43 gene (Strittmatter *et al.*, 1995; Kruger *et al.*, 1998; Sretavan & Kruger, 1998). These GAP-43 'knockout' mice were derived by homologous recombination using embryonal stem cells. Their nervous systems are grossly normal, and embryonic dorsal root ganglion cells isolated from these mice form growth cones and extend axons in culture normally, suggesting that GAP-43 is not important for axonogenesis and growth cone formation *per se*. Most strikingly, however, the pathfinding of retinal ganglion cells through the optic chiasm is abnormal. Many growth cones fail to take the appropriate pathway on reaching the chiasm. This is prima facie evidence that GAP-43 is involved in the intracellular signalling mechanisms used by some guidance cues for pathfinding.

Non-receptor tyrosine kinases and phosphatases

There is good evidence that non-receptor, i.e. soluble or cytoplasmic, tyrosine kinases, particularly those of the *src* family ($p59^{fyn}$, $pp60^{c-src}$, $pp62^{c-yes}$, hereafter referred to as Fyn, Src and Yes, respectively), are important elements in growth cone signal transduction pathways (reviewed in: Bixby & Harris, 1991; Grabham *et al.*, 1996; Maness *et al.*, 1996). All three kinases are highly expressed in the nervous system and are developmentally regulated (Cotton & Brugge, 1983; Levy *et al.*, 1984; Sorge, Levy & Maness, 1984; Schartl & Barnekow, 1984; Fults *et al.*, 1985; Maness, Sorge & Fults, 1986; Steedman & Landreth, 1989). Src expression in the central nervous system is highest during development and the kinase is concentrated in growth cones (Maness *et al.*,

1988; Wiestler & Walter, 1988; Maness, Shores & Ignelzi, 1990; Bixby & Jhabvala, 1993; Helmke & Pfenninger, 1995). There are also high levels of Fyn and Yes in growth cones (Bare et al., 1993; Bixby & Jhabvala, 1993; Helmke & Pfenninger, 1995) and developing axonal tracts (Bare et al., 1993). In PC12 cells (Sobue & Kanda, 1988) and retinal ganglion cells in culture (Maness et al., 1988), Src is highly concentrated in growth cones. Not all non-receptor tyrosine kinases are enriched in growth cones and some are absent (e.g. Lck and Blk; Helmke & Pfenninger, 1995). In Drosophila embryos there is a subset of growing axons that express a non-receptor tyrosine kinase called Abelson (Abl; Gertler et al., 1989) and, although embryos with mutated forms of Abl are relatively normal, double mutants in which both Abl and the cell adhesion molecule fasciclin I (see Chapter 3) are mutated show pathfinding errors and abnormal commissures (Elkins et al., 1990).

Transgenic mice lacking either Src, Fyn or Yes have been generated, but none of these mutants show gross abnormalities in the development of their nervous systems. However, they do show alterations in long-term potentiation and synaptic plasticity in the adult hippocampus (reviewed in Maness et al., 1996). This is in stark contrast to many of the transgenic mice lacking receptor tyrosine kinases. However, the non-receptor tyrosine kinase knockouts do show unexpected abnormalities in other systems, for instance, the Src knockout mouse suffers from osteopetrosis (Soriano et al., 1991). Although there are no gross abnormalities of the nervous systems in non-receptor tyrosine kinase knockouts, there are subtle changes. For instance, when embryonic neurons from mice lacking Fyn are cultured on the cell adhesion molecule NCAM (see Chapter 3), they do not extend axons as rapidly as their wild-type controls (Beggs et al., 1994). Similarly, cerebellar neurons from the Src-minus mutant mice show impaired neurite outgrowth on the cell adhesion molecule L1 (Ignelzi et al., 1994). These experiments have led to the suggestion that NCAM signals through Fyn and L1 through Src (Maness et al., 1996). L1 may also activate another non-receptor tyrosine kinase, Rsk (Wong et al., 1996). Rsk can be co-immunoprecipitated from neuronal cell lines and brain extracts with L1 using antibodies to L1 and the kinase phosphorylates L1 in developing brain (Wong et al., 1996). More significantly, neurite outgrowth on substrate-bound L1 in vitro can be partially inhibited by introducing into cells phosphorylated peptides that correspond to the region on L1 phosphorylated by Rsk (Wong et al., 1996). These peptides presumably act either by competing for the binding of Rsk to L1 or other molecules that bind to the phosphorylated site on L1. However, how these non-receptor tyrosine kinases fit into the intracellular signalling pathway activated by cell adhesion molecules proposed by Doherty and Walsh (see above) is not clear.

A PC12 cell line expressing a constitutively active viral form of Src extends neurites spontaneously, without the need for nerve growth factor induction (Alemá et al., 1985; Rausch et al., 1989), suggesting that Src is required for neurite outgrowth. However, pharmacological experiments with tyrosine kinase inhibitors are contradictory; in some studies, inhibition of neurite outgrowth is seen (Muroya et al., 1992; Ohmichi et al., 1993; Jian, Hidaka &

Schmidt, 1994; Tsukada et al., 1994; Williams et al., 1994a; Worley & Holt, 1996; Oberstar et al., 1997), while in others a stimulation (Bixby & Jhabvala, 1992; Miller, Lee & Maness, 1993; Oberstar et al., 1997). In PC12 cells induced to differentiate and produce neurites by the addition of nerve growth factor, some protein tyrosine kinase inhibitors enhance neurite outgrowth (Miller, Lee & Maness, 1993), while others inhibit it (Hashimoto, 1988; Knusel & Hefti, 1992; Knusel et al., 1992). These contradictory results may be due to a variation in the response of growth cones from different neuronal types or to a lack of specificity of these inhibitors, or a dependency on inhibitor concentration (e.g. Oberstar et al., 1997). In only a few of these studies was the activity of the kinase directly measured to ensure inhibition and, similarly, few studies employ more than one pharmacological inhibitor with a different mechanism of action (see Worley & Holt, 1996).

What are the molecular interactions of the non-receptor tyrosine kinases in growth cones? Src has a Src-homology 2 domain which is known to bind to phosphotyrosine residues in proteins such as tyrosine kinase receptors (see above) and focal adhesion kinase (Koch et al., 1991). There is evidence for a stable association of Src and Fyn with focal adhesion kinase (Cobb et al., 1994). The major phosphotyrosine-containing protein in growth cones is tubulin (Aubry & Maness, 1988) and Src phosphorylates tubulin in growth cones (Matten et al., 1990; Cox & Maness, 1993). In subcellular fractions of isolated growth cone particles, soluble forms of NCAM and L1 inhibit the tyrosine phosphorylation of membrane-bound tubulin (Atashi et al., 1992). This effect is produced by both inhibition of tyrosine kinase activity and stimulation of tyrosine phosphatases (Klinz, Schachner & Maness, 1995). It is not known whether this effect is mediated via the cis-interaction between NCAM/L1 and fibroblast growth factor receptor (see earlier text).

Heterotrimeric G proteins, small GTP-binding proteins and GTPases

Growth cone collapse of chick dorsal root ganglion cell growth cones produced by collapsin-I, a member of the collapsin/semaphorin family of growth cone guidance molecules (see Chapter 3), may be mediated by heterotrimeric G proteins. Biochemical analysis of isolated growth cone particles has shown that heterotrimeric G proteins are present in growth cones (Edmonds et al., 1990). A specific activator of G proteins, mastoparan, mimics the growth cone collapse caused by collapsin-I and the bacterial toxin, pertussis toxin, which is thought to be a specific inactivator of G proteins, blocks growth cone collapse (Igarashi et al., 1993). However, doubt has been cast on this interpretation because Kindt and Lander (1995) have shown that growth cone collapse can be blocked by a form of pertussis toxin that lacks the G protein inactivating activity.

Rho subgroup of the Ras superfamily

Three members of the mammalian Rho subgroup of the Ras superfamily of small GTPases have roles in actin-dependent fibroblast migration: Rho, Rac and Cdc42 (reviewed in: Mackay, Nobes & Hall, 1995; Takai *et al.*, 1995; Ridley, 1996; Symons, 1996; Tapon & Hall, 1997) and recent evidence suggests that they may also be involved in the regulation of neurite growth (reviewed in: Luo, Jan & Jan, 1997; Vancura & Jay, 1998). In fibroblasts, microinjection experiments with dominant-negative forms of Rho proteins and pharmacological inhibition studies have shown that Rho is involved in the regulation of stress fibre formation, Rac in the formation of lamellipodia and that Cdc42 is important for filopodial extension. Exactly how these small GTPases regulate actin dynamics in cells is not clear. The neuroblastoma cell line, N1E-115, has fibroblast-like properties when growing in serum containing medium but differentiates into a neuronal morphology when serum is withdrawn. Microinjection of Cdc42 or Rac1 into N1E-115 neuroblastoma cells induces filopodial and lamellipodial extension, respectively, while microinjection of the *Clostridium botulinum* toxin, C3 transferase, which specifically inactivates Rho, induces neurite outgrowth in the absence of serum withdrawal, suggesting that Rho inactivation, and hence suppression of stress fibre formation, is necessary for neurite outgrowth (Kozma *et al.*, 1997). In contrast, injection of dominant-negative forms of either Rac1 or Cdc42, blocks neurite outgrowth induced by serum withdrawal and co-injection of dominant-negative forms of either Rac1 or Cdc42 with C3 transferase blocks the induction of neurite outgrowth produced by Rho inactivation (Kozma *et al.*, 1997). These experiments suggest that Cdc42 and Rac1 have similar functions in regulating actin filament dynamics in neurite outgrowth in neuroblastoma cells as they do in fibroblast and other cell types.

Serum contains lysophosphatidic acid, which has been shown to cause growth cone collapse and neurite retraction when added to differentiated N1E-115 neuroblastoma cells (Jalink & Moolenaar, 1992; Jalink *et al.*, 1994; Kozma *et al.*, 1997). Lysophosphatidic acid-mediated growth cone collapse is inhibited by microinjecting C3 transferase and, since this inactivates Rho, this suggests that growth cone collapse induced by lysophosphatidic acid involves changes in actin filament dynamics regulated by Rho. Further support for the involvement of Rho comes from the finding that microinjection of Rho or a constitutively active form of Rho into N1E-115 neuroblastoma cells, which have extended neurites following serum withdrawal, also causes growth cone collapse (Kozma *et al.*, 1997). Expression of a dominant-negative form of Rac in PC12 cells inhibits the normal response to nerve growth factor – formation of neurites – but does not inhibit the formation of a neurite produced by touching the cells with a laminin-coated micropipette and withdrawing the micropipette (Lamoureux *et al.*, 1997). Since such mechanically produced neurites depend on microtubule function, this finding suggests that Rac is not involved in regulation of the microtubule cytoskeleton in neurites, but is somehow involved in growth cone tension or attachment to the substratum. Rac may also mediate growth cone collapse produced by collapsin-I (Jin & Strittmatter, 1997).

These *in vitro* experiments showing roles for small GTPases in growth cone function beg the question of whether such molecules function *in vivo*. Genetic experiments in *Drosophila*, *C. elegans* and mice suggest that they do (Luo *et al.*, 1994, 1996; Zipkin, Kindt & Kenyon, 1997; reviewed in Luo, Jan & Jan, 1997). When a constitutively active form of Rac1 is expressed in transgenic mice, Purkinje cells in the cerebellum showed aberrations in the development of dendritic spines and axon extension (Luo *et al.*, 1996).

The intracellular signalling molecules that link membrane receptor activation with these small GTPases are unknown. However, likely candidates include cyclic adenosine $3',5'$-monophosphate (cAMP), since cAMP can negatively regulate RhoA and recent experiments show that intracellular levels of cAMP within growth cones can influence the effects of certain growth cone guidance molecules *in vitro* (Lohof *et al.*, 1992; Zheng *et al.*, 1994; Ming *et al.*, 1997; Song *et al.*, 1997). When cAMP levels in spinal cord neuron growth cones from *Xenopus* are decreased or when cAMP-dependent protein kinase A is inhibited, the normal attractive response of these growth cones to a gradient of acetylcholine, brain-derived neurotrophic factor or netrin-1 (Lohof *et al.*, 1992; Zheng *et al.*, 1994; Ming *et al.*, 1997) is rapidly converted to one of repulsion (Ming *et al.*, 1997; Song *et al.*, 1997). These experiments show that growth cones can respond in diametrically opposite ways to the same guidance cue depending on intracellular levels of signalling molecules and point to a commonality of intracellular mechanisms underlying diverse responses.

5

Synaptogenesis

Introduction

The term 'synapse' was coined by the English physiologist Charles Sherrington from the Greek 'to fasten or clasp together' (Sherrington, 1947). He inferred the existence of a gap between neurons that contacted each other from experiments using the ganglionic blocking agent nicotine, which he painted on to peripheral ganglia. The synapse is a specialised region of close apposition between two cells where intercellular communication occurs (Burns & Augustine, 1995). Sherrington proposed the existence of a synapse from his physiological studies of the effects of nicotine on autonomic ganglia. Synapses form between neurons or between a neuron and an effector cell, such as a muscle or an endocrine cell. At the culmination of pathfinding, the growth cone must select a synaptic partner and form a synapse: a process known as synaptogenesis. In the case of most axonal growth cones, synaptogenesis requires the growth cone to stop growing and to transform into a presynaptic nerve terminal (reviewed in: Rees, 1978; Vaughn, 1989; Garrity & Zipursky, 1995; Haydon & Drapeau, 1995). Although there are presynaptic dendrites in the adult central nervous system, dendritic growth cones develop mainly into postsynaptic sites.

In this chapter, I will review our understanding of the early events in synaptogenesis, particularly the selection of a synaptic partner by the growth cone and the initial stages of the interaction between the growth cone and its synaptic partner leading to the differentiation of the synapse. Later events in synaptogenesis, such as synapse elimination and synaptic plasticity (sprouting or remodelling) in the adult nervous system, are outside the scope of this monograph (reviewed in: Goodman & Shatz, 1993; Hall & Sanes, 1993; Bonhoeffer, 1996; Henderson, 1996; Katz & Schatz, 1996).

Synaptogenesis has been studied most thoroughly at the vertebrate neuromuscular junction, largely because of the accessibility of this synapse to both visual observation and electrophysiological recording and, more recently, to genetic analysis (see below). Consequently, we know far more about synaptogenesis at the neuromuscular junction than at any other synapse (reviewed in: Hall & Sanes, 1993; Bowe & Fallon, 1995; Keshishian *et al.*, 1996; Ruegg, 1996; Wallace, 1996; Sanes, 1997). It is important to remember, however, that the neuromuscular junction differs in structure and physiological properties from synapses between neurons in a number of respects and therefore some aspects of synaptogenesis at the neuromuscular junction may differ from those at synapses between neurons. Whether these differences are fundamental or in the detail has yet to be determined. One prominent feature of synaptogenesis is the requirement of the growth cone and its synaptic partner to communicate with each other. Communication takes place at two levels: local signalling between the growth cone and the postsynaptic cell; and more distant signalling between the growth cone and the nucleus, at the level of gene expression. The nature of these signals and the molecules involved are now being defined.

Morphological differentiation of synapses

There have been numerous electron microscopical studies of synaptogenesis, both *in vivo* and in cell culture (e.g. Glees & Sheppard, 1964; Meller, 1964; Aghajanian & Bloom, 1967; Bunge, Bunge & Peterson, 1967; James & Tresman, 1969; Johnson & Armstrong-James, 1970; Molliver & Van der Loos, 1970; Adinolfi, 1972a,b; Molliver, Kostovic & Van der Loos, 1973; Vaughn *et al.*, 1974; König, Roch & Marty, 1975; Rees *et al.*, 1976; Vaughn & Sims, 1978; Juraska & Fifková, 1979; McGraw & McLaughlin, 1980; reviewed in Vaughn, 1989). These have shown that there is a gradual accumulation of synaptic vesicles and the appearance of elements of the synaptic membrane specialisations, such as presynaptic and postsynaptic densities, as synaptogenesis proceeds. The precise order of the appearance of presynaptic and postsynaptic membrane specialisations varies between different locations. For example, in the embryonic spinal cord of amphibians, the presynaptic density is present before the postsynaptic density (Hayes & Roberts, 1973, 1974), whereas the reverse is true in the olfactory bulb of the mouse (Hinds & Hinds, 1976a,b; see also: Adinolfi, 1972a,b; Blue & Parnevelas, 1983; Krisst & Molliver, 1976). However, there is evidence, that, in widely different parts of the central and peripheral nervous system, immature postsynaptic densities are present in target neurons before growth cones have invaded the neuropil (Stelzner, Martin & Scott, 1973; Westrum, 1975; Hinds & Hinds, 1976a,b; Smolen, 1981; Blue & Parnavelas, 1983; Bähr & Wolff, 1985; Mihailoff & Bourell, 1986).

That these so-called vacant postsynaptic densities become incorporated into synapses and are not lost has not been demonstrated directly, although there is a temporal correlation with their appearance and decline, and synaptogenesis that is consistent with this idea (Blue & Parnavelas, 1983). Furthermore, vacant postsynaptic densities persist if growth cones are prevented from entering the region. This has been demonstrated for vacant postsynaptic densities on the dendritic spines of cerebellar Purkinje cells (reviewed in: Hirano & Dembitzer, 1975; Sotello, 1990). When cerebellar granule cells, the principal afferents to Purkinje cells, are destroyed during development by X-ray irradiation, neurotropic viruses or cytotoxins, such as cycasin, the vacant postsynaptic densities on the Purkinje cells survive into adulthood and become opposed by glial cell processes (Herndon, Margolis & Kilham, 1971; Hirano, Dembitzer & Jones, 1972). In the mouse *weaver* mutant, in which cerebellar granule cells die and are congenitally absent, vacant postsynaptic sites on Purkinje cells are retained into adulthood, also opposed by glial cell processes (Hirano & Dembitzer, 1973, 1974; Llinás, Hillman & Precht, 1973; Rakic & Sidman, 1973a,b,c; Sotello, 1973, 1975; Landis & Reese, 1977). In the cerebellum of these mutant and experimental animals, there are subtle structural differences between vacant postsynaptic densities, and the dendritic spines that carry them, and postsynaptic densities and spines in normal cerebellum, implying that complete morphological maturation requires afferent, parallel fibre input (Hirano *et al.*, 1972). Purkinje neurons also receive synaptic input from other sources, such as the inferior olive via the climbing fibres, and, since these are present in these experimental and mutant animals, it is not clear to what extent the development of vacant postsynaptic sites by Purkinje cells is dependent on some form of synaptic input.

These findings have led to the suggestion that the postsynaptic neuron dictates where on its surface synaptogenesis is to occur and thus the relative position of its synapses. This is an important issue since many neurons receive multiple synaptic inputs from diverse sources and these are often spatially organised into discrete regions of the somatodendritic plasma membrane. Furthermore, these varied inputs may use different neurotransmitters and therefore require different sets of postsynaptic receptors and intracellular signalling pathways. Although there has been much recent progress in determining the molecular composition of postsynaptic densities at adult synapses (reviewed in Kennedy, 1997), the molecular composition of vacant postsynaptic densities has not been determined.

Partial differentiation of the presynaptic element can also occur in the absence of the postsynaptic cell. The cerebellum of the *staggerer* mutant mouse is deficient in Purkinje cells, which die before most granule cells are born (Sidman, Lane & Dickie, 1962). Thus, the synaptic target of the granule cell parallel fibre axons is missing. Despite this, parallel fibres are able to develop axonal varicosities complete with synaptic vesicles and presynaptic membrane specialisations (Sotello, 1973; Sotello & Changeux, 1974). These are opposed to glial cell processes in a mirror image of the situation seen with presynaptic loss. Neurons grown on beads coated with poly-basic amino acids can also develop presynaptic elements in the absence of postsynaptic cells (Burry, 1980).

Acquisition of synaptic properties by growth cones

Several components of the mature synapse appear in the growth cone and the postsynaptic cell before contact is made between the two. They include voltage-gated calcium channels (see Chapter 4), neurotransmitters, and some of the proteins associated with synaptic vesicles and the machinery for vesicle exocytosis (Phelan & Gordon-Weeks, 1992; Osen-Sand et al., 1993; Igarishi et al., 1996). The acquisition of these features of the mature synapse continues progressively and more rapidly after contact has been established. For example, growth cones acquire some of the biochemical machinery for neurotransmission while they are pathfinding, but this is not fully developed until after pathfinding has ceased and synaptogenesis begun. Why this is so is not entirely clear. One possibility is that certain biochemical components of the mature synapse may also serve a function during pathfinding, for example, some components of the biochemical machinery underlying synaptic vesicle exocytosis and recycling may also be used for membrane addition at the growth cone (see below). Also, during target recognition at the end of pathfinding, the growth cone must communicate with its intended partner – an event which necessitates an exchange of signalling molecules. The mechanism of release of such signalling molecules may have common molecular components to signalling at mature synapses. In the interests of parsimony, it may be that neurotransmitters are used both as signalling molecules at mature synapses and to signal to the postsynaptic cell the arrival of the growth cone.

Neurotransmitter release from growth cones

Growth cones contain, and are able to release, neurotransmitters (reviewed in: Lockerbie, 1990). Many neurotransmitters, including γ-amino butyric acid (GABA), glutamate, glycine and serotonin, have been localised in growth cones by immunocytochemistry (Dale et al., 1986, 1987a,b; van Mier et al., 1986; Roberts et al., 1987; Roberts, 1988; Broadie & Bate, 1993a; van den Pol, 1997). In the embryonic spinal cord and hindbrain of Xenopus, neurotransmitters are present within axonal growth cones of some classes of neuron from the moment that they emerge from the cell body. For example, the neuronal cell bodies of developing Kolmer-Agduhr neurons in the spinal cord label strongly with antibodies to GABA, as does their axonal growth cone as it forms and emerges from the cell body (Dale et al., 1987b), whereas glycine-containing commissural interneurons in Xenopus spinal cord first express glycine in their growth cones only after crossing the floor-plate (Dale et al., 1986). Developing hypothalamic neurons in the rat express GABA in both dendritic and axonal growth cones, although apparently at higher levels in the latter (van den Pol, 1997).

The release of endogenous neurotransmitter by growth cones at the point of, or soon after, contacting their target cells was first demonstrated by Cohen (1980). In these experiments, Cohen assayed acetylcholine release from embryonic chick spinal cord growth cones contacting muscle cells in culture by intracellular electrophysiological recording from the muscle cells. For technical reasons he was only able to record stimulus-evoked release but clearly showed that it occurred within 30 minutes of growth cone filopodia contacting a muscle cell. It was not clear from these experiments whether neurotransmitter release was induced by contact between the growth cone and the muscle cell or whether growth cones can release neurotransmitter independently of their target cell. This issue was resolved by the experiments of Young & Poo (1983) and Hume, Role and Fischbach (1983), who enlarged upon Cohen's findings by showing that growth cones can release neurotransmitters independently of the target cell. As a probe for extracellular acetylcholine, they used pipettes tipped with excised patches of embryonic muscle cell membrane containing acetylcholine receptors. They found that growth cones of cultured *Xenopus* spinal cord neurons and chick ciliary ganglion neurons release the neurotransmitter acetylcholine both spontaneously and in response to action potentials in the absence of muscle cells. Neurotransmitter release from growth cones in culture has since been demonstrated by a number of workers by electrophysiological recording from muscle cells or neurons as the growth cone makes contact with them, and it has been confirmed that release occurs within a short time (less than 1 minute) following contact (Xie & Poo, 1986; Buchanan, Sun & Poo, 1989; Evers *et al.*, 1989; Haydon & Zoran, 1989). Some, but not all, of the growth cones of *Helisoma* neurons can also release neurotransmitter immediately on contact with target cells in culture. The B5 neuron is highly promiscuous in culture, and will form synapses with muscle cells and many neuron types. Neurotransmitter is released from the growth cone of B5 neurons within seconds of contacting a cell (Haydon & Kater, 1988; Haydon & Zoran, 1989). In contrast, the B19 neuron is highly selective in its choice of synaptic partner *in vitro* and it takes several hours for its growth cone to achieve evoked neurotransmitter release after contact with its synaptic partner (Haydon & Kater, 1988; Haydon & Zoran, 1989; Zoran, Doyle & Haydon, 1990, 1991; Zoran *et al.*, 1993). Thus, some classes of growth cones are able to release neurotransmitters independently of contact with a postsynaptic cell, while other classes of growth cones release neurotransmitter only after contacting a postsynaptic cell. In both cases, neurotransmitter release from growth cones on to target cells can occur well before any morphological development of the synapse.

What is the purpose of this precocious release of neurotransmitter by growth cones? Since, in some cases, neurotransmitter is present in growth cones from the time that they emerge from the neuronal cell body, neurotransmitters could be involved in pathfinding as well as in synaptogenesis. At the neuromuscular junction, blockade of postsynaptic acetylcholine receptors with D-tubocurarine does not appear to affect synaptogenesis, suggesting that release of acetylcholine by motoneuron growth cones is not important for synaptogenesis (Cohen, 1972). It is not clear whether this is also true of other synapses. In contrast to neurotransmitters, other molecules released from growth cones have been shown to have important roles in synaptogenesis (see later text).

Neurotransmitter storage in growth cones

At mature, chemical synapses, neurotransmitters are stored in synaptic vesicles, and released into the synaptic cleft by synaptic vesicle exocytosis through the presynaptic membrane. Synaptic vesicle exocytosis at synapses is calcium dependent. It is not clear whether growth cones contain synaptic vesicles, or to what extent neurotransmitters in growth cones are stored within vesicles or are in a soluble pool in the cytoplasm. Antibodies to neurotransmitters usually label the entire growth cone including the filopodia and lamellipodia, suggesting that some fraction of the neurotransmitter is free in the cytoplasm because filopodia do not usually contain vesicles (see Chapter 1). Some synaptic vesicle-specific proteins are expressed in growth cones (Phelan & Gordon-Weeks, 1992; Osen-Sand et al., 1993, 1996; Igarashi et al., 1995, 1996; Igarashi, Tagaya & Komiya, 1997), but this may be related to membrane recycling in the growth cone rather than neurotransmitter release (reviewed in Futerman & Banker, 1996). Consistent with this view is the finding that two proteins that are part of the synaptic vesicle fusion complex at mature synapses, SNAP-25 (Osen-Sand et al., 1993) and syntaxin (Igarishi et al., 1996), are required for neurite extension. Growth cones contain vesicles but many of these are distinct from synaptic vesicles: they are larger and more heterogeneous in size (see Chapter 1, Table 1.2; Pfenninger et al., 1992; Pfenninger & Friedman, 1993). However, neurotransmitter release from the growth cones of some neurons is calcium-dependent, which implies vesicular storage. Evoked acetylcholine release from *Xenopus* spinal cord neuron growth cones is dependent on extracellular calcium, as it is at mature synapses, since it can be blocked by cobalt ions (Co^{2+}), which block voltage-gated calcium channels (Xie & Poo, 1986; Sun & Poo, 1987; Evers et al., 1989; Dan, Song & Poo, 1994).

Neurotransmitter receptors

Isolated growth cone particles from neonatal rat forebrain retain osmotic activity and certain physiological functions (reviewed in Lockerbie, 1990). For example, they can take up and release the neurotransmitters GABA and noradrenaline (Gordon-Weeks, Lockerbie & Pearce, 1984, 1995; Lockerbie et al., 1985; Lockerbie & Gordon-Weeks, 1985, 1986; Taylor & Gordon-Weeks, 1989, 1991). They contain endogenous GABA, which is also releasable (Taylor et al., 1990), and express $GABA_A$ receptors (Lockerbie & Gordon-Weeks, 1985; Fukura, Komiya & Igarashi, 1996). In fact it has been known for some time that certain neurotransmitter receptors are expressed by neurons in the developing nervous system before synaptogenesis, for example, GABA receptors (Laurie, Wisden & Seeberg, 1992; Poulter et al., 1992; Walton, Schaffner & Barker, 1993; Chen, Trombley & van den Pol, 1995; reviewed in: Luis de Blas, 1993) and acetylcholine receptors (Perrins & Roberts, 1994; Erskine & McCaig, 1995; reviewed in Zagon & McLaughlin, 1993). Activation of GABA receptors on growth cones can lead to modulation of GABA release

(Lockerbie & Gordon-Weeks, 1985) and affect neurite outgrowth of neurons in culture (Spoerri, 1988; Michler, 1990; Barbin *et al.*, 1993; Behar *et al.*, 1994). $GABA_A$ receptor activation of isolated growth cones from developing rat brain increases protein kinase C phosphorylation of GAP-43 and MARCKS (Fukura *et al.*, 1996) and the growth cones of developing hypothalamic neurons in culture respond to GABA receptor activation with an increase in cytosolic calcium (Obrietan & van den Pol, 1996). In the developing cerebral cortex, GABA may act as a chemoattractant for migrating neurons (Behar *et al.*, 1996).

In the adult nervous system, the $GABA_A$ receptor is a ligand-gated chloride channel whose activation leads to chloride influx and membrane hyperpolarisation; in other words, the $GABA_A$ receptor is inhibitory. There is evidence, however, that the $GABA_A$ receptor in the embryonic nervous system is excitatory and that its activation leads to a chloride efflux and calcium influx (Obata, Oide & Tanaka, 1978; Connor, Tseng & Hockberger, 1987; Ben-Ari *et al.*, 1989; Chen, Trombley & van den Pol, 1996; reviewed in Cherubini, Gaiarsa & Ben-Ari, 1991; Ben-Ari *et al.*, 1997). Functional $GABA_A$ receptors are widely expressed on immature neurons and their growth cones in the developing nervous system and activation of these receptors leads to membrane depolarisation which is sufficient to reach the threshold for sodium action potential generation. The $GABA_A$ receptors expressed in the developing nervous system do not appear to be different from those expressed in the adult nervous system and it seems likely that the chloride *efflux*, and thus membrane depolarisation, occurs because of a higher intracellular chloride concentration in developing neurons.

Molecular mechanisms underlying synaptogenesis at the neuromuscular junction

Far more is known about the molecular mechanisms underlying synaptogenesis at the neuromuscular junction than at any other synapse (reviewed in: Hall & Sanes, 1993; Keshishian & Chiba, 1993; Keshishian *et al.*, 1996; Ruegg, 1996; Wells & Fallon, 1996; Sanes, 1997). This is mainly because of the accessibility of the neuromuscular junction for visual and electrophysiological studies, both in vertebrates and invertebrates and, more recently, the availability of genetic techniques particularly in mouse and *Drosophila*. To what extent synaptogenesis at the neuromuscular junction resembles that at synapses between neurons is not clear since the importance of the differences between the two types of synapse for synaptogenesis is not known. Among the features that are unique to the neuromuscular junction is the presence of a basal lamina in the synaptic cleft; there is no such structure at synapses between neurons. Although continuous with the basal lamina surrounding the muscle and Schwann cells, it differs from the extrajunctional basal lamina in a number of important respects and plays an essential role in synaptogenesis at this synapse (see below).

The behaviour of motoneuron growth cones as they approach and contact muscle cells and the consequent changes in the muscle cell have been thoroughly documented (reviewed in Sanes, 1997; Fig. 5.1). One of the first consequences of contact between the growth cone and muscle cell is a redistribution of neurotransmitter receptors in the muscle cell. Before growth cone contact, neurotransmitter receptors are distributed uniformly throughout the muscle cell plasma membrane (sarcolemma; Fig. 5.1A). At vertebrate neuromuscular junctions, the neurotransmitter is acetylcholine and, on growth cone contact, pre-existing acetylcholine receptors in the muscle sarcolemma begin to aggregate at the site of the future synapse (Fig. 5.1B,C; Anderson & Cohen, 1977; Frank & Fischbach, 1979), probably by a mechanism that traps receptors diffusing in the plane of the sarcolemma (see below). This event marks the beginning of neuromuscular junction formation and will not occur unless a motoneuron growth cone contacts the muscle cell; other neuronal cell types, such as sensory or cerebral cortical neurons, cannot act as a substitute (Anderson & Cohen, 1977; Frank & Fischbach, 1979). A similar phenomenon occurs at invertebrate junctions, for instance, in *Drosophila*, where the neurotransmitter of somatic motoneurons is glutamate (Broadie & Bate, 1993a,b). Vertebrate skeletal muscle cells are multinucleate and all nuclei produce mRNA encoding for the acetylcholine receptors. Following the arrival of the growth cone, extrajunctional receptors begin to disappear, and only those nuclei immediately subjacent to the developing junction continue to produce mRNA encoding receptors and thus direct the synthesis of new receptors (Fig. 5.1C). Acetylcholine receptors are composed of five subunits. In immature muscle cells (myotubes), before the arrival of the motoneuron growth cone, the acetylcholine receptor subunit composition is $\alpha_2\beta\gamma\delta$ and it is this immature, foetal form of the receptor, which accumulates in the postsynaptic sarcolemma following the arrival of the growth cone. The nerve-induced up-regulation of acetylcholine receptor synthesis in subsynaptic nuclei is associated with the production of the ε receptor subunit which appears in mature acetylcholine receptors ($\alpha_2\beta\varepsilon\delta$). Concomitant with receptor aggregation, synapse-specific molecules, such as neuronal agrin (see later text), appear in the basal lamina between the developing nerve terminal and the muscle at the postsynaptic site. Finally, the postsynaptic membrane reorganises to produce its characteristic junctional folds, at the crests of which appears the enzyme that breaks down acetylcholine, acetylcholinesterase, anchored to the basal lamina (Fig. 5.1D). Acetylcholinesterase is synthesised by the muscle cell and secreted into the junctional cleft, where it becomes anchored to the basal lamina. While postsynaptic differentiation is proceeding, the presynaptic terminal is maturing – accumulating synaptic vesicles and developing presynaptic dense projections (Fig. 5.1D).

Some important clues about the molecular mechanisms of synaptogenesis at the neuromuscular junction have come from studies of the adult peripheral nervous system. Motoneuron axons in the adult peripheral nervous system can regenerate new presynaptic terminals if disconnected from their terminal synapses, for example, by severing or crushing the axon. A growth cone

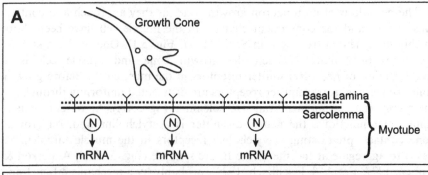

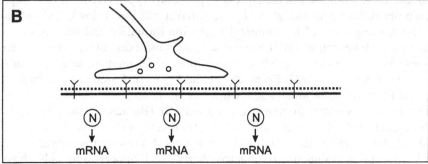

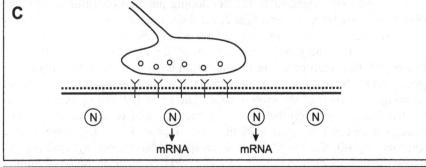

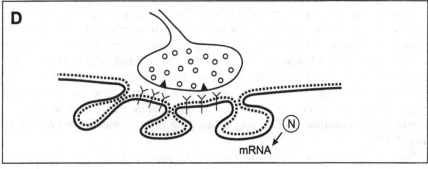

forms at the severed distal end of the axon and, if the conditions are conducive, can grow back to the site of the original muscle cell synapse (end-plate) and form a new synapse (Bennett, McClachlan & Taylor, 1973). Under ideal circumstances, these events can lead to complete recovery of function. Furthermore, if the muscle cell is also destroyed at the time of nerve severance, for instance, by directly injecting local anaesthetics into it, the basal lamina tube surrounding the muscle remains and regenerating axonal growth cones can return to the site of the old basal lamina synaptic region (Sanes, Marshall & McMahan, 1978; Burden, Sargent & McMahan, 1979; Glicksman & Sanes, 1983). Despite the absence of a muscle cell, these growth cones form all the normal structures associated with the pre-synaptic terminal. If the muscle is also allowed to regenerate, the postsynaptic elements are also re-formed at the original end-plate by the regenerated muscle, even if the nerve is prevented from regenerating (Anglister & McMahan, 1985). These observations strongly suggest that there are stably expressed molecules in the adult basal lamina at the neuromuscular junction that instruct the regenerating growth cone and muscle cell to stop growing, and commence the programme for synaptogenesis. It is assumed that regeneration recapitulates development and therefore that there are similar or identical molecules also acting during development.

One of the first events during synaptogenesis is the halting of the advance of the growth cone. Since, at least for regeneration, the basal lamina contains sufficient information to orchestrate synaptogenesis, there must be molecules, specific to the synaptic basal lamina, that have this function. One candidate molecule that might act as a growth cone stop signal in the basal lamina at the neuromuscular junction is laminin (see Chapter 3). Both laminin-3 (s-laminin; $\alpha1\beta2\gamma1$; see Fig. 3.1) and laminin-11 ($\alpha5\beta2\gamma1$) are concentrated at the neuromuscular junction in the basal lamina folds of the synaptic cleft and, unlike the other laminin family members, do not stimulate neurite outgrowth, although they are, significantly, highly adhesive for motoneuron cell bodies, axons and growth cones (Hunter et al., 1989a,b, 1991; Patton et al., 1997; reviewed in: McKerracher et al., 1996; Ruegg, 1996). There is a tripeptide sequence, leucine–arginine–glutamate (LRE in the single letter code) within the $\beta2$ chain, which has been implicated in growth cone adhesion (Hunter et al., 1989b;

Figure 5.1 Drawings illustrating synaptogenesis at the neuromuscular junction. (A) At the time of arrival of the growth cone, acetylcholine receptors (Y symbols in sarcolemma) are synthesised from mRNA produced by all muscle cell (myotube) nuclei (N). (B) On growth cone filopodial contact with the myotube basal lamina, the growth cone stops growing and increases its contact area with the myotube. (C) The first sign of synaptogenesis is the aggregation of acetylocholine receptors in the sarcolemma underlying the growth cone and the concomitant shutdown of acetylcholine mRNA synthesis by myotube nuclei outside the region of the forming neuromuscular junction. During this time, synaptic vesicles begin to accumulate in the growth cone. (D) In the final stages of synaptogenesis, the presynaptic terminal acquires the apparatus for synaptic vesicle exocytosis and the postsynaptic membrane junctional folds form. At the crests of the folds, acetylcholinesterase becomes concentrated in the basal lamina.

Porter, Weis & Sanes, 1995). Transgenic mice in which the β2 gene, and thus laminin-3 and laminin-11, has been knocked out have anatomically and functionally abnormal neuromuscular junctions (Noakes *et al.*, 1995). The nerve terminals show reduced branching, the distribution of synaptic vesicles is abnormal, and the number of pre-synaptic active zones is dramatically reduced. The difference between laminin-3 and laminin-11 is that the former has an α1 chain and the latter an α5 chain. Since α1 is not found in the synaptic basal lamina (Patton *et al.*, 1997), laminin-11 is now the more likely candidate for a role in synaptogenesis. However, interpretation of the β2 gene knockout phenotype is complicated by the compensatory up-regulation of some laminin genes that occurs in these animals (Patton *et al.*, 1997). Despite this it is clear that synapse-specific laminins are important in neuromuscular junction differentiation and probably contribute to the mechanism which arrests growth cone movement. There are, however, other contributors to growth cone arrest and these most likely include agrin and MuSK (see below).

Motoneuron growth cones release factors that induce synaptogenesis

Neurotransmitter receptor aggregation is one of the first signs of synaptogenesis and is induced by the growth cone, either by the release of a factor on to the muscle cell or by a cell–cell-mediated event. Growth cones release neurotransmitter (see earlier text) but there is no evidence that neurotransmitter released on to the muscle membrane from growth cones is responsible for receptor aggregation. For instance, blockade of acetylcholine receptors with D-tubocurarine does not affect synaptogenesis (Cohen, 1972). However, there is clear-cut evidence that motoneuron axons and growth cones produce and release factors on to muscle cells that are important for synaptogenesis. A considerable amount of effort has been expended in trying to identify neuronal factors that regulate acetylcholine receptor expression and organisation at the developing neuromuscular junction (reviewed in: Ruegg, 1996; Ruegg & Bixby, 1998). The aggregation of acetylcholine receptors at the neuromuscular junction appears to be under the influence of agrin (reviewed in Sanes, 1997), whereas up-regulation of synthesis appears to be regulated by neuregulin/ARIA (acetylcholine receptor inducing activity; Jessell, Siegel & Fischbach, 1979) and calcitonin gene-related peptide (New & Mudge, 1986; Peng *et al.*, 1989; Changeux *et al.*, 1992; Falls *et al.*, 1993). The suppression of extrajunctional synthesis of acetylcholine receptors is regulated by electrical activity in the muscle, a fact which has been known for a long time (Merlie & Sanes, 1985). Muscle electrical activity is also important for the development of postsynaptic junctional folds (Brenner, Meier & Widmer, 1983) and the accumulation of acetylcholinesterase in the junctional cleft (Lømo & Slater, 1980).

Agrin induces neurotransmitter receptor aggregation at the synapse

Considerable biochemical and genetic evidence has accumulated to suggest that a neuronal form of agrin is secreted by motoneuron axons and growth cones into the developing synaptic cleft at the neuromuscular junction, where it binds to laminin and organises the aggregation of acetylcholine receptors and probably other junction-specific molecules.

Agrin (Greek *ageirein*, meaning 'to assemble') was first isolated from basal lamina extracts of the electric organs of the marine ray, *Torpedo californica*, which are rich in cholinergic synapses (Nitkin *et al.*, 1987; Smith *et al.*, 1987). Agrin is a large (225 kD) multidomain heparan sulphate proteoglycan with laminin-binding activity. It is secreted by the motoneuron growth cone into the forming synaptic cleft, where it accumulates, presumably by binding to laminin (Magill-Solc & McMahan, 1988; Cohen & Godfrey, 1992). When added to cultures of developing muscle cells (myotubes), soluble agrin causes the clustering of acetylcholine receptors in a dose-dependent manner (Wallace, 1989; Campanelli *et al.*, 1991; Ruegg *et al.*, 1992) and antibodies to agrin inhibit the formation of acetylcholine receptor clusters (Reist, Werle & McMahan, 1992). Although agrin is produced by non-neuronal cells, including muscle cells, alternative splicing of the gene generates a neuron-specific form of the protein. Several lines of evidence suggest that only the neuronal form can cause the aggregation of acetylcholine receptors (reviewed in Ruegg, 1996). After experimental degeneration of the nerve and muscle (see earlier text), agrin remains associated with the basal lamina at the site of the end-plate for several weeks, during which time regeneration of the synapse at the original site can occur (Reist, Magill & McMahan, 1987). These observations have led to a general hypothesis that agrin, secreted by the motoneuron, induces synaptic differentiation at the neuromuscular junction (McMahan, 1990). However, although mutant mice lacking agrin expression by homologous recombination of the gene show reduced acetylcholine receptor aggregation in their skeletal muscle membranes, receptor clustering in these animals is not entirely absent (Gautam *et al.*, 1996). These clusters are distributed throughout the sarcolemma and may correspond to the small clusters of receptor that form in the sarcolemma during normal development. Such receptor clusters are associated with muscle-derived agrin and the $\beta 2$-chain containing laminins. Importantly, cultured muscle cells (myotubes) from mice lacking agrin responded to recombinant agrin with an appropriate aggregation of their acetylcholine receptors (Gautam *et al.*, 1996). Unexpectedly, the agrin knockout mouse also shows defective pre-synaptic nerve terminal differentiation, including defects in axon branching and abnormalities in acetylcholine receptor synthesis from subsynaptic nuclei. Thus, agrin may not be the only molecule involved in receptor clustering and may also influence differentiation of the pre-synaptic nerve terminal. Consistent with this, experiments with recombinant agrin *in vitro* show that it inhibits axon outgrowth of chick ciliary ganglion neurons, but not of sensory neurons, and causes the localised accumulation of synaptic vesicle protein markers (synaptotagmin) at the site of contact between agrin and the axon (Campagna, Ruegg & Bixby, 1995). Finally, expression of neuronal agrin in muscle cells in the extrasynaptic basal lamina by transfection or injection of agrin cDNA into the muscle cell can induce ectopic postsynaptic differentiation, including the aggregation of acetylcholine receptors (Cohen *et al.*, 1997a; Jones *et al.*, 1997; Meier *et al.*, 1997). The muscle form of agrin is not capable of inducing these postsynaptic changes (Meier *et al.*, 1997). These findings imply that activation of the agrin receptor by neuronal agrin released from motoneuron growth cones causes a retrograde signal to pass from the muscle cell to the growth cone, which triggers pre-synaptic differentiation.

How does neuronal agrin produce acetylcholine receptor aggregation? Agrin does not appear to interact directly with the acetylcholine receptor, instead it is likely that a signal transduction mechanism is involved in which there is a substantial amplification factor (reviewed in Bowe & Fallon, 1995). This latter aspect is important because the number of acetylcholine receptors at the neuromuscular junction (more than 10,000 per square micrometre) far outnumber the agrin molecules. The consensus view is that agrin binds to a multicomponent receptor complex in the sarcolemma at the neuromuscular junction that signals through tyrosine kinase phosphorylation to produce cross-linking of the acetylcholine receptor to the cytoskeleton. Although all the components of the complex have yet to be identified, they include a transmembrane, muscle-specific kinase (MuSK) and an intracellular linker protein, receptor-associated protein at synapse (rapsyn).

MuSK and rapsyn

MuSK is a transmembrane receptor tyrosine kinase in the sarcolemma at the synapse that interacts indirectly with agrin, possibly through an agrin receptor, which has yet to be identified (DeChiara *et al.*, 1996; Glass *et al.*, 1996). As myoblasts exit the cell cycle and fuse to form multinucleated myotubes, MuSK expression is up-regulated and, as maturation proceeds, becomes concentrated at the postsynaptic membrane. Addition of neuronal agrin to myotubes in culture causes a rapid phosphorylation of MuSK, whereas no MuSK phosphorylation is observed when agrin is added to fibroblast cell lines expressing MuSK (DeChiara *et al.*, 1996). This suggests that some intermediary factor, presumably the agrin receptor, is necessary for signalling, and is present in muscle cells but not fibroblast cell lines. Agrin, operating through MuSK is involved in the phosphorylation of the β-subunit of the acetylcholine receptor (Wallace, Qu & Huganir, 1991; Wallace, 1994; Meier, Perez & Wallace, 1995; Ferns, Deiner & Hall, 1996), an essential prerequisite for receptor clustering. Transgenic mice lacking MuSK protein have a phenotype which is similar to that of the agrin knockout, which is consistent with MuSK being downstream of the agrin-signalling pathway (DeChiara *et al.*, 1996). MuSK knockout mice do not form neuromuscular junctions, and there is dramatically impaired differentiation of the postsynaptic site, including acetylcholine receptor aggregation (DeChiara *et al.*, 1996). In the MuSK knockout, as with the agrin knockout, the range of defects seen extend beyond that of postsynaptic receptor aggregation and implicate this molecule in other aspects of synaptogenesis. In both knockouts, axon terminals extend more widely than normal throughout the muscle, which implicates these molecules in the growth cone arrest mechanism.

The intracellular, peripheral membrane protein rapsyn (also known as 43 kD protein) is also thought to be a component of the intracellular signalling pathway for the agrin-mediated aggregation of acetylcholine receptors (Gautam et al., 1996; Gillespie et al., 1996). Like agrin, rapsyn was first isolated from synapses of the electric organ of the marine ray Torpedo californica (Neubig et al., 1979; Frail et al., 1988). Co-expression studies with the acetylcholine receptor in heterologous cell lines indicates that rapsyn can aggregate acetylcholine receptors (Froehner et al., 1990; Phillips et al., 1991). When acetylcholine receptors are expressed in cell lines, they appear uniformly distributed throughout the cell plasma membrane. If cells expressing acetylcholine receptors are fused with cells expressing rapsyn, the receptors become aggregated in the fused cell plasma membrane, and rapsyn associates with the receptor aggregates, suggesting that rapsyn functions to aggregate acetylcholine receptors. Confirmation of a role for rapsyn in aggregating acetylcholine receptors has come from examining mice in which the gene for rapsyn has been deleted (Gautam et al., 1995). Homozygous mutant mice lacking rapsyn protein die soon after birth, and have movement and breathing defects. Most significantly, rapsyn-deficient mice fail to cluster acetylcholine receptors at their neuromuscular junctions. This is not due to a failure to express acetylcholine receptors, since the number of receptors is about 60% higher in mutant mice compared to the wild type. Furthermore, myotubes cultured from mutant mice failed to show receptor clustering when exposed to agrin, unlike normal myotubes (see earlier text). Up-regulation of acetylcholine receptor synthesis by subsynaptic nuclei is not affected in the rapsyn-deficient mice and presynaptic differentiation in rapsyn-deficient mice proceeds relatively normally.

Despite the structural differences between the neuromuscular junction and inter-neuronal synapses, there may be a conservation of the developmental mechanisms between the two types of synapse. Agrin and receptor tyrosine kinases related to MuSK have been found in the developing central nervous system (Rupp et al., 1991; Masiakowski & Carroll, 1992; Hoch et al., 1993; Thomas et al., 1993; Kröger, Horton & Honig, 1996).

References

Aberle, H., Schwartz, H., & Kemler, R. (1996). Cadherin–catenin complex: protein interactions and their implications for cadherin function. *Journal of Cell Biochemistry* **61**, 514–523.

Abosch, A., & Lagenaur, C. (1993). Sensitivity of neurite outgrowth to microfilament disruption varies with adhesion molecule substrate. *Journal of Neurobiology* **24**, 344–355.

Ackerman, S.L., Kozak, L.P., Przyborski, S.A., Rund, L.A., Boyer, B.B., & Knowles, B.B. (1997). The mouse rostral cerebellar malformation gene encodes an UNC-5-like protein. *Nature* **386**, 838–842.

Acklin, S.E., & Nicholls, J.G. (1990). Intrinsic and extrinsic factors influencing properties and growth patterns of identified leech neurons in culture. *Journal of Neuroscience* **10**, 1082–1090.

Adams, R.H., Betz, H., & Püschel, A.W. (1996). A novel class of murine semaphorins with homology to thrombospondin is differentially expressed during early embryogenesis. *Mechanisms of Development* **57**, 33–45.

Adinolfi, A.M. (1972a). Morphogenesis of synaptic junctions in layers I and II of the somatic sensory cortex. *Experimental Neurology* **34**, 372–382.

Adinolfi, A.M. (1972b). The organization of paramembranous densities during postnatal maturation of synaptic junctions in the cerebral cortex. *Experimental Neurology* **34**, 383–393.

Adler, R., Jerden, J., & Hewitt, A.T. (1985). Responses of cultured neural retinal cells to substratum-bound laminin and other extracellular matrix molecules. *Developmental Biology* **112**, 100–114.

Aghajanian, G.K., & Bloom, F.E. (1967). The formation of synaptic junctions in developing rat brain: A quantitative electron microscopic study. *Brain Research* **6**, 716–727.

Ahmad, F.J., & Baas, P.W. (1995). Microtubules released from the neuronal centrosome are transported into the axon. *Journal of Cell Science* **108**, 2761–2769.

Ahmad, F.J., Joshi, H.C., Centonze, V.E., & Baas, P.W. (1994). Inhibition of microtubule nucleation at the neuronal centrosome compromises axon growth. *Neuron* **12**, 271–280.

Ahmed, Z., & Connor, J.A. (1988). Calcium regulation by and buffering capacity of molluscan neurons during calcium transients. *Cell Calcium* **9**, 57–70.

Aigner, L., & Caroni, P. (1993). Depletion of 43–kD growth associated protein in primary sensory neurons leads to diminished formation and spreading of growth cones. *Journal of Cell Biology* **123**, 417–429.

Aigner, L., & Caroni, P. (1995). Absence of persistent spreading, branching, and adhesion in GAP-43-depleted growth cones. *Journal of Cell Biology* **128**, 647–660.

Aigner, L., Arber, S., Kapfhammer, J.P., Laux, T., Schneider, C., Botteri, F., Brenner, H.R., & Caroni, P. (1995). Overexpression of the neural growth-associated protein GAP-43 induces nerve sprouting in the adult nervous system of transgenic mice. *Cell* **83**, 269–278.

Akers, R.F., & Routtenberg, A. (1985). Brain protein phosphorylates a 47 Mr protein (F1) directly related to synaptic plasticity. *Brain Research* **334**, 147–151.

Akers, R.M., Mosher, D.F., & Lilien, J.E. (1981). Promotion of retinal neurite ougrowth by substratum-bound fibronectin. *Developmental Biology* **86**, 179–188.

Alemá, S.P., Casalbore, E., Agostini, E., & Tagó, F. (1985). Differentiation of PC12 pheochromocytoma cells induced by v-*src* oncogene. *Nature* **316**, 557–559.

Aletta, J.M., & Greene, L.A. (1988). Growth cone configuration and advance: A time-lapse study using video-enhanced differential interference contrast microscopy. *Journal of Neuroscience* **8**, 1425–1435.

Aletta, J.M., Lewis, S.A., Cowan, N.J., & Greene, L.A. (1988). Nerve growth factor regulates both the phosphorylation and steady-state levels of microtubule associated protein 1.2 (MAP 1.2). *Journal of Cell Biology* **106**, 1573–1581.

Alexander, K.A., Cimler, B.M., Meier, K.E., & Storm, D.R. (1987). Regulation of calmodulin binding to P-57. *Journal of Biological Chemistry* **262**, 6108–6113.

Alexander, K.A., Wakim, B.T., Doyle, G.S., Walsh, K.A., & Storm, D.R. (1988). Identification and characterization of the calmodulin-binding domain of neuromodulin, a neurospecific calmodulin-binding protein. *Journal of Biological Chemistry* **263**, 7544–7549.

Al-Ghaith, L.K., & Lewis, J.H. (1982). Pioneer growth cones in virgin mesenchyme: An electron-microscope study in the developing chick wing. *Journal of Embryology and Experimental Morphology* **68**, 149–160.

Allen, R.D. (1985). New observations on cell architecture and dynamics by video-enhanced contrast optical microscopy. *Annual Review of Biophysical Chemistry* **14**, 265–290.

Al-Mohanna, F., Cave, J., & Bolsover, S.R. (1992). A narrow window of intracellular calcium concentration is optimal for neurite outgrowth in rat sensory neurons. *Developmental Brain Research* **70**, 287–290.

Aloyo, V., Zwiers, H., & Gispen, W.H. (1983). Phosphorylation of B-50 by calcium-activated, phospholipid-dependent protein kinase and B-50 protein kinase. *Journal of Neurochemistry* **41**, 649–653.

Anderson, M.J., & Cohen, M.W. (1977). Nerve induced and spontaneous redistribution of acetylcholine receptors on cultured muscle cells. *Journal of Physiology* **268**, 757–773.

Andreason, T.J., Luetje, C.W., Heidman, W., & Storm, D.R. (1983). Purification of a novel calmodulin binding protein from bovine cerebral cortex. *Biochemistry* **22**, 4615–4618.

Anglister, L., & McMahan, U.J. (1985). Basal lamina directs acetylcholinesterase accumulation at synaptic sites in regenerating msucle. *Journal of Cell Biology* **101**, 735–743.

Anglister, L., Farber, I.C., Shahar, A., & Grinvald, A. (1982). Localization of voltage-sensitive calcium channels along developing neurites: Their possible role in regulating neurite elongation. *Developmental Biology* **94**, 351–365.

Apel, E.D., Byford, M.F., Au, D., Walsh, K.A., & Storm, D.R. (1990). Identification of the protein kinase C phosphorylation site in neuromodulin. *Biochemistry (U.S.A.)* **29**, 2330–2335.

Argarana, C.E., Barra, H.S., & Caputto, R. (1978). Release of [^{14}C]-tyrosine from tubilinyl-[^{14}C]-tyrosine by brain extract. Separation of a carboxypeptidase from tubulin tyrosine ligase. *Molecular Cell Biochemistry* **19**, 17–22.

Argiro, V., Bunge, M.B., & Johnson, M.I. (1984). Correlation between growth cone form and movement and their dependence on neuronal age. *Journal of Neuroscience* **4**, 3051–3062.

Argiro, V., Bunge, M.E., & Johnson, M.I. (1985). A quantitative study of growth cone filopodial extension. *Journal of Neuroscience Research* **13**, 149–162.

Arndt, K., & Redies, C. (1996). Restricted expression of R-cadherin by brain nuclei and neural circuits of the developing chick brain. *Journal of Comparative Neurology* **373**, 373–399.

Arni, S., Keilbaugh, S.A., Ostermeyer, A.G., & Brown, D.A. (1998). Association of GAP-43 with detergent-resistant membranes requires two palmitoylated cysteine residues. *Journal of Biological Chemistry* **273**, 28478–28485.

Arregui, C., Busciglio, J., Caceres, A., & Barra, H.S. (1991). Tyrosinated and detyrosinated microtubules in axonal processes of cerebellar macroneurons grown in culture. *Journal of Neuroscience Research* **28**, 171–181.

Arregui, C., Carbonetto, S., & McKerracher, L. (1994). Characterization of neural cell adhesion sites: Point contacts are the sites of interaction between integrins and the cytoskeleton in PC12 cells. *Journal of Neuroscience* **14**, 6967–6977.

Asai, D.J., Thompson, W.C., Wilson, L., Dresden, C.F., Schulman, H., & Purich, D.L. (1985). Microtubule-associated proteins (MAPs): A monoclonal antibody to MAP1 decorates microtubules in vitro but stains stress fibres and not microtubules in vivo. *Proceedings of the National Academy of Sciences U.S.A.* **81**, 5613–5617.

Atashi, J.R., Klinz, S.G., Ingraham, C.A., Matten, W.T., Schachner, M., & Maness, P.F. (1992). Neural cell adhesion molecules modulate tyrosine phosphorylation of tubulin in nerve growth cone membranes. *Neuron* **8**, 831–842.

Attardi, D.G., & Sperry, R.W. (1963). Preferential selection of central pathways by regenerating optic fibres. *Experimental Neurology* **7**, 46–64.

Aubry, M., & Maness, P.F. (1988). Developmental regulation of protein tyrosine phosphorylation in rat brain. *Journal of Neuroscience Research* **21**, 473–479.

Augusti-Tocco, G., & Sato, G. (1969). Establishment of functional clonal lines of neurones from mouse neuroblastoma. *Proceedings of the National Academy of Sciences U.S.A.* **64**, 311–315.

Avila, J., Domínguez, J., & Díaz-Nido, J. (1994). Regulation of microtubule dynamics by microtubule-associated protein expression and phosphorylation during neuronal development. *International Journal of Developmental Biology* **38**, 13–25.

Baas, P., & Ahmad, F. (1992). The plus ends of stable microtubules are the exclusive nucleating structures for microtubules in the axon. *Journal of Cell Biology* **116**, 1231–1241.

Baas, P., & Brown, A. (1997). Slow axonal transport: The polymer transport model. *Trends in Cell Biology* **7**, 380–384.

Baas, P.W. (1997). Microtubules and axonal growth. *Current Opinion in Cell Biology* **9**, 29–36.

Baas, P.W., & Black, M.M. (1990). Individual microtubules in the axon consist of domains that differ in both composition and stability. *Journal of Cell Biology* **111**, 495–509.

Baas, P.W., & Joshi, H.C. (1992). Gamma-tubulin distribution in the neuron: implications for the origins of neuritic microtubules. *Journal of Cell Biology* **119**, 171–178.

Baas, P.W., White, L.A., & Heidemann, S.R. (1987). Microtubule polarity reversal accompanies regrowth of amputated neurites. *Proceedings of the National Academy of Sciences U.S.A.* **84**, 5272–5276.

Baas, P.W., Deitch, J.S., Black, M.M., & Banker, G.A. (1988). Polarity orientation of microtubules in hippocampal neurons: Uniformity in the axon and nonuniformity in the dendrite. *Proceedings of the National Academy of Sciences U.S.A.* **85**, 8335–8339.

Baas, P.W., Pienkowski, T.P., & Kosik, K.S. (1991). Processes induced by tau expression in Sf9 cells have an axon-like microtubule organization. *Journal of Cell Biology* **115**, 1333–1344.

Baetge, E.E., & Hammang, J.P. (1991). Neurite outgrowth in PC12 cells deficient in GAP-43. *Neuron* **6**, 21–30.

Baetge, E.E., Hammang, J.P., Gribkoff, V.K., & Meiri, K.E. (1992). The role of GAP-43 in the molecular regulation of axon outgrowth and electrical excitability. *Perspectives on Developmental Neurobiology* **1**, 21–28.

Bähr, S., & Wolff, J.R. (1985). Postnatal development of axosomatic synapses in the rat visual cortex: Morphogenesis and quantitative evaluation. *Journal of Comparative Neurology* **233**, 405–420.

Baier, H., & Bonhoeffer, F. (1992). Axon guidance by gradients of a target-derived component. *Science* **255**, 472–475.

Baier, H., & Klostermann, S. (1994). Axon guidance and growth cone collapse *in vitro*. In *Axon Growth and Guidance in Vitro*, ed. P.R. Gordon-Weeks, pp. 96–105. New York: Academic Press.

Baldwin, T.J., Walsh, F.S., & Doherty, P. (1996). The role of protein tyrosine kinases in transducing signals from cell adhesion molecules to promote neurite outgrowth. In *Nerve Growth and Guidance*, ed. C.D. McCaig, pp. 21–34. London: Portland Press.

Bamburg, J.R., & Bernstein, B.W. (1991). Actin and actin-binding proteins in neurons. In *The Neuronal Cytoskeleton*, ed. R.D. Burgoyne, pp. 121–160. New York: Wiley-Liss, Inc.

Bamburg, J.R., Bray, D., & Chapman, K. (1986). Assembly of microtubules at the tip of growing axons. *Nature* **321**, 788–790.

Bantlow, C.E., Zachleder, T., & Schwab, M.E. (1990). Oligodendrocytes arrest neurite growth by contact inhibition. *Journal of Neuroscience* **10**, 3837–3848.

Bantlow, C.E., Schmidt, M.F., Hassinger, T.D., Schwab, M.E., & Kater, S.B. (1993). Role of intracellular calcium in NI-35–evoked collapse of neuronal growth cones. *Science* **259**, 80–83.

Barbin, G., Pollard, H., Gaiarsa, J.L., & Ben-Ari, Y. (1993). Involvement of GABA_A receptors in the outgrowth of cultured hippocampal neurons. *Neuroscience Letters* **152**, 150–154.

Barde, Y.-A. (1989). Trophic factors and neuronal survival. *Neuron* **2**, 1525–1534.

Barde, Y.-A., & Thoenen, H. (1980). Physiology of nerve growth factor. *Physiology Reviews* **60**, 1284–1335.

Bare, D.J., Lauder, J.M., Wilkie, M.B., & Maness, P.F. (1993). p59fyn in rat brain is localized in developing axonal tracts and subpopulations of adult neurons and glia. *Oncogene* **8**, 1429–1436.

Baron van Evercooren, A., Kleinman, H.K., Ohno, S., Marganos, P., Schwartz, J.P., & Dubois-Dalcq, M. (1982). Nerve growth factor, laminin, and fibronectin promote neurite growth in human fetal sensory ganglion cultures. *Journal of Neuroscience Research* **8**, 179–193.

Barra, H.S., Rodriguez, J.A., Arce, C.A., & Caputto, R. (1973). A soluble preparation from rat brain that incorporates into its own proteins [^{14}C]-arginine by a ribonuclease-sensitive system and [^{14}C]-tyrosine by a ribonuclease-insensitive system. *Journal of Neurochemistry* **20**, 97–108.

Bartsch, S., Bartsch, U., Dorries, U., Faissner, A., Weller, A., Ekblom, P., & Schachner, M. (1992). Expression of tenascin in the developing and adult cerebellar cortex. *Journal of Neuroscience* **12**, 736–749.

Bartsch, U., Faissner, A., Trotter, J., Dörries, U., Bartsch, S., Mohajeri, H., & Schachner, M. (1994). Tenascin demarks the boundaries between myelinated and non-myelinated retinal ganglion cell axons of the adult mouse. *Journal of Neuroscience* **14**, 4756–4768.

Barylko, B., Wagner, M.C., Reizes, O., & Albanesi, J.P. (1992). Purification and characterization of a mammalian myosin I. *Proceedings of the National Academy of Sciences U.S.A.* **89**, 490–494.

Bastiani, M.J., & Goodman, C.S. (1984). Neuronal growth cones – specific interactions mediated by filopodial insertion and induction of coated vesicles. *Proceedings of the National Academy of Sciences U.S.A.* **81**, 1849–1853.

Bastiani, M., Pearson, K.G., & Goodman, C.S. (1984a). From embryonic fascicles to adult tracts: Organization of neuropile from a developmental perspective. *Journal of Experimental Biology* **112**, 45–64.

Bastiani, M.J., Raper, J.A., & Goodman, C.S. (1984b). Pathfinding by neuronal growth cones in grasshopper embryos. III. Selective affinity of the G growth cone for the P cells within the A/P fascicle. *Journal of Neuroscience* **4**, 2311–2328.

Bastiani, M.J., Doe, C.Q., Helfand, S.L., & Goodman, C.S. (1985). Neuronal specificity and growth cone guidance in grasshopper and *Drosophila* embryos. *Trends in Neurosciences* **6**, 257–266.

Bastiani, M.J., Harrelson, A.L., Snow, P.M., & Goodman, C.S. (1987). Expression of fasciclin I and II glycoproteins on subsets of axon pathways during neuronal development in the grasshopper. *Cell* **48**, 745–755.

Bastmeyer, M., & O'Leary, D.D.M. (1996). Dynamics of target recognition by interstitial axon branching along developing cortical axons. *Journal of Neuroscience* **16**, 1450–1459.

Bastmeyer, M., & Stuermer, C.A.O. (1993). Behaviour of fish retinal growth cones encountering chick caudal tectal membranes – a time-lapse study on growth cone collapse. *Journal of Neurobiology* **24**, 37–50.

Bastmeyer, M., Ott, H., Leppert, C.A., & Stuermer, C.A. (1995). Fish E587 glycoprotein, a member of the L1 family of cell adhesion molecules, participates in axonal fasciculation and the age-related order of ganglion cell axons in the goldfish retina. *Journal of Cell Biology* **130**, 969–976.

Bate, C.M. (1976). Pioneer neurones in an insect embryo. *Nature* **260**, 54–56.

Bedlack, R.S., Jr, Wei, M.-D., & Loew, L.M. (1992). Localized membrane depolarizations and localized calcium influx during electric field-guided neurite growth. *Neuron* **9**, 393–403.

Beggs, H.E., Soriano, P., & Maness, P.F. (1994). NCAM-dependent neurite outgrowth is inhibited in neurons from Fyn-minus mice. *Journal of Cell Biology* **127**, 825–833.

Behar, O., Golden, J.A., Mashimo, H., Schoen, F.J., & Fishman, M.C. (1996). Semaphorin III is needed for normal patterning and growth of nerves, bones and heart. *Nature* **383**, 525–528.

Behar, T.N., Schaffner, A.E., Colton, C.A., Somogyi, R., Olah, Z., Lehel, C., & Barker, J.L. (1994). GABA-induced chemokinesis and NGF-induced chemotaxis of embryonic spinal cord neurons. *Journal of Neuroscience* **14**, 29–38.

Ben-Ari, Y., Cherubini, R., Corradetti, R., & Gaiarsa, J. (1989). Giant synaptic potentials in immature rat CA3 hippocampal neurons. *Journal of Physiology* **416**, 303–325.

Ben-Ari, Y., Khazipov, R., Leinekugel, X., Caillard, O., & Gaiarsa, J.-L. (1997). GABA$_A$, NMDA and AMPA receptors: A developmentally regulated 'menage à trois'. *Trends in the Neurosciences* **20**, 523–529.

Bennett, M.R., McClachlan, E.M., & Taylor, R.S. (1973). The formation of synapses in reinnervated mammalian striated muscle. *Journal of Physiology* **233**, 481–500.

Benowitz, L.I., & Lewis, E.R. (1983). Increased transport of 44,000- to 49,000-dalton acidic proteins during regeneration of the goldfish optic nerve: A two-dimensional gel analysis. *Journal of Neuroscience* **3**, 2153–2163.

Benowitz, L.I., & Routtenberg, A. (1987). A membrane phosphoprotein associated with neural development, axonal regeneration, phospholipid metabolism and synaptic plasticity. *Trends in Neurosciences* **10**, 527–532.

Benowitz, L.I., & Routtenberg, A. (1997). GAP-43: an intrinsic determinant of neuronal development and plasticity. *Trends in Neurosciences* **20**, 84–91.

Benowitz, L.I., Shashoua, V.E., & Yoon, M. (1981). Specific changes in rapidly transported proteins during regeneration of the goldfish optic nerve. *Journal of Neuroscience* **1**, 300–307.

Bentley, D., & Caudy, M. (1984). Navigational substrates for peripheral pioneer growth cones: Limb-axis polarity cues, limb-segment boundaries, and guidepost neurons. *Cold Spring Harbor Symposium of Quantitative Biology* **48**, 573–585.

Bentley, D., & Keshishian, H. (1982). Pathfinding by peripheral pioneer neurons in grasshoppers. *Science* **218**, 1082–1088.

Bentley, D., & O'Connor, T.P. (1991). Guidance and steering of peripheral pioneer growth cones in grasshopper embryos. In *The Nerve Growth Cone*, ed. P.C. Letourneau, S.B. Kater & E.R. Macagno, pp. 265–282. New York: Raven Press.

Bentley, D., & O'Connor, T.P. (1994). Cytoskeletal events in growth cone steering. *Current Opinion in Neurobiology* **4**, 43–48.

Bentley, D., & Toroian-Raymond, A. (1986). Disoriented pathfinding by pioneer growth cones deprived of filopodia by cytochalasin treatment. *Nature* **323**, 712–715.

Bentley, D., Guthrie, P.B., & Kater, S.B. (1991). Calcium ion distribution in nascent pioneer axons and coupled preaxonogenesis neurons in situ. *Journal of Neuroscience* **11**, 1300–1308.

Bershadsky, A.D., & Vasiliev, J.M. (1988). *Cytoskeleton*. New York: Plenum Press.

Bieber, A.J., Snow, P.M., Hortsch, M., Patel, N.H., Jacobs, J.R., Traquina, Z.R., Schilling, J., & Goodman, C.S. (1989). *Drosophila* neuroglian: A member of the immunoglobulin superfamily with extensive homology to the vertebrate neural adhesion molecule L1. *Cell* **59**, 447–460.

Birks, R.I., Mackey, M.C., & Weldon, P.R. (1972). Organelle formation from pinocytotic elements in neurites of cultured sympathetic ganglia. *Journal of Neurocytology* **1**, 311–340.

Bixby, J.L., & Harris, W.A. (1991). Molecular mechanisms of axon growth and guidance. *Annual Review of Cell Biology* **7**, 117–159.

Bixby, J.L., & Jhabvala, P. (1992). Inhibition of tyrosine phosphorylation potentiates substrate-induced neurite outgrowth. *Journal of Neurobiology* **23**, 468–480.

Bixby, J.L., & Jhabvala, P. (1993). Tyrosine phosphorylation in early embryonic growth cones. *Journal of Neuroscience* **13**, 3421–3432.

Bixby, J.L., & Spitzer, N.C. (1984). Early differentiation of vertebrate spinal neurons in the absence of Ca^{2+} and Na^+ influx. *Developmental Biology* **106**, 89–96.

Bixby, J.L., & Zhang, R. (1990). Purified N-cadherin is a potent substrate for the rapid induction of neurite outgrowth. *Journal of Cell Biology* **110**, 1253–1260.

Bixby, J.L., Pratt, R.S., Lilien, J., & Riechardt, L.F. (1987). Neurite outgrowth on muscle cell surfaces involves extracellular matrix receptors as well as Ca^{2+}-dependent and -independent cell adhesion molecules. *Proceedings of the National Academy of Sciences U.S.A.* **84**, 2555–2559.

Bixby, J.L., Lilien, J., & Reichardt, L.F. (1988). Identification of the major proteins that promote neuronal process outgrowth on Schwann cells. *Journal of Cell Biology* **107**, 353–361.

Black, M.M., Keyser, P., & Sobel, E. (1986). Interval between the synthesis and assembly of cytoskeletal proteins in cultured neurons. *Journal of Neuroscience.* **6**, 1004–1012.

Black, M.M., Baas, P.W., & Humphries, S. (1989). Dynamics of α-tubulin deacetylation in intact neurons. *Journal of Neuroscience* **9**, 358–368.

Black, M.M., Slaughter, T., & Fischer, I. (1994). Microtubule-associated protein 1b (MAP1b) is concentrated in the distal region of growing axons. *Journal of Neuroscience* **14**, 857–870.

Black, M.M., Slaughter, T., Moshiach, S., Obrocka, M., & Fischer, I. (1996). Tau is enriched on dynamic microtubules in the distal region of growing axons. *Journal of Neuroscience* **16**, 3601–3619.

Blaustein, M.P. (1988). Calcium transport and buffering in neurons. *Trends in Neurosciences* **11**, 438–443.

Blue, M.E., & Parnavelas, J.G. (1983). The formation and maturation of synapses in the visual cortex of the rat. I. Qualitative analysis. *Journal of Neurocytology* **12**, 599–616.

Bodian, D. (1966). Development of the fine structure of spinal cord in monkey fetuses. *Bulletin of The Johns Hopkins Hospital* **119**, 129–149.

Bodian, D., Melby, E.C., Jr, & Taylor, N. (1968). Development of fine structure of spinal cord in monkey fetuses. II. Pre-reflex period to period of long intersegmental reflexes. *Journal of Comparative Neurology* **133**, 113–166.

Boisseau, S., Nedelec, J., Poirier, V., Rougon, G., & Simonneau, M. (1991). Analysis of high PSA N-CAM expression during mammalian spinal cord and peripheral nervous system development. *Development* **112**, 69–82.

Bolsover, S., & Silver, R.A. (1991). Artifacts in calcium measurement: Recognition and remedies. *Trends in Cell Biology* **1**, 71–74.

Bolsover, S.R., & Spector, I. (1986). Measurements of calcium transients in the soma, neurite and growth cone of single cultured neurons. *Journal of Neuroscience* **6**, 1934–1940.

Bolz, J., Novak, N., Götz, M., & Bonhoeffer, T. (1990). Formation of target-specific neuronal projections in organotypic slice cultures from rat visual cortex. *Nature* **346**, 359–362.

Bonhoeffer, T. (1996). Neurotrophins and activity-dependent development of the neocortex. *Current Opinion in Neurobiology* **6**, 119–126.

Borgens, R.B., Roederer, E., & Cohen, M.J. (1981). Enhanced spinal cord regeneration in lamprey by applied electric fields. *Science* **213**, 611–617.

Bovolenta, P., & Dodd, J. (1990). Guidance of commissural growth cones at the floor plate in the embryonic rat spinal cord. *Development* **109**, 435–447.

Bovolenta, P., & Mason, C. (1987). Growth cone morphology varies with position in the developing mouse visual pathway from retina to first targets. *Journal of Neuroscience* **7**, 1447–1460.

Bowe, M.A., & Fallon, J.R. (1995). The role of agrin in synapse formation. *Annual Review of Neuroscience* **18**, 443–462.

Boyne, L.J., Martin, K., Hockfield, S., & Fischer, I. (1995). Expression and distribution of phosphorylated MAP1B in growing axons of cultured hippocampal neurons. *Journal of Neuroscience Research* **40**, 439–450.

Bozyczko, D., & Horwitz, A.F. (1986). The participation of a putative cell surface receptor for laminin and fibronectin in peripheral neurite extension. *Journal of Neuroscience* **6**, 1241–1251.

Brambilla, R., & Klein, R. (1995). Telling axons where to grow: A role for Eph receptor tyrosine kinases in guidance. *Molecular and Cellular Neuroscience* **6**, 487–495.

Braun, J., & Stent, G. (1989). Axon outgrowth along segmental nerves in the leech. I. Identification of candidate guidance cells. *Developmental Biology* **132**, 471–485.

Bray, D. (1970). Surface movements during the growth of single explanted neurons. *Proceedings of the National Academy of Science U.S.A.* **65**, 905–910.

Bray, D. (1973a). Branching patterns of individual sympathetic neurons in culture. *Journal of Cell Biology* **56**, 702–712.

Bray, D. (1973b). Model for membrane movements in the neural growth cone. *Nature* **244**, 93–96.

Bray, D. (1979). Mechanical tension produced by nerve cells in tissue culture. *Journal of Cell Science* **37**, 391–410.

Bray, D. (1982). Filopodial contraction and growth cone guidance. In *Cell Behaviour*, ed. R. Bellair, A. Curtis & G. Dunn, pp. 299–317. Cambridge: Cambridge University Press.

Bray, D. (1984). Axonal growth in response to experimentally applied mechanical tension. *Developmental Biology* **102**, 379–389.

Bray, D. (1987). Growth cones: do they pull or are they pushed? *Trends in Neurosciences* **10**, 431–434.

Bray, D. (1992). *Cell Movements*. New York & London: Garland Publishing, Inc., pp. 1–406.

Bray, D. (1996). The dynamics of growing axons. *Current Biology* **6**, 241–243.

Bray, D. (1997). The riddle of slow transport an introduction. *Trends in Cell Biology* **7**, 379.

Bray, D., & Bunge, M.B. (1981). Serial analysis of microtubules in cultured rat sensory neuron axons. *Journal of Neurocytology* **10**, 589.

Bray, D., & Chapman, K. (1985). Analysis of microspike movements on the neuronal growth cone. *Journal of Neuroscience* **5**, 3204–3213.

Bray, D., & White, J.G. (1988). Cortical flow in animal cells. *Science* **239**, 833–888.

Bray, D., Thomas, C., & Shaw, G. (1978). Growth cone formation in cultures of sensory neurons. *Proceedings of the National Academy of Sciences U.S.A.* **75**, 5226–5229.

Bray, D., Wood, P., & Bunge, R.P. (1980). Selective fasciculation of nerve fibers in culture. *Experimental Cell Research* **130**, 241–250.

Brenner, H.R., Meier, T., & Widmer, B. (1983). Early action of nerve determines motor endplate differentiation in rat muscle. *Nature* **305**, 536–537.

Brenner, S. (1974). The genetics of *Caenorhabditis elegans*. *Genetics* **77**, 71–94.

Bridgman, P.C., & Dailey, M.E. (1989). The organization of myosin and actin in rapid frozen nerve growth cones. *Journal of Cell Biology* **108**, 95–109.

Bridgman, P.C., Rochlin, M.W., Lewis, A.K., & Evans, L.L. (1994). Contributions of multiple forms of myosin to nerve outgrowth. *Progress in Brain Research* **103**, 99–107.

Brittis, P.A., & Silver, J. (1995). Multiple factors govern intraretinal axon guidance: A time-lapse study. *Molecular and Cellular Neuroscience* **6**, 413–432.

Brittis, P.A., Canning, D.R., & Silver, J. (1992). Chondroitin sulphate as a regulator of neuronal patterning in the retina. *Science* **255**, 733–736.

Brittis, P.A., Lemmon, V., Rutishauser, U., & Silver, J. (1995). Unique changes of ganglion cell growth cone behaviour following cell adhesion molecule perturbations: A time-lapse study of the living retina. *Molecular and Cellular Neurosciences* **6**, 433–449.

Brittis, P.A., Silver, J., Walsh, F.S., & Doherty, P. (1996). Fibroblast growth factor receptor function is required for the orderly projection of ganglion cell axons in the developing mammalian retina. *Molecular and Cellular Neuroscience* **8**, 120–128.

Broadie, K., & Bate, M. (1993a). Innervation directs receptor synthesis and localisation in *Drosophila* embryo synaptogenesis. *Nature* **361**, 350–353.

Broadie, K.S., & Bate, M. (1993b). Development of the embryonic neuromuscular synapse of *Drosophila melanogaster. Journal of Neuroscience* **13**, 144–166.

Brown, A., Li, Y., Slaughter, T., & Black, M.M. (1993). Composite microtubules of the axon: Quantitative analysis of tyrosinated and acetylated tubulin along individual axonal microtubules. *Journal of Cell Science* **104**, 339–352.

Brugg, B., & Matus, A. (1988). PC12 cells express juvenile microtubule-associated proteins during nerve growth factor-induced neurite outgrowth. *Journal of Cell Biology* **107**, 643–650.

Brugg, B., Reddy, D., & Matus, A. (1993). Attenuation of microtubule-associated protein 1B expression by antisense oligodeoxynucleotides inhibits initiation of neurite outgrowth. *Neuroscience* **52**, 489–496.

Brückner, K., Pasquale, E.B., & Klein, R. (1997). Tyrosine phosphorylation of transmembrane ligands for Eph receptors. *Science* **275**, 1640–1643.

Brümmendorf, T., & Rathjen, F.G. (1996). Structure/function relationships of axon-associated adhesion receptors of the immunoglobulin superfamily. *Current Opinion in Neurobiology* **6**, 584–593.

Brümmendorf, T., Wolff, J.M., Frank, R., & Rathjen, F.G. (1989). Neural cell recognition molecule F11: Homology with fibronectin type III and immunoglobulin type C domains. *Neuron* **2**, 1351–1361.

Brümmendorf, T., Hubert, M., Treubert, U., Leuschner, R., Tarnok, A., & Rathjen, F.G. (1993). The axonal recognition molecule F11 is a multifunctional protein: Specific domains mediate interactions with Ng-CAM and restrictin. *Neuron* **10**, 711–727.

Buchanan, J., Sun, Y., & Poo, M.-M. (1989). Studies of nerve–muscle interactions in *Xenopus* cell culture: Morphology of early functional contacts. *Journal of Neuroscience* **9**, 1540–1554.

Bunge, M.B. (1973). Fine structure of nerve fibers and growth cones of isolated sympathetic neurons in culture. *Journal of Cell Biology* **56**, 713–735.

Bunge, M.B. (1977). Initial endocytosis of peroxidase or ferritin by growth cones of cultured nerve cells. *Journal of Neurocytology* **6**, 407–439.

Bunge, M.B., Bunge, R.P., & Peterson, E.R. (1967). The onset of synapse formation in spinal cord cultures as studied by electron microscopy. *Brain Research* **6**, 728–749.

Bunge, M.B., Johnson, M.I., & Argiro, V.J. (1983). Studies of regenerating nerve fibers and growth cones. In *Spinal Cord Reconstruction*, ed. C.C. Kao, R.P. Bunge & P.J. Reier, pp. 99–120. New York: Raven Press.

Burden, S.J., Sargent, P.B., & McMahan, U.J. (1979). Acetylcholine receptors in regenerating muscle accumulate at original synaptic sites in the absence of nerve. *Journal of Cell Biology* **82**, 412–425.

Burden-Gulley, S.M., & Lemmon, V. (1996). L1, N-cadherin, and laminin induce distinct distribution patterns of cytoskeletal elements in growth cones. *Cell Motility and the Cytoskeleton* **35**, 1–23.

Burden-Gulley, S.M., Payne, H.R., & Lemmon, V. (1995). Growth cones are actively influenced by substrate-bound adhesion molecules. *Journal of Neuroscience* **15**, 4370–4381.

Burmeister, D.W., & Goldberg, D.J. (1988). Micropruning: The mechanism of turning of *Aplysia* growth cones at substrate borders *in vitro*. *Journal of Neuroscience* **8**, 3151–3159.

Burns, F.R., von Kannen, S., Guy, L., Raper, J.A., Kamholz, J., & Chang, S. (1991). DM-GRASP, a novel immunoglobulin superfamily axonal surface protein that supports neurite extension. *Neuron* **7**, 209–220.

Burns, M.E., & Augustine, G.J. (1995). Synaptic structure and function: Dynamic organization yields architectural precision. *Cell* **83**, 187–194.

Burridge, K., Fath, K., Kelly, T., Nuckolls, G., & Turner, D.C. (1988). Focal adhesions: Transmembrane junctions between the extracellular matrix and the cytoskeleton. *Annual Review of Cell Biology* **4**, 487–525.

Burrows, M.T. (1911). The growth of tissues of the chick embryo outside the animal body, with special reference to the nervous system. *Journal of Experimental Zoology* **10**, 63–84.

Burry, R.W. (1980). Formation of apparent presynaptic elements in response to polybasic compounds. *Brain Research* **184**, 85–98.

Bush, M.S., & Gordon-Weeks, P.R. (1994). Distribution and expression of developmentally regulated phosphorylation epitopes on MAP 1B and neurofilament proteins in the developing rat spinal cord. *Journal of Neurocytology* **23**, 682–698.

Bush, M.S., Goold, R.G., Moya, F., & Gordon-Weeks, P.R. (1996a). An analysis of an axonal gradient of phosphorylated MAP 1B in cultured sensory neurons. *European Journal of Neuroscience* **8**, 235–248.

Bush, M.S., Tonge, D.A., Woolf, C., & Gordon-Weeks, P.R. (1996b). Expression of a developmentally-regulated, phosphorylated isoform of MAP 1B in regenerating axons of the sciatic nerve. *Neuroscience* **73**, 553–563.

Bush, M., Eagles, P.A.M., & Gordon-Weeks, P.R. (1996c). Neuronal cytoskeleton. In *Treatise on the Cytoskeleton, III Cytoskeleton in Specialized Tissues*, ed. J.E. Hesketh & I.F. Pryme, pp. 185–227. Connecticut: JAI Press.

Cáceres, A., & Kosik, K.S. (1990). Inhibition of neurite polarity by tau antisense oligonucleotides in primary cerebellar neurons. *Nature* **343**, 461–463.

Cáceres, A., Banker, G., Steward, O., Binder, L., & Payne, M. (1984). MAP2 is located to the dendrites of hippocampal neurons which develop in culture. *Developmental Brain Research* **13**, 314–318.

Cáceres, A., Potrebic, S., & Kosik, K.S. (1991). The effect of tau antisense oligonucleotides on neurite formation of cultured cerebellar macroneurons. *Journal of Neuroscience* **11**, 1515–1523.

Cáceres, A., Mautino, J., & Kosik, K.S. (1992). Suppression of MAP2 in cultured cerebellar macroneurons inhibits minor neurite formation. *Neuron* **9**, 607–618.

Calof, A.L., & Lander, A.D. (1991). Relationship between neuronal migration and cell-substratum adhesion: Laminin and merosin promote olfactory neuronal migration but are anti-adhesive. *Journal of Cell Biology* **115**, 779–794.

Calof, A.L., Campanero, M.R., O'Rear, J.J., Yurchenko, P.D., & Lander, A.D. (1994). Domain-specific activation of neuronal migration and neurite outgrowth-promoting activities of laminin. *Neuron* **13**, 117–130.

Calvert, R., & Anderton, B.H. (1985). A microtubule associated protein MAP 1 which is expressed at elevated levels during development of rat cerebellum. *EMBO Journal* **4**, 1171–1176.

Campagna, J.A., Ruegg, M.A., & Bixby, J.L. (1995). Agrin is a differentiation-inducing 'stop signal' for motoneurons. *Neuron* **15**, 1365–1374.

Campanelli, J.T., Hoch, W., Rupp, F., Kreiner, T., & Scheller, R.H. (1991). Agrin mediates cell contact-induced acetylcholine receptor clustering. *Cell* **67**, 909–916.

Campenot, R.B. (1977). Local control of neurite development by nerve growth factor. *Proceedings of the National Academy of Sciences U.S.A.* **74**, 4516–4519.

Campenot, R.B., & Dracker, D.D. (1989). Growth of sympathetic nerve fibers in culture does not require extracellular calcium. *Neuron* **3**, 733–743.

Campenot, R.B., Lund, K., & Senger, D.L. (1996). Delivery of newly synthesized tubulin to rapidly growing distal axons of rat sympathetic neurons in compartmented cultures. *Journal of Cell Biology* **135**, 701–709.

Carbonetto, S., Gruver, M.M., & Turner, D. (1983). Nerve fibre growth in culture on fibronectin, collagen and glycosaminoglycan substrates. *Journal of Neuroscience* **3**, 2324–2335.

Carlier, M.-F. (1989). Role of nucleotide hydrolysis in the dynamics of actin filaments and microtubules. *International Review of Cytology* **115**, 139–170.

Carter, S.B. (1967). Haptotaxis and the mechanism of cell motility. *Nature* **213**, 256–260.

Caudy, M., & Bentley, D. (1986). Pioneer growth cone morphologies reveal proximal increases in substrate affinity within leg segments of grasshopper embryos. *Journal of Neuroscience* **6**, 364–379.

Caviness, V.S., Jr (1976). Patterns of cell and fiber distribution in the neocortex of the reeler mutant mouse. *Journal of Comparative Neurology* **170**, 435–448.

Caviness, V.S., Jr & Yorke, C.H., Jr (1976). Interhemispheric neocortical connections of the corpus callosum in the reeler mutant mouse: A study based on anterograde and retrograde methods. *Journal of Comparative Neurology* **170**, 449–460.

Challacombe, J.F., Snow, D.M., & Letourneau, P.C. (1996a). Role of the cytoskeleton in growth cone motility and axonal elongation. *Seminars in the Neurosciences* **8**, 67–80.

Challacombe, J.F., Snow, D.M., & Letourneau, P.C. (1996b). Actin filament bundles are required for microtubule reorientation during growth cone turning to avoid an inhibitory guidance cue. *Journal of Cell Science* **109**, 2031–2040.

Challacombe, J.F., Snow, D.M., & Letourneau, P.C. (1997). Dynamic microtubule ends are required for growth cone turning to avoid an inhibitory guidance cue. *Journal of Neuroscience* **17**, 3085–3095.

Chamley, J.H., Goller, I., & Burnstock, G. (1973). Selective growth of sympathetic nerve fibers to explants of normally densely innervated autonomic effector organs in tissue culture. *Developmental Biology* **31**, 362–379.

Chan, S.S., Zheng, H., Su, M.W., Wilk, R., Killeen, M.T., Hedgecock, E.M., & Culotti, J.G. (1996). UNC-40, a *C. elegans* homolog of DCC (Deleted in Colorectal

Cancer), is required in motile cells responding to UNC-6 netrin cues. *Cell* **87**, 187–195.

Chang, C.M., & Goldman, R.D. (1973). The localization of actin-like fibres in cultured neuroblastoma cells as revealed by heavy meromyosin binding. *Journal of Cell Biology* **57**, 867–874.

Chang, H.Y., Takei, K., Sydor, A.M., Born, T., Rusnak, F., & Jay, D.G. (1995). Asymmetric retraction of growth cone filopodia following focal inactivation of calcineurin. *Nature* **376**, 686–690.

Chang, S., Rathjen, F.G., & Raper, J.A. (1987). Extension of neurites on axons is impaired by antibodies against specific neural cell surface glycoproteins. *Journal of Cell Biology* **104**, 355–362.

Changeux, J.P., Duclert, A., & Sekine, S. (1992). Calcitonin gene-related peptides and neuromuscular interactions. *Annals of the New York Academy of Sciences* **657**, 361–378.

Chapman, E., Au, D., Alexander, K., Nicolson, T., & Storm, D. (1991). Characterization of the calmodulin binding domain of GAP-43. *Journal of Biological Chemistry* **266**, 207–213.

Chen, C., Trombley, P.Q., & van den Pol, A.N. (1995). GABA receptors precede glutamate receptors in hypothalamic development: Differential regulation by astrocytes. *Journal of Neurophysiology* **74**, 1473–1484.

Chen, C., Trombley, P.Q., & van den Pol, A.N. (1996). Excitatory action of GABA in developing hypothalamic neurons. *Journal of Physiology* **494**, 451–464.

Chen, H., Chédotal, A., He, Z., Goodman, C.S., & Tessier-Lavigne, M. (1997). Neuropilin-2, a novel member of the neuropilin family, is a high affinity receptor for the semaphorins sema E and sema IV but not sema III. *Neuron* **19**, 547–559.

Chen, J., Kanai, Y., Cowan, N.J., & Hirokawa, N. (1992). Projection domains of MAP2 and tau determine spacings between microtubules in dendrites and axons. *Nature* **360**, 674–677.

Chen, J.S., & Levi-Montalcini, R. (1970). Axonal growth from insect neurons in glia-free cultures. *Proceedings of the National Academy of Sciences U.S.A.* **66**, 32–39.

Cheney, R.E., Oshea, M.K., Heuser, J.E., Coelho, M.V., Wolenski, J.S., Espreafico, E.M., Forscher, P., Larson, R.E., & Mooseker, M. (1993). Brain myosin-V is a two-headed unconventional myosin with motor activity. *Cell* **75**, 13–23.

Cheng, H.-J., Nakamoto, M., Bergemann, A.D., & Flanagan, J.G. (1995). Complementary gradients in expression and binding of ELF-1 and Mek-4 in development of the topographical retinotectal projection map. *Cell* **82**, 371–381.

Cheng, N., & Sahyoun, N. (1988). The growth cone cytoskeleton. Glycoprotein association, calmodulin binding, and tyrosine/serine phosphorylation of tubulin. *Journal of Biological Chemistry* **263**, 3935–3942.

Cheng, T.P.O., & Reese, T.S. (1985). Polarized compartmentalization of organelles in growth cones from developing optic tectum. *Journal of Cell Biology* **101**, 1473–1480.

Cheng, T.P.O., & Reese, T.S. (1987). Recycling of plasmalemma in chick tectal growth cones. *Journal of Neuroscience* **7**, 1752–1759.

Cheng, T.P.O., & Reese, T.S. (1988). Compartmentalization of anterogradely and retrogradely transported organelles in axons and growth cones from chick optic tectum. *Journal of Neuroscience* **8**, 3190–3199.

Cheng, T.P.O., Murakami, N., & Elizinga, M. (1993). Localization of myosin IIB at the leading edge of growth cones from rat dorsal root ganglion cells. *FEBS Letters* **311**, 91–94.

Chenney, R., & Mooseker, M. (1992). Unconventional myosins. *Current Opinion in Cell Biology* **4**, 27–35.

Cherubini, R., Gaiarsa, J., & Ben-Ari, Y. (1991). GABA: An excitatory transmitter in early postnatal life. *Trends in the Neurosciences* **14**, 515–519.

Chiba, A., Snow, P.M., Keshishian, H., & Hotta, Y. (1995). Fasciclin III as a synaptic recognition molecule in *Drosophila*. *Nature* **374**, 166–168.

Chien, C.-B., Rosenthal, D.E., Harris, W.A., & Holt, C.E. (1993). Navigational errors made by growth cones without filopodia in the embryonic *Xenopus* brain. *Neuron* **11**, 237–251.

Chiquet-Ehrismann, R. (1995). Inhibition of cell adhesion by anti-adhesive molecules. *Current Opinion in Cell Biology* **7**, 715–719.

Chitnis, A.B., & Kuwada, J.Y. (1991). Elimination of a brain tract increases errors in pathfinding by follower growth cones in the zebrafish embryo. *Neuron* **7**, 277–285.

Chiu, A.Y., & Sanes, J.R. (1984). Differentiation of basal lamina in synaptic and extra-synaptic portions of embryonic rat muscle. *Developmental Biology* **103**, 456–467.

Cimler, B.M., Giebelhaus, D.H., Wakim, B.T., Storm, D.R., & Moon, R.T. (1987). Characterization of murine cDNAs encoding P-57, a neural-specific calmodulin-binding protein. *Journal of Biological Chemistry* **262**, 12158–12163.

Clark, E.A., & Brugge, J.S. (1995). Integrins and signal transduction pathways: The road taken. *Science* **268**, 233–239.

Cobb, B.S., Schaller, M.D., Leu, T.-H., & Parsons, J.T. (1994). Stable association of $pp60^{c\text{-}src}$ and $pp59^{fyn}$ with the focal adhesion-associated protein tyrosine kinase, $pp125^{FAK}$. *Molecular Biology of the Cell* **14**, 147–155.

Coggins, P.J., & Zwiers, H. (1989). Evidence for a single protein kinase C-mediated phosphorylation site in rat brain protein B-50. *Journal of Neurochemistry* **53**, 1895–1901.

Cohan, C.S., Haydon, P.G., & Kater, S.B. (1985). Single channel activity differs in growing and non-growing growth cones of isolated neurons of *Helisoma*. *Journal of Neuroscience Research* **13**, 285–306.

Cohan, C.S., Connor, J.A., & Kater, S.B. (1987). Electrically and chemically mediated increases in intracellular calcium in neuronal growth cones. *Journal of Neuroscience* **7**, 3588–3599.

Cohen, I., Rimer, M., Lømo, T., & McMahan, U.J. (1997a). Agrin-induced postsynaptic-like apparatus in skeletal muscle fibers in vivo. *Molecular and Cellular Neuroscience* **9**, 237–253.

Cohen, J., & Johnson, A.R. (1991). Differential effects of laminin and merosin on neurite outgrowth by developing RGCs. *Journal of Cell Science* **15**, 1–7.

Cohen, J., Burne, J.F., Winter, J., & Bartlett, J. (1986). Retinal ganglion cells lose response to laminin with maturation. *Nature* **322**, 465–467.

Cohen, J., Burne, J.F., McKinlay, C., & Winter, J. (1987). The role of laminin and the laminin/fibronectin receptor complex in the outgrowth of retinal ganglion cell axons. *Developmental Biology* **122**, 407–418.

Cohen, J., Nurcombe, V., Jeffrey, P., & Edgar, D. (1989). Developmental loss of functional laminin receptors on retinal ganglion cells is regulated by their target tissue, the optic tectum. *Development* **107**, 381–387.

Cohen, M.W. (1972). The development of neurotransmitter connections in the presence of d-tubocurarine. *Brain Research* **41**, 457–463.

Cohen, M.W., & Godfrey, E.W. (1992). Early appearance of and neuronal contribution to agrin-like molecules at embryonic frog nerve-muscle synapses formed in culture. *Journal of Neuroscience* **12**, 2982–2992.

Cohen, N.R., Taylor, J.S.H., Scott, L.B., Guillary, R.W., Soriano, P., & Furley, A.J.W. (1997b). Errors in corticospinal axon guidance in mice lacking the neural cell adhesion molecule L1. *Current Biology* **8**, 26–33.

Cohen, S.A. (1980). Early nerve-muscle synapses *in vitro* release transmitter over postsynaptic membrane having low acetylcholine sensitivity. *Proceedings of the National Academy of Sciences U.S.A.* **77**, 644–648.

Colamarino, S.A., & Tessier-Lavigne, M. (1995). The axonal chemoattractant Netrin-1 is also a chemorepellent for trochlear motor axons. *Cell* **81**, 621–629.

Cole, G.J., & McCabe, C.F. (1991). Identification of a developmentally regulated keratan sulphate proteoglycan that inhibits cell adhesion and neurite outgrowth. *Neuron* **7**, 1007–1018.

Condic, M.L., & Bentley, D. (1989a). Removal of the basal lamina *in vivo* reveals growth cone–basal lamina adhesive interactions and axonal tension in grasshopper embryos. *Journal of Neuroscience* **9**, 2678–2686.

Condic, M.L., & Bentley, D. (1989b). Pioneer growth cone adhesion *in vivo* to boundary cells and neurons after enzymatic removal of basal lamina in grasshopper embryos. *Journal of Neuroscience* **9**, 2687–2696.

Condic, M.L., & Bentley, D. (1989c). Pioneer neuron pathfinding from normal and ectopic locations *in vivo* after removal of the basal lamina. *Neuron* **3**, 427–439.

Condic, M.L., & Letourneau, P.C. (1997). Ligand-induced changes in integrin expression regulate neuronal adhesion and neurite outgrowth. *Nature* **389**, 852–856.

Connolly, J.L., Seeley, P.J., & Greene, L.A. (1985). Regulation of growth cone morphology by nerve growth factor: A comparative study by scanning electron microscopy. *Journal of Neuroscience Research* **13**, 183–198.

Connor, J., Tseng, H., & Hockberger, P. (1987). Depolarization- and transmitter-induced changes in intracellular Ca^{2+} of rat cerebellar granule cells in explant cultures. *Journal of Neuroscience* **7**, 1384–1400.

Connor, J.A. (1986). Digital imaging of free calcium changes and of spatial gradients in growing processes in single, mammalian central nervous system cells. *Proceedings of the National Academy of Sciences U.S.A.* **83**, 6179–6183.

Cooper, J.A. (1987). Effects of cytochalasin and phalloidin on actin. *Journal of Cell Biology* **105**, 1473–1478.

Costero, I., & Pomerat, C.M. (1951). Cultivation of neurons from the adult human cerebral and cerebellar cortex. *American Journal of Anatomy* **89**, 405–467.

Cotton, P.C., & Brugge, J.S. (1983). Neural tissues express high levels of the cellular *src* gene product pp60$^{c\text{-}src}$. *Molecular and Cellular Biology* **3**, 1157–1162.

Coughlin, M.D. (1975). Target organ stimulation of parasympathetic nerve growth in the developing mouse submandibular gland. *Developmental Biology* **43**, 140–158.

Cox, M.E., & Maness, P.F. (1993). Tyrosine phosphorylation of α-tubulin is an early response to NGF and pp60$^{v\text{-}src}$ in PC12 cells. *Journal of Molecular Neuroscience* **4**, 63–72.

Cox, E.C., Müller, B., & Bonhoeffer, F. (1990). Axonal guidance in the chick visual system: posterior tectal membranes induce collapse of growth cones from the temporal retina. *Neuron* **2**, 31–37.

Craig, A.M., Jareb, M., & Banker, G. (1992). Neuronal polarity. *Current Opinion in Neurobiology* **2**, 602–606.

Craig, A.M., Wyborski, R.J., & Banker, G. (1995). Preferential addition of newly synthesized membrane protein at axonal growth cones. *Nature* **375**, 592–594.

Cramer, L.P. (1997). Molecular mechanisms of actin-dependent retrograde flow in lamellipodia of motile cells. *Frontiers in Bioscience* **2**, 260–270.

Cremer, H., Lange, R., Christoph, A., Plomann, M., Vopper, G., Roes, J., Brown, R., Baldwin, S., Kraemer, P., Scheff, S., Barthels, D., Rajewsky, K., & Wille, W. (1994). Inactivation of the N-CAM gene in mice results in size reduction of the olfactory bulb and deficits in spatial learning. *Nature* **367**, 455–459.

Cremer, H., Chazal, G., Goridis, C., & Represa, A. (1996). NCAM is essential for axonal growth and fasciculation in the hippocampus. *Molecular and Cellular Neuroscience* **8**, 323–335.

Crossin, K.L., Hoffman, S., Grumet, S., Thiery, J.-P., & Edelman, G.M. (1986). Site-restricted expression of cytotactin during development of the chick embryo. *Journal of Cell Biology* **102**, 1917–1930.

Crossin, K.L., Hoffman, S., Tan, S.-S., & Edelman, G.M. (1989). Cytotactin and its proteoglycan ligand mark structural and functional boundaries in somatosensory cortex of the early postnatal mouse. *Developmental Biology* **136**, 381–392.

Crossin, K.L., Prieto, A., Hoffman, S., Jones, F.S., & Friedlander, D. (1990). Expression of adhesion molecules and the establishment of boundaries during embryonic and neural development. *Experimental Neurology* **109**, 6–18.

Crossland, W.J., Cowan, W.M., Rogers, L.A., & Kelly, J.P. (1974). Specification of the retino-tectal projection in the chick. *Journal of Comparative Neurology* **155**, 127–164.

Culotti, J.G. (1994). Axon guidance mechanisms in *Caenorhabditis elegans*. *Current Opinion in Genetics and Development* **4**, 587–595.

Culotti, J.G., & Kolodkin, A.L. (1996). Functions of netrins and semaphorins in axon guidance. *Current Opinion in Neurobiology* **6**, 81–88.

Cunningham, B.A., Hemperly, J.J., Murray, B.A., Prediger, E.A., Brackenbury, R., & Edelman, G.M. (1987). Neural cell adhesion molecule: structure, immunoglobulin-like domains, cell surface modulation and alternative RNA splicing. *Science* **236**, 799–806.

Curtis, R., Hardy, R., Reynolds, R., Spruce, B.A., & Wilkin, G.P. (1991). Down-regulation of GAP-43 during oligodendrocyte development and lack of expression by astrocytes *in vitro:* Implications for macroglial differentiation. *European Journal of Neuroscience* **3**, 876–886.

Curtis, R., Stewart, H.J.S., Hall, S.M., Wilkin, J.P., Mirsky, R., & Jessen, K.R. (1992). GAP-43 is expressed by non-myelin forming Schwann cells of the peripheral nervous system. *Journal of Cell Biology* **116**, 1455–1464.

Cypher, C., & Letourneau, P.C. (1991). Identification of cytoskeletal, focal adhesion and cell adhesion proteins in growth cone particles isolated from developing chick brain. *Journal of Neuroscience Research* **30**, 259–265.

Cypher, C., & Letourneau, P.C. (1992). Growth cone motility. *Current Opinion in Cell Biology* **4**, 4–7.

da Cunha, A., & Vitkovic, L. (1990). Regulation of immunoreactive GAP-43 expression in rat cortical macroglia is cell type specific. *Journal of Cell Biology* **111**, 209–215.

Dahme, M., Bartsch, U., Martini, R., Anliker, B., Schachner, M., & Mantei, N. (1997). Disruption of the mouse L1 gene leads to malformation of the nervous system. *Nature Genetics* **17**, 346–349.

Dai, J., & Sheetz, M. (1995a). Axon membrane flows from the growth cone to the cell body. *Cell* **83**, 693–701.

Dai, J., & Sheetz, M. (1995b). Mechanical properties of neuronal growth cone membrane studied by tether formation with laser optical tweezers. *Biophysical Journal* **68**, 988–996.

Dailey, M.E., & Bridgman, P.C. (1989). Dynamics of the endoplasmic reticulum and other membrane organelles in growth cones of cultured neurons. *Journal of Neuroscience* **9**, 1897–1909.

Dailey, M.E., & Bridgman, P.C. (1991). Structure and organization of membrane organelles along distal microtubule segments in growth cones. *Journal of Neuroscience Research* **30**, 242–258.

Dailey, M.E., & Bridgman, P.C. (1993). Vacuole dynamics in growth cones: Correlated EM and video observations. *Journal of Neuroscience Research* **13**, 3375–3393.

Dale, N., Ottersen, O.P., Roberts, A., & Storm-Mathisen, J. (1986). Inhibitory neurons of a motor pattern generator in *Xenopus* revealed by antibodies to glycine. *Nature* **324**, 255–257.

Dale, N., Roberts, A., Ottersen, O.P., & Storm-Mathisen, J. (1987a). The morphology and distribution of 'Kolmer-Agduhr' cells, a class of cerebrospinal fluid-contacting neurons revealed in the frog embryo spinal cord by GABA immunocytochemistry. *Proceedings of the Royal Society Series B* **232**, 193–203.

Dale, N., Roberts, A., Ottersen, O.P., & Storm-Mathisen, J. (1987b). The development of a population of spinal cord neurons and their axonal projections revealed by GABA immunocytochemistry. *Proceedings of the Royal Society Series B* **232**, 205–215.

Damsky, C.H., & Werb, Z. (1992). Signal transduction by integrin receptors for extracellular matrix: cooperative processing of extracellular information. *Current Opinion in Cell Biology* **4**, 772–781.

Dan, Y., Song, H.J., & Poo, M.-M. (1994). Evoked neuronal secretion of false neurotransmitters. *Neuron* **13**, 909–917.

Daniels, M. (1972). Colchicine inhibiton of nerve fiber formation *in vitro*. *Journal of Cell Biology* **53**, 164–176.

Daniels, M.P. (1975). The role of microtubules in the growth and stabilization of nerve fibers. *Annals of the New York Academy of Sciences* **253**, 535

Davenport, R.W. (1996). Functional domains and intracellular signalling: clues to growth cone dynamics. In *Nerve Growth and Guidance*, ed. C.D. McCaig, pp. 55–75. London: Portland Press.

Davenport, R.W., & Kater, S.B. (1992). Local increases in intracellular calcium elicit local filopodial responses in *Helisoma* neuronal growth cones. *Neuron* **9**, 405–416.

Davenport, R.W., & McCaig, C.D. (1993). Hippocampal growth cone responses to focally applied electric fields. *Journal of Neurobiology* **24**, 89–100.

Davenport, R.W., Dou, P., Rehder, V., & Kater, S.B. (1993). A sensory role for neuronal growth cone filopodia. *Nature* **361**, 721–723.

Davies, A.M. (1987). Molecular and cellular aspects of patterning sensory neurone connections in the vertebrate nervous system. *Development* **101**, 185–208.

Davies, A.M., Bandtlow, C., Heumann, R., Korsching, S., Rohrer, H., & Thoenen, H. (1987). The site and timing of nerve growth factor (NGF) synthesis in developing skin in relation to its innervation by sensory neurons and their expression of NGF receptors. *Nature* **326**, 353–363.

Davies, J.A., Cook, G.M.W., Stern, C.D., & Keynes, R.J. (1990). Isolation from chick somites of a glycoprotein fraction that causes collapse of dorsal root ganglion growth cones. *Neuron* **2**, 11–20.

Davis, L., Dou, P., DeWit, M., & Kater, S.B. (1992). Protein synthesis within neuronal growth cones. *Journal of Neuroscience* **12**, 4867–4877.

DeChiara, T.M., Bowen, D.C., Valenzuela, D.M., Simmons, M.V., Poueymirou, W.T., Thomas, S., Kinetz, E., Compoton, D.L., Rojas, E., Park, J.S., Smith, C.,

DiStefano, P.S., Glass, D.J., Burden, S.J., & Yancopoulos, G.D. (1996). The receptor tyrosine kinase MuSK is required for neuromuscular junction formation *in vivo*. *Cell* **85**, 501–512.

Deiner, M.S., Kennedy, T.E., Fazeli, A., Serafini, T., & Tessier-Lavigne, M. (1997). Netrin-1 and DCC mediate axon guidance locally at the optic disc: Loss of function leads to optic nerve hypoplasia. *Neuron* **19**, 575–589.

Deitch, J.S., & Banker, G.A. (1993). An electron microscopic analysis of hippocampal neurons developing in culture: Early stages in the emergence of polarity. *Journal of Neuroscience* **13**, 4301–4315.

de la Torre, J.R., Höpker, V.H., Ming, G.-I., Poo, M.-M., Tessier-Lavigne, M., Hemmati-Brivanlou, A., & Holt, C.E. (1997). Turning of retinal growth cones in a netrin-1 gradient mediated by the netrin receptor DCC. *Neuron* **19**, 1211–1224.

Del Cerro, M.P., & Snider, R.S. (1968). Studies on the developing cerebellum. Ultrastructure of the growth cones. *Journal of Comparative Neurology* **133**, 341–362.

Deloulme, J.-C., Thierry, J., Au, D., Storm, D.R., Sensenbrenner, M., & Baudier, J. (1990). Neuromodulin (GAP-43): A neuronal protein kinase C substrate is also present in 0–2A glial cell lineage. Characterisation of neuromodulin in secondary cultures of oligodendrocytes and comparison with the neuronal antigen. *Journal of Cell Biology* **111**, 1559–1569.

Dennerll, T.J., Joshi, H.C., Steel, V.L., Buxbaum, R.E., & Heidemann, S.R. (1988). Tension and compression in the cytoskeleton of PC12 neurites II: Quantitative measurements. *Journal of Cell Biology* **107**, 665–674.

Dennerll, T.J., Lamoureux, P., Buxbaum, R.E., & Heidemann, S.R. (1989). The cytomechanics of axonal elongation and retraction. *Journal of Cell Biology* **109**, 3073–3083.

Dent, E.W., & Meiri, K.F. (1992). GAP-43 phosphorylation is dynamically regulated in individual growth cones. *Journal of Neurobiology* **23**, 1037–1053.

Dent, E.W., & Meiri, K.F. (1998). Distribution of phosphorylated GAP-43 (neuromodulin) in growth cones directly reflects growth cone behavior. *Journal of Neurobiology* **35**, 287–299.

Díaz-Nido, J., Serrano, L., Méndez, E., & Avila, J. (1988). A casein kinase II-related activity is involved in phosphorylation of microtubule-associated protein MAP1B during neuroblastoma cell differentiation. *Journal of Cell Biology* **106**, 2057–2065.

Díaz-Nido, J., Hernandez, M.A. & Avila, J. (1990). Microtubule proteins in neuronal cells. In *Microtubule Proteins*, ed. J. Avila, pp. 193–257. Boca Raton, Florida: CRC Press, Inc.

Díaz-Nido, J., Armas-Portela, R., Correas, I., Domínguez, J.E., Montejo, E., & Avila, J. (1991). Microtubule protein phosphorylation in neuroblastoma cells and neurite growth. *Journal of Cell Science* **15**, 51–59.

Dinsmore, J.H., & Solomon, F. (1991). Inhibition of MAP2 expression affects both morphological and cell division phenotypes of neuronal differentiation. *Cell* **64**, 817–826.

DiPaolo, G., Lutjens, R., OsenSand, A., Sobel, A., Catsicas, S., & Grenningloh, G. (1997). Differential distribution of stathmin and SCG10 in developing neurons. *Journal of Neuroscience Research* **50**, 1000–1009.

DiTella, M., Feiguin, F., Morfini, G., & Cáceres, A. (1994). Microfilament-associated growth cone component depends upon tau for its intracellular localization. *Cell Motility and the Cytoskeleton* **29**, 117–130.

DiTella, M.C., Feiguin, F., Carri, N., Kosik, K.S., & Cáceres, A. (1996). MAP-1B/TAU functional redundancy during laminin-enhanced axonal growth. *Journal of Cell Science* **109**, 467–477.

Dodd, J., & Jessell, T.M. (1988). Axon guidance and the patterning of neuronal projections in vertebrates. *Science* **242**, 692–699.

Dodd, J., Morton, S.B., Karagorgeos, D., Yamamoto, M., & Jessell, T.M. (1988). Spacial regulation of axonal glycoprotein expression on subsets of embryonic spinal neurons. *Neuron* **1**, 105–116.

Doherty, P., & Walsh, F. (1994). Signal transduction events underlying neurite outgrowth stimulated by cell adhesion molecules. *Current Opinion in Neurobiology* **4**, 49–55.

Doherty, P., & Walsh, F.S. (1996). CAM-FGF receptor interactions: a model for axonal growth. *Molecular and Cellular Neuroscience* **8**, 99–111.

Doherty, P., Fruns, M., Seaton, P., Dickson, G., Barton, C.H., Sears, T.A., & Walsh, F.S. (1990a). A threshold effect of the major isoforms of NCAM on neurite outgrowth. *Nature* **343**, 464–466.

Doherty, P., Cohen, J., & Walsh, F.S. (1990b). Neurite outgrowth in response to transfected N-CAM changes during development and is modulated by polysialic acid. *Neuron* **5**, 209–219.

Doherty, P., Ashton, S.V., Moore, S.E., & Walsh, F.S. (1991a). Morphoregulatory activities of N-CAM and N-cadherin can be accounted for by G-protein dependent activation of L- and N-type neuronal calcium channels. *Cell* **67**, 21–33.

Doherty, P., Rowett, L.H., Moore, S.E., Mann, D.A., & Walsh, F.S. (1991b). Neurite outgrowth in response to transfected NCAM and N-cadherin reveals fundamental differences in neuronal responsiveness to CAMs. *Neuron* **6**, 247–258.

Doherty, P., Rimon, G., Mann, D.A., & Walsh, F.S. (1992a). Alternative splicing of the cytoplasmic domain of neural cell adhesion molecule alters its ability to act as a substrate for neurite outgrowth. *Journal of Neurochemistry* **58**, 2338–2341.

Doherty, P., Moolenaar, C.E.C.K., Ashton, S.A., Michalides, R.J.A.M., & Walsh, F.S. (1992b). Use of the VASE exon down regulates the neurite growth promoting activity of NCAM 140. *Nature* **356**, 791–793.

Doherty, P., Williams, E.J., & Walsh, F.S. (1995). A soluble chimeric form of the L1 glycoprotein stimulates neurite outgrowth. *Neuron* **14**, 1–20.

Doherty, P., Smith, P., & Walsh, F.S. (1997). Shared cell adhesion molecule (CAM) homology domains point to CAMs signalling via FGF receptors. *Perspectives on Developmental Neurobiology* **4**, 157–168.

Dorries, U., Taylor, J., Xiao, Z., Lochter, A., Montag, D., & Schachner, M. (1996). Distinct effects of recombinant tenascin-C domains on neuronal cell adhesion, growth cone guidance, and neuronal polarity. *Journal of Neuroscience Research* **43**, 420–438.

Dotti, C.G., Banker, G.A., & Binder, L.I. (1987). The expression and distribution of the microtubule-associated proteins tau and microtubule-associated protein 2 in hippocampal neurons in the rat *in situ* and in cell culture. *Neuroscience* **23**, 121–130.

Dotti, C.G., Sullivan, C.A., & Banker, G.A. (1988). The establishment of polarity by hippocampal neurons in culture. *Journal of Neuroscience* **8**, 1454–1468.

Dräger, U.C. (1981). Observations on the organization of the visual cortex in the reeler mutant mouse. *Journal of Comparative Neurology* **201**, 555–570.

Drazba, J., & Lemmon, V. (1990). The role of cell adhesion molecules in neurite outgrowth on Müller cells. *Developmental Biology* **138**, 82–93.

Drazba, J., Liljelund, P., Smith, C., Payne, R., & Lemmon, V. (1997). Growth cone interactions with purified cell and substrate adhesion molecules visualized by interference reflection microscopy. *Developmental Brain Research* **100**, 183–197.

Drescher, U., Kremoser, C., Handwerker, C., Löschinger, J., Noda, M., & Bonhoeffer, F. (1995). In vitro guidance of retinal ganglion cell axons by RAGS, a 25 kDa tectal protein related to ligands for Eph receptor tyrosine kinases. *Cell* **82**, 359–370.

Drubin, D.G., Feinstein, S.C., Shooter, E.M., & Kirschner, M.W. (1985). Nerve growth factor-induced neurite outgrowth in PC12 cells involves the coordinate induction of microtubule assembly and assembly-promoting factors. *Journal of Cell Biology* **101**, 1799–1807.

Ebendal, T. (1976). The relative roles of contact inhibition and contact guidance in orientation of axons extending on aligned collagen fibrils *in vitro*. *Experimental Cell Research* **98**, 159–169.

Ebendal, T., & Jacobson, C.-O. (1977). Tissue explants affecting extension and orientation of axons in cultured chick embryo ganglia. *Experimental Cell Research* **105**, 379–387.

Edelman, G.M., & Crossin, K.L. (1991). Cell adhesion molecules: implications for a molecular histology. *Annual Review of Biochemistry* **60**, 155–190.

Edelmann, W., Zervas, M., Costello, P., Roback, L., Fischer, I., Hammarback, J.A., Cowan, J., Davies, P., Wainer, B., & Kucherlapati, R. (1996). Neuronal abnormalities in microtubule-associated protein 1B mutant mice. *Proceedings of the National Academy of Sciences U.S.A.* **93**, 1270–1275.

Edgar, D., Timpl, R., & Thoenen, H. (1984). The heparin binding domain of laminin is responsible for its effect on neurite outgrowth and neuronal survival. *EMBO Journal* **3**, 1463–1488.

Edgar, M., Simpson, S.B., & Singer, M. (1970). The growth and differentiation of the regenerating spinal cord of the lizard, *Anolis carolinensis*. *Journal of Morphology* **131**, 131–152.

Edmonds, B.T., Moomaw, C.R., Hsu, J.T., Slaughter, C., & Ellis, L. (1990). The p38 and p34 polypeptides of the growth cone particle membranes are the alpha- and beta- subunits of G proteins. *Developmental Brain Research* **56**, 131–136.

Edson, K., Weisshaar, B., & Matus, A. (1993a). Actin depolymerisation induces process formation on MAP 2-transfected non-neuronal cells. *Development* **117**, 689–700.

Edson, K.J., Lim, S.-S., Borisy, G.G., & Letourneau, P.C. (1993b). FRAP analysis of the stability of the microtubule population along the neurites of chick sensory neurons. *Cell Motility and the Cytoskeleton* **25**, 59–72.

Edwards, R.A., & Bryan, J. (1995). Fascins, a family of actin bundling proteins. *Cell Motility and the Cytoskeleton* **32**, 1–9.

Egar, M., Simpson, S.B., & Singer, M. (1970). The growth and differentiation of the regenerating spinal cord of the lizard, *Anolis carolinensis*. *Journal of Morphology* **131**, 131–152.

Ehrig, K., Leivo, I., Argraves, W.S., Ruoslahti, E., & Engvall, E. (1990). Merosin, a tissue-specific basement membrane protein, is a laminin-like protein. *Proceedings of the National Academy of Sciences U.S.A.* **87**, 3264–3268.

Ehrlich, Y., & Routtenberg, A. (1974). Cyclic AMP regulates phosphorylation of three protein components of rat cerebral cortex membranes for thirty minutes. *FEBS Letters* **45**, 237–243.

Eisen, J.S. (1992). Development of motoneuronal identity in the Zebrafish. In *Determinants of Neuronal Identity*, ed. M. Shankland & E.R. Macagno, pp. 469–496. New York: Academic Press.

Eisen, J.S., Pike, S.H., & Debu, B. (1989). The growth cones of identified motoneurons in embryonic zebrafish select appropriate pathways in the absence of specific cellular interactions. *Neuron* **2**, 1097–1104.

Elkins, T., Zinn, K., McAllister, L., Hoffmann, F.M., & Goodman, C.S. (1990). Genetic analysis of a *Drosophila* neural cell adhesion molecule: interaction of fasciclin I and Abelson tyrosine kinase mutations. *Cell* **60**, 565–575.

Engvall, E., Davis, G.E., Dickerson, K., Ruoslahti, E., Varon, S., & Manthorpe, M. (1986). Mapping of domains in human laminin using monoclonal antibodies: Localization of the neurite-promoting site. *Journal of Cell Biology* **103**, 2457–2465.

Engvall, E., Earwicker, D., Day, A., Muir, D., Manthorpe, M., & Paulsson, M. (1992). Merosin promotes cell attachment and neurite outgrowth and is a component of the neurite-promoting factor of RN22 Schwannoma cells. *Experimental Cell Research* **198**, 115–123.

Eph Nomenclature Committee (1997). Unified nomenclature for Eph family receptors and their ligands, the ephrins. *Cell* **90**, 403–404.

Erickson, H.P. (1993). Tenascin-C, tenascin-R, and tenascin-X – a family of talented proteins in search of functions. *Current Opinion in Cell Biology* **5**, 869–876.

Ernsberger, U., & Rohrer, H. (1994). Neurotrophins and neurite outgrowth in the peripheral nervous system. *Developmental Biology* **5**, 403–410.

Erskine, L., & McCaig, C.D. (1995). Growth cone neurotransmitter receptor activation modulates electric field-guided nerve growth. *Developmental Biology* **171**, 330–339.

Erskine, L., Stewart, R., & McCaig, C.D. (1995). Electric field-directed growth and branching of cultured frog nerves: Effects of aminoglycosides and polycations. *Journal of Neurobiology* **26**, 523–536.

Esmaeli-Azad, B., McCarty, J.H., & Feinstein, S.C. (1994). Sense and antisense transfection analysis of tau function: Tau influences net microtubule assembly, neurite outgrowth and neurite stability. *Journal of Cell Science* **107**, 869–879.

Espreafico, E.M., Cheney, R.E., Matteoli, M., Nascimento, A.A., DeCamilli, P.V., Larson, R.E., & Mooseker, M.S. (1992). Primary structure and cellular localization of chicken brain myosin-V (p190), an unconventional myosin with calmodulin light chains. *Journal of Cell Biology* **119**, 1541–1557.

Estable, C., Acosta-Ferreira, W., & Sotelo, J.R. (1957). An electron microscope study of the regenerating nerve fibres. *Zeitschrift für Zellforschung und Mikroskopische Anatomie* **46**, 387–400.

Evans, L.L., Hammer, J., & Bridgman, P.C. (1997). Subcellular localization of myosin V in nerve growth cones and outgrowth from *dilute-lethal* neurons. *Journal of Cell Science* **110**, 439–449.

Evers, J., Laser, M., Sun, Y., Xie, Z., & Poo, M.-M. (1989). Studies of nerve-muscle interactions in *Xenopus* cell culture: Analysis of early synaptic currents. *Journal of Neuroscience* **9**, 1523–1539.

Eyer, J., & Peterson, A. (1994). Neurofilament-deficient axons and perikaryal aggregates in viable transgenic mice expressing a neurofilament-β-galactosidase fusion protein. *Neuron* **12**, 389–405.

Faissner, A. (1997). The tenascin gene family in axon growth and guidance. *Cell and Tissue Research* **290**, 331–341.

Faissner, A., & Kruse, J. (1990). J1/tenascin is a repulsive substrate for central nervous system neurons. *Neuron* **5**, 627–637.

Faissner, A., & Steindler, D.A. (1995). Boundaries and inhibitory molecules in developing neural tissues. *Glia* **13**, 233–254.

Faivre-Bauman, A., Puymirat, J., Loudes, J., Barret, A., & Tixier-Vidal, A. (1984). Laminin promotes attachment and neurite elongation of fetal hypothalamic neurons grown in serum-free medium. *Neuroscience Letters* **44**, 83–89.

Faivre-Sarrailh, C., Lena, J.Y., Had, L., Vignes, M., & Lindberg, U. (1993). Location of profilin at presynaptic sites in the cerebellar cortex; implication for the regulation of the actin-polymerization state during axonal elongation and synaptogenesis. *Journal of Neurocytology* **22**, 1060–1072.

Falls, D.L., Rosen, K.M., Corfas, G., Lane, W.S., & Fischbach, G.D. (1993). ARIA, a protein that stimulates acetylcholine receptor synthesis, is a member of the Neu ligand family. *Cell* **72**, 801–815.

Fan, J., & Raper, J.A. (1995). Localized collapsing cues can steer growth cones without inducing their full collapse. *Neuron* **14**, 263–274.

Fan, J., Mansfield, S.G., Redmond, T., Gordon-Weeks, P.R., & Raper, J. (1993). The organization of F-actin and microtubules in growth cones exposed to a brain-derived collapsing factor. *Journal of Cell Biology* **121**, 867–878.

Fawcett, J.W. (1993). Growth cone collapse: too much of a good thing? *Trends in Neurosciences* **16**, 165–167.

Fawcett, J.W., Rokos, J., & Bakst, I. (1989). Oligodendrocytes repel axons and cause growth cone collapse. *Journal of Cell Science* **92**, 93–100.

Fawcett, J.W., Mathews, G., Housden, E., Goedert, M., & Matus, A. (1995). Regenerating sciatic nerve axons contain the adult rather than the embryonic pattern of microtubule associated proteins. *Neuroscience* **61**, 789–804.

Fazeli, A., Dickinson, S.L., Hermiston, M.L., Tighe, R.V., Steen, R.G., Small, C.G., Stoeckli, E.T., Keino-Masu, K., Masu, M., Rayburn, H., Simons, J., Bronson, R.T., Gordon, J.I., Tessier-Lavigne, M., & Weinberg, R.A. (1997). Phenotype of mice lacking functional *Deleted in colorectal cancer (Dcc)* gene. *Nature* **386**, 796–804.

Feiner, L., Koppel, A.M., Kobayashi, H., & Raper, J.A. (1997). Secreted chick semaphorins bind recombinant neuropilin with similar affinites but bind different subsets of neurons in situ. *Neuron* **19**, 539–545.

Feldman, E.L., Axelrod, D., Schwartz, M., Heacock, A.M., & Agranoff, B.W. (1981). Studies on the localization of newly added membrane in growing neurites. *Journal of Neurobiology* **12**, 591–598.

Ferns, M., Deiner, M., & Hall, Z. (1996). Agrin-induced acetylcholine-receptor clustering requires tyrosine phosphorylation. *Journal of Cell Biology* **132**, 937–944.

Ferreira, A., & Cáceres, A. (1989). The expression of acetylated microtubules during axonal and dendritic growth in cerebellar macroneurons which develop *in vitro*. *Developmental Brain Research* **49**, 204–213.

Ferreira, A., Busciglio, J., & Càceres, A. (1989). Microtubule formation and neurite growth in cerebellar macroneurons which develop *in vitro*: Evidence for the involvement of the microtubule-associated proteins MAP-1a, HMW-MAP-2 and tau. *Developmental Brain Research* **34**, 9–31.

Ferreira, A., Kincaid, R., & Kosik, K. (1993). Calcineurin is associated with the cytoskeleton of cultured neurons and has a role in the acquisition of polarity. *Molecular Biology of the Cell* **4**, 1225–1238.

Fields, R.D., Guthrie, P.B., Russell, J.T., Kater, S.B., Malhotra, B.S., & Nelson, P.G. (1993). Accommodation of mouse DRG growth cones to electrically induced collapse: Kinetic analysis of calcium transients and set-point theory. *Journal of Neurobiology* **24**, 1080–1098.

Fischer, I., & Romano-Clarke, G. (1990). Changes in microtubule-associated protein MAP1B phosphorylation during rat brain development. *Journal of Neurochemistry* **55**, 328–333.

Fischer, I., & Romano-Clarke, G. (1991). Association of microtubule associated protein (MAP 1B) with growing axons in cultured hippocampal neurons. *Molecular and Cellullar Neurosciences* **2**, 39–51.

Fitzgerald, M., Reynolds, M.L., & Benowitz, L.I. (1991). GAP-43 expression in the developing rat lumbar spinal cord. *Neuroscience* **41**, 187–199.

Fitzgerald, M., Kwiat, G.C., Middleton, J., & Pini, A. (1993). Ventral spinal cord inhibition of neurite outgrowth from embryonic rat dorsal root ganglia. *Development* **117**, 1377–1384.

Flenniken, A.M., Gale, N.W., Yancopoulos, G.D., & Wilkinson, D.G. (1996). Distinct and overlapping expression patterns of ligands for Eph-related receptor tyrosine kinases during mouse embryogenesis. *Developmental Biology* **179**, 382–401.

Forsberg, E., Hirsch, E., Fröhlich, L., Meyer, M., Ekblom, P., Aszodi, A., Werner, S., & Fässler, R. (1996). Skin wounds and severed nerves heal normally in mice lacking tenascin-C. *Proceedings of the National Academy of Sciences U.S.A.* **93**, 6594–6599.

Forscher, P., & Smith, S.J. (1988). Actions of cytochalasins on the organization of actin filaments and microtubules in a neuronal growth cone. *Journal of Cell Biology* **107**, 1505–1516.

Forscher, P., Kaczmarek, L.K., Buchanan, J., & Smith, S.J. (1987). Cyclic AMP induces changes in distribution and transport of organelles within growth cones of *Aplysia* bag cell neurons. *Journal of Neuroscience* **7**, 3600–3611.

Forscher, P., Lin, C.-H., & Thompson, C. (1992). Novel form of growth cone motility involving site-directed actin filament assembly. *Nature* **357**, 515–518.

Forssman, J. (1898). Ueber die Ursachen, welche die Wachstumsrichtung der peripheren Nervenfasern bei der Regeneration bestimmen. *(Zeigler's) Beitrage zur Pathologischen Anatomie und zur Allgemeinen Pathologie* **24**, 56–100.

Forssman, J. (1900). Zur Kenntnis des Neurotropismus. *(Zeigler's) Beitrage zur Pathologischen Anatomie und zur Allgemeinen Pathologie* **27**, 407–430.

Fox, G.Q., Pappas, G.D., & Purpura, D.P. (1976). Fine structure of growth cones in medullary raphe nuclei in the postnatal cat. *Brain Research* **101**, 411–425.

Frail, D.E., McLaughlin, L.L., Mudd, J., & Merlie, J.P. (1988). Identification of the mouse muscle 43,000-Dalton acetylcholine receptor-associated protein (RAPsyn) by cDNa cloning. *Journal of Biological Chemistry* **263**, 15602–15607.

Frank, E., & Fischbach, G.D. (1979). Early events in neuromuscular junction formation in vitro: Induction of acetylcholine receptor clusters in the postsynaptic membrane and morphology of newly formed synapses. *Journal of Cell Biology* **83**, 143–158.

Fredette, B.J., & Ranscht, B. (1994). T-cadherin expression delineates specific regions of the developing motor axon–hindlimb projection pathway. *Journal of Neuroscience* **14**, 7331–7346.

Fredette, B.J., Miller, J., & Ranscht, B. (1996). Inhibiton of motor axon growth by T-cadherin substrata. *Development* **122**, 3163–3171.

Freeman, J.A., Manis, P.B., Snipes, G.J., Mayes, B.N., Samson, P.C., Wikswo, J.P., Jr, & Freeman, D.B. (1985). Steady growth cone currents revealed by a novel circularly vibrating probe: A possible mechanism underlying neurite growth. *Journal of Neuroscience Research* **13**, 257–283.

Friedman, G.C., & O'Leary, D.D.M. (1996). Eph receptor tyrosine kinases and their ligands in neural development. *Current Opinion in Neurobiology* **6**, 127–133.

Froehner, S.C., Luetje, C.W., Scotland, P.B., & Patrick, J. (1990). The postsynaptic 43K protein clusters muscle nicotinic acetylcholine receptors in *Xenopus* oocytes. *Neuron* 5, 403–410.

Fryer, H.J.L., & Hockfield, S. (1996). The role of polysialic acid and other carbohydrate polymers in neural structural plasticity. *Current Opinion in Neurobiology* 6, 113–118.

Fukura, H., Komiya, Y., & Igarashi, M. (1996). Signaling pathway downstream of GABA$_A$ receptor in the growth cone. *Journal of Neurochemistry* 67, 1426–1434.

Fults, D.W., Towle, A.C., Lauder, J.M., & Maness, P.F. (1985). pp60$^{c\text{-}src}$ in the developing cerebellum. *Molecular and Cellular Biology* 5, 27–32.

Furley, A.J., Morton, S.B., Manalo, D., Karagogeos, D., Dodd, J., & Jessell, T.M. (1990). The axonal glycoprotein TAG-1 is an immunoglobulin superfamily member with neurite outgrowth-promoting activity. *Cell* 81, 157–170.

Futerman, A., & Banker, G. (1996). The economics of neurite outgrowth – the addition of new membrane to growing axons. *Trends in the Neurosciences* 19, 144–149.

Gähwiler, B.H., & Brown, D.A. (1985). Functional innervation of cultured hippocampal neurones by cholinergic afferents from co-cultured septal explants. *Nature* 313, 577–579.

Gähwiler, B.H., & Hefti, F. (1984). Guidance of acetylcholinesterase-containing fibres by target tissue in co-cultured brain slices. *Neuroscience* 13, 681–689.

Gale, N.W., & Yancopoulos, G.D. (1997). Ephrins and their receptors: A repulsive topic? *Cell and Tissue Research* 290, 227–241.

Gale, N.W., Holland, S.J., Valenzuela, D.M., Flenniken, A., Pan, L., Henkemeyer, M., Strebhardt, K., Hirai, H., Wilkinson, D.G., & Pawson, T. (1996a). Eph receptors and ligands comprise two major specificity subclasses, and are reciprocally compartmentalized during embryogenesis. *Neuron* 17, 9–19.

Gale, N.W., Flenniken, A., Compton, D.C., Jenkins, N., Copeland, N.G., Gilbert, D.J., Davis, S., Wilkinson, D.G., & Yancopoulos, G.D. (1996b). Elk-L3, a novel transmembrane ligand for the Eph family of receptor tyrosine kinases, expressed in embryonic floor plate, roof plate and hindbrain segments. *Oncogene* 13, 1343–1352.

Gallo, G., Lefcort, F.B., & Letourneau, P.C. (1997). The trkA receptor mediates growth cone turning toward a localized source of nerve growth factor. *Journal of Neuroscience* 17, 5445–5454.

Gao, P.P., Zhang, J.-H., Yokoyama, M., Racey, B., Dreyfus, C.F., Black, I.B., & Zhou, R. (1996). Regulation of topographic projection in the brain: Elf-1 in the hippocamposeptal system. *Proceedings of the National Academy of Sciences U.S.A.* 93, 11161–11166.

García-Alonso, L., Fetter, R.D., & Goodman, C.S. (1996). Genetic analysis of Laminin A in *Drosophila*: Extracellular matrix containing Laminin A is required for ocellar axon pathfinding. *Development* 122, 2611–2621.

Gard, D.L., & Kirschner, M.W. (1985). A polymer dependent increase in phosphorylation of β-tubulin accompanies differentiation of a mouse neuroblastoma cell line. *Journal of Cell Biology* 100, 764–774.

Garner, C.C., Tucker, R.P., & Matus, A. (1988). Selective localization of messenger RNA for cytoskeletal protein MAP2 in dendrites. *Nature* 336, 674–677.

Garner, C.C., Matus, A., Anderton, B., & Calvert, R. (1989). Microtubule-associated proteins MAP 5 and MAP 1x: Closely related components of the neuronal cytoskeleton with different cytoplasmic distribution in the developing brain. *Molecular Brain Research* 5, 85–92.

Garner, C.C., Garner, A., Huber, G., Kozak, C., & Matus, A. (1990). Molecular cloning of MAP 1 (MAP 1A) and MAP 5 (MAP 1B): Identification of distinct genes and their differential expression in developing brain. *Journal of Neurochemistry* **55**, 146–154.

Garrity, P.A., & Zipursky, S.L. (1995). Neuronal target recognition. *Cell* **83**, 177–185.

Garyantes, T.K., & Regehr, W.G. (1992). Electrical activity increases growth cone calcium but fails to inhibit neurite outgrowth from rat sympathetic neurons. *Journal of Neuroscience* **12**, 96–103.

Gautam, M., Noakes, P.G., Mudd, J., Nichol, M., Chu, G.C., Sanes, J.R., & Merlie, J.P. (1995). Failure of postsynaptic specialization to develop at neuromuscular junctions of rapsyn-deficient mice. *Nature* **377**, 232–236.

Gautam, M., Noakes, P.G., Moscoso, L.M., Rupp, F., Scheller, R.H., Merlie, J.P., & Sanes, J.R. (1996). Defective neuromuscular synaptogenesis in agrin-deficient mice. *Cell* **85**, 525–536.

Geiger, B., & Ayalon, O. (1992). Cadherins. *Annual Review of Cell Biology* **8**, 307–332.

Gennarini, G., Cibelli, G., Rougon, G., Mattei, M.-G., & Goridis, C. (1989). The mouse neuronal cell surface protein F3: A phosphatidylinositol-anchored member of the immunoglobulin superfamily related to chicken contactin. *Journal of Cell Biology* **109**, 775–788.

Gennarini, G., Durbec, P., Boned, A., Rougon, G., & Goridis, C. (1991). Transfected F3/F11 neuronal cell surface protein mediates intercellular adhesion and promotes neurite outgrowth. *Neuron* **6**, 595–606.

Gertler, F.B., Bennett, R.L., Clark, M.J., & Hoffmann, F.M. (1989). *Drosophila abl* tyrosine kinase in embryonic CNS axons: A role in axiogenesis is revealed through dosage-sensitive interactions with *disabled*. *Cell* **58**, 103–113.

Ghosh, A., Antonini, A., McConnell, S.K., & Shatz, C.J. (1990). Requirement for subplate neurons in the formation of thalamocortical connections. *Nature* **347**, 179–181.

Gillespie, S.K.H., Balasubramanian, S., Fung, E.T., & Huganir, R.L. (1996). Rapsyn clusters and activates the synapse-specific receptor tyrosine kinase MuSK. *Neuron* **16**, 953–962.

Gimona, M., Vandekerckhove, J., Goethals, M., Herzog, M., Lando, Z., & Small, J.V. (1994). β-actin specific monoclonal antibody. *Cell Motility and the Cytoskeleton* **27**, 108–116.

Glass, D.J., Bowen, D.C., Stitt, T.N., Radziejewski, C., Bruno, J., Ryan, T.E., Gies, D.R., Shah, S., Mattsson, K., Burden, S.J., DiStefano, P.S., Valenzuela, D.M., DeChiara, T.M., & Yancopoulos, G.D. (1996). Agrin acts via a MuSK receptor complex. *Cell* **85**, 1–20.

Glees, P., & Sheppard, B.L. (1964). Electron microscopical studies of the synapse in the developing chick spinal cord. *Zeitschrift für Zellforschung und mikroskopische Anatomie* **62**, 356–362.

Glicksman, M.A., & Sanes, J.R. (1983). Differentiation of motor nerve terminals formed in the absence of muscle fibres. *Journal of Neurocytology* **12**, 666–677.

Godement, P. (1994). Axonal pathfinding at the optic chiasm and at decision regions: Control of growth cone motility, guidance, and cellular contacts. *Seminars in Developmental Biology* **5**, 381–389.

Godement, P., Salaun, J., & Mason, C.A. (1990). Retinal axons pathfinding in the optic chiasm: Divergence of crossed and uncrossed fibers. *Neuron* **5**, 173–186.

Godement, P., Wang, L.C., & Mason, C.A. (1994). Retinal axon divergence in the optic chiasm: dynamics of growth cone behaviour at the midline. *Journal of Neuroscience* **14**, 7024–7039.

Goldberg, D.J. (1988). Local role of Ca^{++} in formation of veils in growth cones. *Journal of Neuroscience* **8**, 2596–2605.

Goldberg, D.J., & Burmeister, D.W. (1986). Stages in axon formation: Observations of growth of *Aplysia* axons in culture using video-enhanced contrast-differential interference contrast microscopy. *Journal of Cell Biology* **103**, 1921–1931.

Goldberg, D.J., & Burmeister, D.W. (1988). Growth cone movement. *Trends in Neurosciences* **11**, 257–258.

Goldberg, D.J., & Burmeister, D.W. (1989). Looking into growth cones. *Trends in Neurosciences* **12**, 503–506.

Goldstein, M.N., & Pinkel, D. (1957). Long-term tissue culture of neuroblastomas. *Journal of the National Cancer Institute* **20**, 675–689.

Gomez, T.M., & Letourneau, P.C. (1994). Filopodia initiate choices made by sensory neuron growth cones at laminin/fibronectin borders *in vitro*. *Journal of Neuroscience* **14**, 5959–5972.

Gomez, T.M., Snow, D.M., & Letourneau, P.C. (1995). Characterization of spontaneous calcium transients in nerve growth cones and their effect on growth cone migration. *Neuron* **14**, 1233–1246.

Gomez, T.M., Roche, F.K., & Letourneau, P.C. (1996). Chick sensory neuronal growth cones distinguish fibronectin from laminin by making substratum contacts that resemble focal contacts. *Journal of Neuroscience* **29**, 18–34.

Gonzalez, M.D., & Silver, J. (1994). Axon-glia interactions regulate ECM patterning in the postnatal rat olfactory bulb. *Journal of Neuroscience* **14**, 6121–6131.

Gonzalez, M.D., Malemud, C.J., & Silver, J. (1993). Role of astroglial extracellular matrix in the formation of rat olfactory bulb glomeruli. *Experimental Neurology* **123**, 91–105.

Gonzalez-Agosti, C., & Solomon, F. (1996). Response of radixin to perturbations of growth cone morphology and motility in chick sympathetic neurons in vitro. *Cell Motility and the Cytoskeleton* **34**, 122–136.

Goodman, C.S. (1996). Mechanisms and molecules that control growth cone guidance. *Annual Review of Neuroscience* **19**, 341–377.

Goodman, C., & Shatz, C. (1993). Developmental mechanisms that generate precise patterns of neuronal connectivity. *Cell* **72** (suppl.), 77–98.

Goodman, C.S., & Tessier-Lavigne, M. (1997). Molecular mechanisms of axon guidance and target recognition. In *Molecular and Cellular Approaches to Neural Development*, ed. W.M. Cowan, T. Jessell & S.L. Zipursky, pp. 108–178. New York: Oxford University Press.

Goodman, C., Bastiani, M., Doe, C.Q., Lac, S.d., Helfand, S.L., Kuwada, J.Y., & Thomas, J.B. (1984). Cell recognition during neuronal development. *Science* **225**, 1271–1279.

Goodman, C.S., Grenningloh, G., & Bieber, A.J. (1991). Molecular genetics of neural cell adhesion molecules in *Drosophila*. In *The Nerve Growth Cone*, ed. P.C. Letourneau, S.B. Kater & E.R. Macagno, pp. 283–303. New York: Raven Press.

Gordon-Weeks, P.R. (1987a). The cytoskeletons of isolated neuronal growth cones. *Neuroscience* **21**, 977–989.

Gordon-Weeks, P.R. (1987b). Isolation of synaptosomes, growth cones and their subcellular components. In *Neurochemistry, A Practical Approach*, ed. H. Batchelard P. Turner, pp. 1–24. London and Washington: IRL Press.

Gordon-Weeks, P.R. (1988a). The ultrastructure of the neuronal growth cone: New insights from subcellular fractionation and rapid freezing studies. *Electron Microscopy Reviews* 1, 201–219.

Gordon-Weeks, P.R. (1988b). The ultrastructure of noradrenergic and cholinergic neurons in the autonomic nervous system. In *Handbook of Chemical Neuroanatomy*, ed. A. Björklund, T. Hökfelt & C. Owman, pp. 117–142. Amsterdam: Elsevier.

Gordon-Weeks, P.R. (1991). Evidence for microtubule capture by filopodial actin filaments in growth cones. *NeuroReport* 2, 573–576.

Gordon-Weeks, P.R. (1993). Organization of microtubules in axonal growth cones: a role for microtubule-associated protein MAP 1B. *Journal of Neurocytology* 22, 717–725.

Gordon-Weeks, P.R. (1997). MAPs in growth cones. In *Brain Microtubule Associated Proteins: Modifications in Disease*, ed. J. Avila, K.S. Kosik & R. Brandt, pp. 53–72. New York: Harwood Academic Publishers.

Gordon-Weeks, P.R., & Lang, R.D.A. (1988). The α-tubulin of the growth cone is predominantly in the tyrosinated form. *Developmental Brain Research* 42, 156–160.

Gordon-Weeks, P.R., & Lockerbie, R.O. (1984). Isolation and partial characterization of neuronal growth cones from neonatal rat forebrain. *Neuroscience* 13, 119–136.

Gordon-Weeks, P.R., & Mansfield, S.G. (1992). Assembly of microtubules in growth cones: The role of microtubule-associated proteins. In *The Nerve Growth Cone*, ed. S.B. Kater, P.C. Letourneau & E.R. Macagno, pp. 55–64. New York: Raven Press.

Gordon-Weeks, P.R., Lockerbie, R.O., & Pearce, B. (1984). Uptake and release of [^{3}H] GABA by growth cones isolated from neonatal rat forebrain. *Neuroscience Letters* 52, 205–210.

Gordon-Weeks, P.R., Lockerbie, R.O., & Pearce, B. (1985). Growth cones isolated from developing rat forebrain: Uptake and release of GABA and noradrenaline. *Developmental Brain Research* 21, 265–275.

Gordon-Weeks, P.R., Mansfield, S.G., & Curran, I. (1989a). Direct visualisation of the soluble pool of tubulin in the neuronal growth cone: Immunofluorescence studies following taxol polymerisation. *Developmental Brain Research* 49, 305–310.

Gordon-Weeks, P.R., Giffin, N., Weekes, C.S.E., & Barben, C. (1989b). Transient expression of laminin immunoreactivity in the developing rat hippocampus. *Journal of Neurocytology* 18, 451–463.

Gordon-Weeks, P.R., Mansfield, S.G., Alberto, C., Johnstone, M., & Moya, F. (1993). Distribution and expression of a phosphorylation epitope on MAP 1B that is transiently expressed in growing axons in the developing rat nervous system. *European Journal of Neuroscience* 5, 1302–1311.

Gorgels, T.G.M.F. (1991). Outgrowth of the pyramidal tract in the rat cervical spinal cord: Growth cone ultrastructure and guidance. *Journal of Comparative Neurology* 306, 95–116.

Gorgels, T.G.M.F., Oestreicher, A.B., De Kort, E.J.M., & Gispen, W.H. (1987). Immunocytochemical distribution of the protein kinase C substrate B-50 (GAP43) in developing rat pyramidal tract. *Neuroscience Letters* 83, 59–64.

Gorgels, T.G.M.F., Campagne, M.V., Oestreicher, A.B., Gribnau, A.A.M., & Gispen, W.H. (1989). B50/GAP-43 is localized at the cytoplasmic side of the plasma-membrane in developing and adult rat pyramidal tract. *Journal of Neuroscience* 9, 3861–3869.

Goshima, Y., Ohsako, S., & Yamauchi, T. (1993). Overexpression of Ca^{2+}/calmodulin-dependent protein kinase II in Neuro2a and NG108–15 neuroblastoma cell lines promotes neurite outgrowth and growth cone motility. *Journal of Neuroscience* **13**, 559–567.

Goshima, Y., Nakamura, F., Strittmatter, P., & Strittmatter, S.M. (1995). Collapsin-induced growth cone collapse mediated by an intracellular protein related to UNC-33. *Nature* **376**, 509–514.

Goslin, K., & Banker, G. (1989). Experimental observations on the development of polarity by hippocampal neurons in culture. *Journal of Cell Biology* **108**, 1507–1516.

Goslin, K., Schreyer, D.J., Skene, J.H.P., & Banker, G. (1988). Development of neuronal polarity: GAP-43 distinguishes axonal from dendritic growth cones. *Nature* **336**, 672–674.

Goslin, K., Birgbauer, E., Banker, G., & Solomon, F. (1989). The role of cytoskeleton in organizing growth cones: A microfilament-associated growth cone component depends upon microtubules for its localization. *Journal of Cell Biology* **109**, 1621–1631.

Gottman, K., & Lux, H.D. (1990). Low and high voltage activated calcium conductances in electrically excitable growth cones of chick dorsal root gangloin neurones. *Neuroscience Letters* **110**, 34–39.

Götz, M., Scholze, A., Clement, A., Joester, A., Schütte, K., Wigger, F., Frank, R., Spiess, E., Ekblom, P., & Faissner, A. (1996). Tenascin-C contains distinct adhesive, anti-adhesive, and neurite outgrowth promoting sites for neurons. *Journal of Cell Biology* **132**, 681–699.

Grabham, P.W., & Goldberg, D.J. (1997). Nerve growth factor stimulates the accumulation of $\beta 1$ integrin at the tips of filopodia in the growth cones of sympathetic neurons. *Journal of Neuroscience* **17**, 5455–5465.

Grabham, P.W., Wu, D.-Y., & Goldberg, D.J. (1996). Protein tyrosine phosphorylation in the growth cone. In *Nerve Growth and Guidance*, ed. C.D. McCaig, pp. 7–19. London: Portland Press.

Green, P.J., Walsh, F.S., & Doherty, P. (1996). Promiscuity of fibroblast growth factor receptors. *Bioessays* **18**, 639–646.

Greene, L.A., & Tischler, A.S. (1976). Establishment of a noradrenergic clonal cell line of rat adrenal pheochromocytoma cells which respond to nerve growth factor. *Proceedings of the National Academy of Sciences U.S.A.* **73**, 2424–2428.

Greene, L.A., Liem, R.K.H., & Shelanski, M.L. (1983). Regulation of a high-molecular weight microtubule-associated protein in PC12 cells by nerve growth factor. *Journal of Cell Biology* **96**, 76–83.

Grenningloh, G., & Goodman, C. (1992). Pathway recognition by neuronal growth cones: Genetic analysis of neural cell adhesion molecules in *Drosophila*. *Current Opinion in Neurobiology* **2**, 42–47.

Grenningloh, G., Rehm, E.J., & Goodman, C.S. (1991). Genetic analysis of growth cone guidance in *Drosophila*: Fasciclin II functions as a neuronal recognition molecule. *Cell* **67**, 45–57.

Grierson, J.P., Petroski, R.E., Ling, D.S.F., & Geller, H. (1990). Astrocyte topography and tenascin/cytotactin expression: Correlation with the ability to support neuritic outgrowth. *Developmental Brain Research* **55**, 11–19.

Griffin, J.W., Price, D.L., Drachman, D.B., & Morris, J. (1981). Incorporation of axonally transported glycoproteins into axolemma during nerve regeneration. *Journal of Cell Biology* **88**, 205–214.

Grinvald, A., & Farber, I. (1981). Optical recording of Ca^{++} action potentials from growth cones of cultured neurons using a laser microbeam. *Science* **212**, 1164–1169.

Grumet, M., Hoffman, S., Crossin, K.L., & Edelman, G.M. (1985). Cytotactin, an extracellular matrix protein of neural and non-neural tissues that mediates glia-neuron interaction. *Proceedings of the National Academy of Sciences U.S.A.* **82**, 8075–8079.

Grumet, M., Mauro, V., Burgoon, M.P., Edelman, G.M., & Cunningham, B.A. (1991). Structure of a new nervous system glycoprotein, Nr-CAM, and its relationship to subgroups of neural cell adhesion molecules. *Journal of Cell Biology* **113**, 1399–1412.

Grunwald, G.B. (1993). The structural and functional analysis of cadherin calcium-dependent cell adhesion molecules. *Current Opinion in Cell Biology* **5**, 797–805.

Grynkiewicz, G., Poenie, M., & Tsien, R.Y. (1985). A new generation of Ca^{2+} indicators with greatly improved fluorescence properties. *Journal of Biological Chemistry* **260**, 3440–3450.

Gundersen, R.W. (1985). Sensory neurite growth cone guidance by substrate adsorbed nerve growth factor. *Journal of Neuroscience Research* **13**, 199–212.

Gundersen, R.W. (1987). Response of sensory neurites and growth cones to patterned substrata of laminin and fibronectin *in vitro*. *Developmental Biology* **121**, 423–431.

Gundersen, R.W. (1988). Interference reflection microscopic study of dorsal root growth cones on different substrates: Assessment of growth cone-substrate contacts. *Journal of Neuroscience Research* **21**, 298–306.

Gundersen, R.W., & Barrett, J.N. (1979). Neuronal chemotaxis; chick dorsal root axons turn towards high concentrations of nerve growth factor. *Science* **206**, 1079–1080.

Gundersen, R.W., & Barrett, J.N. (1980). Characterization of the turning response of dorsal root neurites toward nerve growth factor. *Journal of Cell Biology* **87**, 546–554.

Gundersen, G.G., Kalnoski, M.H., & Bulinski, J.C. (1984). Distinct populations of microtubules: tyrosinated and nontyrosinated alpha-tubulins are distributed differently *in vivo*. *Cell* **38**, 779–789.

Guthrie, P.B., Lee, R.E., & Kater, S.B. (1989). A comparison of neuronal growth cone and cell body membrane: Electrophysiological and ultrastructural properties. *Journal of Neuroscience* **9**, 3596–3605.

Guthrie, S., & Lumsden, A. (1992). Motor neuron pathfinding following rhombomere reversals in the chick embryo hindbrain. *Development* **114**, 663–673.

Guthrie, S., & Pini, A. (1995). Chemorepulsion of developing motor axons by the floor plate. *Neuron* **14**, 1117–1130.

Halegoua, S. (1987). Changes in the phosphorylation and distribution of vinculin during nerve growth factor induced neurite outgrowth. *Developmental Biology* **121**, 97–104.

Halfter, W. (1993). A heparan sulfate proteoglycan in developing avian axonal tracts. *Journal of Neuroscience* **13**, 2863–2873.

Hall, D.E., Neugebauer, K.M., & Reichardt, L.F. (1987). Embryonic neural retinal cell response to extracellular matrix proteins: Developmental changes and effects of the Cell Substratum Attachment antibody (CSAT). *Journal of Cell Biology* **104**, 623–634.

Hall, S.G., & Bieber, A.J. (1997). Mutations in the *Drosophila* neuroglian cell adhesion molecule affect motor neuron pathfinding and peripheral nervous system patterning. *Journal of Neurobiology* **32**, 325–340.

Hall, Z.W., & Sanes, J.R. (1993). Synaptic structure and development: the neuromuscular junction. *Cell* **10**, 99–121.

Halloran, M.C., & Kalil, K. (1994). Dynamic behaviours of growth cones extending in the corpus callosum of living cortical brain slices observed with video microscopy. *Journal of Neuroscience* **14**, 2161–2177.

Halpern, M.E., Chiba, A., Johansen, J., & Keshishian, H. (1991). Growth cone behaviour underlying the development of stereotypic synaptic connections in *Drosophila* embryos. *Journal of Neuroscience* **11**, 3227–3238.

Hamelin, M., Zhou, Y., Su, M.-W., Scott, I.M., & Culotti, J.G. (1993). Expression of the UNC-5 guidance receptor in the touch neurons of *C. elegans* steers their axons dorsally. *Nature* **364**, 327–330.

Hammarback, J.A., & Letourneau, P.C. (1986). Neurite extension across regions of low cell-substratum adhesivity – implications for the guidepost hypothesis of axonal pathfinding. *Developmental Biology* **117**, 655–662.

Hammarback, J.A., Palm, S.L., Furcht, L.T., & Letourneau, P.C. (1985). Guidance of neurite outgrowth by pathways of substratum-adsorbed laminin. *Journal of Neuroscience Research* **13**, 213–220.

Hammer III, J.A. (1991). Novel myosins. *Trends in Cell Biology* **1**, 50–56.

Hammer III, J.A. (1994). The structure and function of unconventional myosins: A review. *Journal of Muscle Research and Cell Motility* **15**, 1–10.

Hanemaaijer, R., & Ginzburg, I. (1991). Involvement of mature tau isoforms in the stabilization of neurites in PC12 cells. *Journal of Neuroscience Research* **30**, 163–171.

Hantaz-Ambroise, D., & Trautmann, A. (1989). Effects of calcium ion on neurite outgrowth of rat spinal cord neurons *in vitro*: The role of non-neuronal cells in regulating neurite sprouting. *International Journal of Developmental Neuroscience* **7**, 591–602.

Hantaz-Ambroise, D., Vigny, M., & Koenig, J. (1987). Heparan-sulfate proteoglycan and laminin mediate two different types of neurite outgrowth. *Journal of Neuroscience* **7**, 2293–2304.

Harada, A., Oguchi, K., Okabe, S., Kuno, J., Terada, S., Ohshima, T., Sato-Yoshitake, R., Takei, Y., Noda, T., & Hirokawa, N. (1994). Altered microtubule organization in small-calibre axons of mice lacking *tau* protein. *Nature* **369**, 488–491.

Harrelson, A.L., & Goodman, C.S. (1988). Growth cone guidance in insects: Fasciclin II is a member of the immunoglobulin superfamily. *Science* **242**, 700–708.

Harris, R., Sabatelli, L.M., & Seeger, M.A. (1996). Guidance cues at the *Drosophila* CNS midline: Identification and characterization of two *Drosophila* Netrin/UNC-6 homologs. *Neuron* **17**, 217–228.

Harris, W.A. (1986). Homing behaviour of axons in the embryonic vertebrate brain. *Nature* **320**, 266–269.

Harris, W.A. (1989). Local positional cues in the neuroepithelium guide retinal axons in embryonic *Xenopus* brain. *Nature* **339**, 218–221.

Harris, W.A., Holt, C.E., & Bonhoeffer, F. (1987). Retinal axons with and without their somata, growing to and arborizing in the tectum of *Xenopus* embryos: A time-lapse video of single fibres *in vivo*. *Development* **101**, 123–133.

Harrison, R.G. (1907). Observations on the living developing nerve fiber. *Anatomical Record* **1**, 116–118.

Harrison, R.G. (1910). The outgrowth of the nerve fiber as a mode of protoplasmic movement. *Journal of Experimental Zoology* **9**, 787–848.

Harrison, R.G. (1914). The reaction of embryonic cells to solid structures. *Journal of Experimental Zoology* **17**, 521–544.

Harrison, R.G. (1935). On the origin and development of the nervous system studied by the methods of experimental embryology. *Proceedings of the Royal Society of London Series B* **118**, 155–196.

Hashimoto, S. (1988). K-252a, a potent protein kinase inhibitor, blocks nerve growth factor-induced neurite outgrowth and changes in the phosphorylation of proteins in PC12h cells. *Journal of Cell Biology* **107**, 1531–1539.

Hassankhani, A., Steinhelper, M.E., Soonpaa, M.H., Katz, E.B., Taylor, D.A., Andrade-Rozental, A., Factor, S.M., Steinberg, J.J., Field, L.J., & Federoff, H.J. (1995). Overexpression of NGF within the heart of transgenic mice causes hyperinnervation, cardiac enlargement, and hyperplasia of ectopic cells. *Developmental Biology* **169**, 309–321.

Hasty, D.L., & Hay, E.D. (1978). Freeze-fracture studies of the developing cell surface-II. Particle-free membrane blisters on glutaraldehyde-fixed corneal fibroblasts are artifacts. *Journal of Cell Biology* **78**, 756–768.

Hawrot, E. (1980). Cultured synpathetic neurones: Effects of cell-derived and synthetic substrate on survival and development. *Developmental Biology* **74**, 136–151.

Haydon, P.G., & Drapeau, P. (1995). From contact to connection: early events during synaptogenesis. *Trends in the Neurosciences* **18**, 196–201.

Haydon, P.G., & Kater, S.B. (1988). The differential regulation of the formation of chemical and electrical connections in *Helisoma*. *Journal of Neurobiology* **19**, 636–655.

Haydon, P.G., & Zoran, M.J. (1989). Formation and modulation of chemical connections: Evoked acetylcholine release from growth cones and neurites of specific identified neurons. *Neuron* **2**, 1483–1490.

Haydon, P.G., McCobb, D.P., & Kater, S.B. (1984). Serotonin selectively inhibits growth cone motility and the synaptogenesis of specific identified neurons. *Science* **226**, 561–564.

Haydon, P.G., Cohan, C.S., McCobb, D.P., Miller, R.H., & Kater, S.B. (1985). Neuron-specific growth cone properties as seen in identified neurons of *Helisoma*. *Journal of Neuroscience Research* **13**, 135–147.

Hayes, B.P., & Roberts, A. (1973). Synaptic junction development in the spinal cord of an amphibian embryo: An electron microscope study. *Zeitschrift für Zellforschung und Mikroskopische Anatomie Abteilung Histochemie* **137**, 251–269.

Hayes, B.P., & Roberts, A. (1974). The distribution of synapses along the spinal cord of an amphibian embryo: An electron microscope study of junction development. *Cell and Tissue Research* **153**, 227–244.

He, Q., Dent, E.W., & Meiri, K.F. (1997). Modulation of actin filament behaviour by GAP-43 (Neuromodulin) is dependent on the phosphorylation status of serine 41, the protein kinase C site. *Journal of Neuroscience* **17**, 3515–3524.

He, Z., & Tessier-Lavigne, M. (1997). Neuropilin is a receptor for the axonal chemorepellant semaphorin III. *Cell* **90**, 739–751.

Hedgecock, E.M., Culotti, J.G., & Hall, D.H. (1990). The *unc-5*, *unc-6*, and *unc-40* genes guide circumferential migrations of pioneer axons and mesodermal cells on the epidermis in *C. elegans*. *Neuron* **4**, 61–85.

Heffner, C.D., Lumsden, A.G.S., & O'Leary, D.D.M. (1990). Target control of collateral extension and directional axon growth in the mammalian brain. *Science* **247**, 217–220.

Heidemann, S.R., & Buxbaum, R.E. (1994). Mechanical tension as a regulator of axonal development. *NeuroToxicology* **15**, 95–108.

Heidemann, S.R., Landers, J.M., & Hamborg, M.A. (1981). Polarity orientation of axonal microtubules. *Journal of Cell Biology* **91**, 661–665.

Heidemann, S.R., Lamoureux, P., & Buxbaum, R.E. (1990). Growth cone behaviour and production of traction force. *Journal of Cell Biology* **111**, 1949–1957.

Heidemann, S.R., Lamoureux, P., & Buxbaum, R.E. (1991). On the cytomechanics and fluid dynamics of growth cone motility. In *Nerve Cell Biology*, ed. D. Bray, N. Holder, R. Keynes, A. Lumsden & H. Perry, pp. 35–44. Cambridge: The Company of Biologists Ltd.

Helfand, S.L., Smith, G.A., & Wessells, N.K. (1976). Survival and development in culture of dissociated parasympathetic neurons from ciliary ganglia. *Developmental Biology* **50**, 541–547.

Helmke, S., & Pfenninger, K.H. (1995). Growth cone enrichment and cytoskeletal association of non-receptor tyrosine kinases. *Cell Motility and the Cytoskeleton* **30**, 194–207.

Henderson, C.E. (1996). Role of neurotrophic factors in neuronal development. *Current Opinion in Neurobiology* **6**, 64–70.

Henkemeyer, M., Marengere, L.E.M., McGlade, J., Olivier, J.P., Conion, R.A., Holmyard, D.P., Letwin, K., & Pawson, T. (1994). Immunolocalization of the Nuk receptor tyrosine kinase suggests roles in segmental patterning of the brain and axonogenesis. *Oncogene* **9**, 1001–1014.

Henkemeyer, M., Orioli, D., Henderson, J.T., Saxton, T.M., Roder, J., Pawson, T., & Klein, R. (1996). Nuk controls pathfinding of commissural axons in the mammalian central nervous system. *Cell* **86**, 35–46.

Hens, J.H., Benfenati, F., Nielander, H.B., Valtorta, F., Gispen, W.H., & De Graan, P.N.E. (1993). B-50/GAP-43 binds to actin filaments without affecting actin polymerization and filament organization. *Journal of Neurochemistry* **61**, 1530–1533.

Herndon, R.M., Margolis, G., & Kilham, L. (1971). The synaptic organization of malformed cerebellum induced by perinatal infection with feline panleukopenia virus (PLV). *Journal of Neuropathology and Experimental Neurology* **30**, 557–580.

Hess, D.T., Patterson, S.I., Smith, D.S., & Skene, J.H.P. (1993). Neuronal growth cone collapse and inhibition of protein fatty acylation by nitric oxide. *Nature* **366**, 562–565.

Hibbard, E. (1965). Orientation and directed growth of Mauthner's cell axons from duplicated vestibular nerve roots. *Experimental Neurology* **13**, 289–301.

Hinds, J.W., & Hinds, P.L. (1972). Reconstruction of dendritic growth cones in neonatal mouse olfactory bulb. *Journal of Neurocytology* **1**, 169–187.

Hinds, J.W., & Hinds, P.L. (1976a). Synapse formation in the mouse olfactory bulb. I. Quantitative studies. *Journal of Comparative Neurology* **169**, 15–40.

Hinds, J.W., & Hinds, P.L. (1976b). Synapse formation in the mouse olfactory bulb. II. Morphogenesis. *Journal of Comparative Neurology* **169**, 41–62.

Hinkle, L., McCaig, C.D., & Robinson, K.R. (1981). The direction of growth of differentiating neurons and myoblasts from frog embryos in an applied electric field. *Journal of Physiology (London)* **314**, 121–135.

Hinnen, R., & Monard, D. (1980). Involvement of calcium ions in neuroblastoma neurite extension. In *Control Mechanisms in Animal Cells*, ed. L. Jimener de Ausa, pp. 315–323. New York: Raven Press.

Hirano, A., & Dembitzer, H.M. (1973). Cerebellar alterations in the weaver mouse. *Journal of Cell Biology* **56**, 478–486.

Hirano, A., & Dembitzer, H.M. (1974). Observations on the development of the weaver mouse cerebellum. *Journal of Neuropathology and Experimental Neurology* **33**, 354–364.

Hirano, A. & Dembitzer, H.M. (1975). Aberrant development of the Purkinje cell dendritic spine. In *Advances in Neurology*, ed. G.W. Kreutzberg, pp. 353–360. New York: Raven Press.

Hirano, A., Dembitzer, H.M., & Jones, M. (1972). An electron microscopic study of cycasin-induced cerebellar alterations. *Journal of Neuropathology and Experimental Neurology* **31**, 113–125.

Hirokawa, N. (1991). Molecular architecture and dynamics of the neuronal cytoskeleton. In *The Neuronal Cytoskeleton*, ed. R.D. Burgoyne, pp. 5–74. New York: Wiley–Liss, Inc.

Hirokawa, N. (1994). Microtubule organization and dynamics dependent on microtubule-associated proteins. *Current Opinion in Cell Biology* **6**, 74–81.

Hirokawa, N., Terada, S., Funakoshi, T., & Takeda, S. (1997). Slow axonal transport: The subunit transport model. *Trends in Cell Biology* **7**, 384–388.

Hoch, W., Ferns, M., Campanelli, J.T., Hall, Z.W., & Scheller, R.H. (1993). Developmental regulation of highly active alternatively spliced forms of agrin. *Neuron* **11**, 479–490.

Hoffman, S., Sorkin, B.C., White, P.C., Brackenbury, R., & Mailhammer, U. (1982). Chemical characterization of a neural cell adhesion molecule purified from embryonic brain membranes. *Journal of Biological Chemistry* **257**, 7720–7729.

Hoffman, S., Crossin, K.L., & Edelman, G.M. (1988). Molecular forms, binding functions, and developmental expression patterns of cytotactin and cytotactin-binding proteoglycan, an interactive pair of extracellular matrix molecules. *Journal of Cell Biology* **106**, 519–532.

Holash, J.A., & Pasquale, E.B. (1995). Polarized expression of the receptor protein tyrosine kinase Cek5 in the developing avian visual system. *Developmental Biology* **172**, 683–693.

Holland, S.J., Gale, N.W., Mbamalu, G., Yancopoulos, G.D., Henkemeyer, M., & Pawson, T. (1996). Bidirectional signalling through the EPH-family receptor Nuk and its transmembrane ligands. *Nature* **383**, 722–725.

Hollenbeck, P.J. (1989). The transport and assembly of the axonal cytoskeleton. *Journal of Cell Biology* **108**, 223–227.

Hollyday, M. (1983). Development of motor innervation of chick limbs. In *Development and Regeneration*, ed. J. Fallon & A. Calpan, pp. 183–193. New York: Alan R. Liss.

Holt, C.E. (1989). A single-cell analysis of early retinal ganglion cell differentiation in *Xenopus*: From soma to axon tip. *Journal of Neuroscience* **9**, 3123–3145.

Holt, C.E., & Harris, W.A. (1983). Order in the initial retinotectal map in *Xenopus:* A new technique for labelling growing nerve fibers. *Nature* **301**, 150–152.

Honig, M.G., & Burden, S.M. (1993). Growth cones respond in diverse ways upon encountering neurites in cultures of chick dorsal root ganglia. *Developmental Biology* **156**, 454–472.

Honig, M.G., & Kueter, J. (1995). The expression of cell adhesion molecules on the growth cones of chick cutaneous and muscle sensory neurons. *Developmental Biology* **167**, 563–583.

Honig, M.G., & Rutishauser, U.S. (1996). Changes in the segmental pattern of sensory neuron projections in the chick hindlimb under conditions of altered cell adhesion molecule function. *Developmental Biology* **175**, 325–337.

Hopkins, J., Ford-Holevinski, T., McCoy, J., & Agranoff, B. (1985). Laminin and optic nerve regeneration in the goldfish. *Journal of Neuroscience* **5**, 3030–3038.

Horstkorte, R., Schachner, M., Magyar, J.P., Vorherr, T., & Schmitz, B. (1993). The fourth immunoglobulin-like domain of NCAM contains a carbohydrate recognition domain for oligomannosidic glycans implicated in association with L1 and neurite outgrowth. *Journal of Cell Biology* **121**, 1409–1421.

Hortsch, M. (1996). The L1 family of neural cell adhesion molecules: Old proteins performing new tricks. *Neuron* **17**, 587–593.

Horwitz, S.B. (1992). Mechanism of action of taxol. *Trends in Pharmacological Sciences* **13**, 134–136.

Hotary, K.B., & Robinson, K.R. (1992). Evidence of a role for endogenous electrical fields in chick embryo development. *Development* **114**, 985–996.

Hoyle, G.W., Mercer, E.H., Palmiter, R.D., & Brinster, R.L. (1993). Expression of NGF in sympathetic neurons leads to excessive axon outgrowth from ganglia but decreased terminal innervation within tissues. *Neuron* **10**, 1019–1034.

Hughes, A. (1953). The growth of embryonic neurites. A study on cultures of chick neural tissues. *Journal of Anatomy* **87**, 150–162.

Hughes, A. (1968). *Aspects of Neural Ontogeny*. New York: Academic Press.

Hume, R.I., Role, L.W., & Fischbach, G.D. (1983). Acetylcholine release from growth cones detected with patches of acetylcholine receptor rich membranes. *Nature* **305**, 632–634.

Hunt, R.K. & Cowan, W.M. (1990). The chemoaffinity hypothesis: An appreciation of Roger W. Sperry's contributions to developmental biology. In *Brain Circuits and Functions of the Mind*, ed. C. Trevarthen, pp. 19–74. Cambridge: Cambridge University Press.

Hunter, D.D., Shah, V., Merlie, J.P., & Sanes, J.R. (1989a). A laminin-like adhesive protein concentrated in the synaptic cleft of the neuromuscular junction. *Nature* **338**, 229–234.

Hunter, D.D., Porter, B.E., Bulock, J.W., Adams, S.P., Merlie, J.P., & Sanes, J.R. (1989b). Primary sequence of the motor neuron-selective adhesive site in the synaptic basal lamina protein S-laminin. *Cell* **59**, 905–913.

Hunter, D.D., Cashman, N., Morris-Valero, R., Bulock, J.W., Adams, S.P., & Sanes, J.R. (1991). An LRE (leucine–arginine–glutamate)-dependent mechanism for adhesion to S-laminin. *Journal of Neuroscience* **11**, 3960–3971.

Hunter, D.D., Llinas, R., Ard, M., Merlie, J.P., & Sanes, J.R. (1992). Expression of s-laminin and laminin in the developing rat central nervous system. *Journal of Comparative Neurology* **323**, 238–251.

Husmann, K., Faissner, A., & Schachner, M. (1992). Tenascin promotes cerebellar granule cell migration and neurite outgrowth by different domains in the fibronectin type III repeats. *Journal of Cell Biology* **116**, 1475–1486.

Hynes, R.O. (1992). Integrins: versatility, modulation, and signaling in cell adhesion. *Cell* **69**, 11–25.

Hynes, R.O., & Lander, A.D. (1992). Contact and adhesive specificities in the associations, migrations, and targeting of cells and axons. *Cell* **68**, 303–322.

Igarashi, M., & Komiya, Y. (1991). Subtypes of protein kinase C in isolated nerve growth cones: only type II is associated with the membrane skeleton from growth cones. *Biochemical and Biophysical Research Communications* **178**, 751–757.

Igarashi, M., Strittmatter, S.M., Vartanian, T., & Fishman, M.C. (1993). Mediation by G proteins of signals that cause collapse of growth cones. *Science* **259**, 77–79.

Igarashi, M., Kozaki, S., Terakawa, S., & Komiya, Y. (1995). Presynaptic proteins in growth cones. *Cell Structure and Function* **20**, 566

Igarashi, M., Kozaki, S., Terakawa, S., Kawano, S., Ide, C., & Komiya, Y. (1996). Growth cone collapse and inhibition of neurite growth induced by *botulinum* neurotoxin C1: a *t*-SNARE is involved in axonal growth. *Journal of Cell Biology* **134**, 205–215.

Igarashi, M., Tagaya, M., & Komiya, Y. (1997). The soluble *N*-ethylmaleimide-sensitive factor attached protein receptor complex in growth cones: Molecular aspects of the axon terminal development. *Journal of Neuroscience* **17**, 1460–1470.

Ignelzi, M.A., Jr., Miller, D.R., Soriano, P., & Maness, P.F. (1994). Impaired neurite outgrowth of src-minus cerebellar neurons on the cell adhesion molecule L1. *Neuron* **12**, 873–884.

Ingvar, S. (1920). Reactions of cells to galvanic current in tissue cultures. *Proceedings of the American Society for Experimental Biology and Medicine* **17**, 198–199.

Inouye, A., & Sanes, J.R. (1997). Lamina-specific connectivity in the brain: Regulation by N-cadherin, neurotrophins, and glycoconguates. *Science* **276**, 1428–1431.

Isenberg, G., & Small, J.V. (1978). Filamentous actin, 100 Å filaments and microtubules in neuroblastoma cells. Their distribution in relation to sites of movement and neuronal transport. *Cytobiologie* **16**, 326–344.

Isenberg, G., Rieske, E., & Kreutzberg, G.W. (1977). Distribution of actin and tubulin in neuroblastoma cells. *European Journal of Cell Biology* **15**, 382–389.

Ishii, N., Wadsworth, W.G., Stern, B.D., Culotti, J.G., & Hedgecock, E.M. (1992). UNC-6, a laminin-related protein, guides cell and pioneer axon migrations in *C. elegans*. *Neuron* **9**, 873–881.

Ivins, J.K., & Pittman, R.N. (1992). Cellular and molecular influences on neurite outgrowth. In *Development, Regeneration and Plasticity of the Autonomic Nervous System*, ed. I.A. Hendry & C.E. Hill, pp. 139–178. Switzerland: Harwood Academic Publishers.

Ivins, J.K., Raper, J.A., & Pittman, R.N. (1991). Intracellular calcium levels do not change during contact mediated collapse of chick DRG growth cone structure. *Journal of Neuroscience* **11**, 1597–1608.

Iwai, Y., Usui, T., Hirano, S., Steward, R., Takeichi, M., & Uemura, T. (1997). Axon patterning requires DN-cadherin, a novel neuronal adhesion receptor, in the *Drosophila* embryonic CNS. *Neuron* **19**, 77–89.

Jacobs, J.R., & Goodman, C.S. (1989). Embryonic development of axon pathways in the *Drosophila* CNS: II. Behaviour of pioneer growth cones. *Journal of Neuroscience* **9**, 2412–2424.

Jacobs, J.R., & Stevens, J.K. (1986). Changes in the organization of the neuritic cytoskeleton during NGF-activated differentiation of PC12 cells: A serial electron-microscopic study of the development and control of neurite shape. *Journal of Cell Biology* **103**, 895–906.

Jacobson, K., & Wojcieszyn, J. (1984). The translational mobility of substances within the cytoplasmic matrix. *Proceedings of the National Academy of Sciences U.S.A.* **81**, 6747–6751.

Jacobson, M. (1991). *Developmental Neurobiology*, 3rd edn. New York: Plenum Press.

Jacobson, M., & Huang, S. (1985). Neurite outgrowth traced by means of horseradish peroxidase inherited from neuronal ancestral cells in frog embryos. *Developmental Biology* **110**, 102–113.

Jacobson, R.D., Virág, I., & Skene, J.H.P. (1986). A protein associated with axon growth, GAP-43, is widely distributed and developmentally regulated in rat CNS. *Journal of Neuroscience* **6**, 1843–1855.

Jaffe, L.F., & Nuccitelli, R. (1977). Electrical controls of development. *Annual Review of Biophysics* **6**, 445–476.

Jaffe, L.F., & Poo, M.-M. (1979). Neurites grow faster towards the cathode than the anode in a steady field. *Journal of Experimental Zoology* **209**, 115–128.

Jalink, K., & Moolenaar, W.H. (1992). Thrombin receptor activation causes rapid neural cell rounding and neurite retraction independent of classic second messengers. *Journal of Cell Biology* **118**, 411–419.

Jalink, K., Eichholtz, T., Postma, F.R., van Corven, E.J., & Moolenaar, W.H. (1993). Lysophosphatidic acid induces neuronal shape changes via a novel, receptor-mediated signaling pathway: Similarity to thrombin action. *Cell Growth and Differentiation* **4**, 247–255.

Jalink, K., Vancorven, E.J., Hengeveld, T., Morii, N., Narumiya, S., & Moolenaar, W.H. (1994). Inhibition of lysophosphatidate-induced and thrombin-induced neurite retraction and neuronal cell rounding by ADP-ribosylation of the small GTP-binding protein Rho. *Journal of Cell Biology* **126**, 801–810.

James, D.W., & Tresman, R.L. (1969). An electron microscopic study of the de novo formation of neuromuscular junctions in tissue culture. *Zeitschrift für Zellforschung und Mikroskopische Anatomie Abteilung Histochemie* **100**, 126–140.

Jap Tjoen San, E.R.A., Schmidt-Michels, M., Oestreicher, A.B., Gispen, W.H., & Schotman, P. (1992). Inhibition of nerve growth factor-induced B-50/GAP-43 expression by antisense oligomers interferes with neurite outgrowth of PC12 cells. *Biochemical and Biophysical Research Communications* **187**, 839–846.

Jay, D.G. (1988). Selective destruction of protein function by chromophore-assisted laser inactivation. *Proceedings of the National Academy of Sciences U.S.A.* **85**, 5454–5458.

Jay, D.G. (1996). Molecular mechanisms of directed growth cone motility. *Perspectives on Developmental Neurobiology* **4**, 137–145.

Jay, D.G., & Keshishian, H. (1990). Laser inactivation of fasciclin I disrupts axon adhesion of grasshopper pioneer neurons. *Nature* **348**, 548–550.

Jaye, M., Schlessinger, J., & Dionne, C.A. (1992). Fibroblast growth factor receptor tyrosine kinases: Molecular analysis and signal transduction. *Biochimica et Biophysica Acta* **1135**, 185–199.

Jessell, T.M., Siegel, R.E., & Fischbach, G.D. (1979). Induction of acetylcholine receptors on cultured skeletal muscle by a factor extracted from brain and spinal cord. *Proceedings of the National Academy of Sciences U.S.A.* **76**, 5397–5401.

Jian, X., Hidaka, H., & Schmidt, J.T. (1994). Kinase requirement for retinal growth cone motility. *Journal of Neurobiology* **25**, 1310–1328.

Jin, Z., & Strittmatter, S.M. (1997). Rac1 mediates collapsin-1 induced growth cone collapse. *Journal of Neuroscience* **17**, 6256–6263.

Jockusch, H., & Jockusch, B.M. (1981). Structural proteins in the growth cone of cultured spinal cord neurons. *Experimental Cell Research* **131**, 345–352.

Jockusch, H., Jockusch, B.M., & Burger, M.M. (1979). Nerve fibers and their interactions with non-neural cells visualized in immunofluorescence. *Journal of Cell Biology* **80**, 629–641.

Johnson, M.H., Maro, B., & Takeichi, M. (1986). The role of cell adhesion in the synchronization and orientation of polarization in 8–cell mouse blastomeres. *Journal of Embryology and Experimental Morphology* **93**, 239–255.

Johnson, R., & Armstrong-James, M. (1970). Morphology of superficial postnatal cerebral cortex with special references to synapses. *Zeitschrift für Zellforschung und mikroskopische Anatomie* **110**, 540–558.

Johnston, A.R., & Gooday, D.J. (1991). *Xenopus* temporal retinal neurites collapse on contact with glial cells from caudal tectum *in vitro*. *Development* **113**, 409–417.

Johnston, R.N., & Wessells, N.K. (1980). Regulation of the elongation of nerve fibers. *Current Topics in Developmental Biology* **16**, 165–206.

Jones, F.S., Hoffman, S., Cunningham, B.A., & Edelman, G.M. (1989). A detailed model of cytotactin: Protein homologies, alternative RNA splicing, and binding regions. *Proceedings of the National Academy of Sciences U.S.A.* **86**, 1905–1909.

Jones, G., Meier, T., Lichtsteiner, M., Witzemann, V., Sakmann, B., & Brenner, H.R. (1997). Induction by agrin of ectopic and functional postsynaptic-like membrane in innervated muscle. *Proceedings of the National Academy of Sciences U.S.A.* **94**, 2654–2659.

Jordan, M.A., & Wilson, L. (1990). Kinetic analysis of tubulin exchange at microtubule ends at low vinblastine concentrations. *Biochemistry* **29**, 2730–2739.

Joshi, H.C. (1994). Microtubule organizing centers and gamma-tubulin. *Current Opinion in Cell Biology* **6**, 55–62.

Joshi, H.C., & Baas, P.W. (1993). A new perspective on microtubules and axon growth. *Journal of Cell Biology* **121**, 1191–1196.

Joshi, H.C., Chu, D., Buxbaum, R.E., & Heidemann, S.R. (1985). Tension and compression in the cytoskeleton of PC 12 neurites. *Journal of Cell Biology* **101**, 697–705.

Joshi, H.C., Baas, P., Chu, D.T., & Heidemann, S.R. (1986). The cytoskeleton of neurites after microtubule depolymerization. *Experimental Cell Research* **163**, 233–245.

Joshi, H.C., Palacios, M.J., McNamara, L., & Cleveland, D.W. (1992). γ-Tubulin is a centrosomal protein required for cell cycled-dependent microtubule nucleation. *Nature* **356**, 80–83.

Juliano, R.L., & Haskill, S. (1993). Signal transduction from the extracellular matrix. *Journal of Cell Biology* **120**, 577–585.

Juraska, J.M., & Fifková, E. (1979). An electron microscope study of the early postnatal development of the visual cortex of the hooded rat. *Journal of Comparative Neurology* **183**, 257–268.

Kadmon, G., Kowitz, A., Altevogt, P., & Schachner, M. (1990a). The neural cell adhesion molecule N-CAM enhances L1-dependent cell–cell interactions. *Journal of Cell Biology* **110**, 193–208.

Kadmon, G., Kowitz, A., Altevogt, P., & Schachner, M. (1990b). Functional cooperation between the neural cell adhesion molecules L1 and N-CAM is carbohydrate dependent. *Journal of Cell Biology* **110**, 209–218.

Kaethner, R.J., & Stuermer, A.O. (1992). Dynamics of terminal arbor formation and target approach of retinotectal axons in living zebrafish embryos: A time-lapse study of single axons. *Journal of Neuroscience* **12**, 3257–3271.

Kalil, K., & Skene, J.H.P. (1986). Elevated synthesis of an axonally transported protein correlates with axon outgrowth in normal and injured pyramidal tracts. *Journal of Neuroscience* **6**, 2563–2570.

Kamiguchi, H., & Lemmon, V. (1997). Neural cell adhesion molecule L1: Signaling pathways and growth cone motility. *Journal of Neuroscience Research* **49**, 1–8.

Kanai, Y., Takemura, R., Oshima, T., Mori, H., Ihara, Y., Yanagisawa, M., Masaki, T., & Hirokawa, N. (1989). Expression of multiple tau isoforms and microtubule bundle formation in fibroblasts transfected with a single tau cDNA. *Journal of Cell Biology* **109**, 1173–1184.

Kapfhammer, J.P., & Raper, J.A. (1987a). Collapse of growth cone structure on contact with specific neurites in culture. *Journal of Neuroscience* **7**, 201–212.

Kapfhammer, J.P., & Raper, J.A. (1987b). Interactions between growth cones and neurites growing from different neural tissues in culture. *Journal of Neuroscience* **7**, 1595–1600.

Kapfhammer, J.P., Grunewald, B.E., & Raper, J.A. (1986). The selective inhibition of growth cone extension by specific neurites in culture. *Journal of Neuroscience* **6**, 2527–2534.

Kater, S.B., & Mills, L.R. (1991). Regulation of growth cone behaviour by calcium. *Journal of Neuroscience* **11**, 891–899.

Kater, S.B., Mattson, M.P., Cohan, C.S., & Connor, J. (1988). Calcium regulation of the neuronal growth cone. *Trends in Neurosciences* **11**, 315–321.

Katz, F., Ellis, L., & Pfenninger, K.H. (1985). Nerve growth cone isolated from fetal rat brain. III. Calcium-dependent protein phosphorylation. *Journal of Neuroscience* **5**, 1402–1411.

Katz, L.C., & Schatz, C.J. (1996). Synaptic activity and the construction of cortical circuits. *Science* **274**, 1133–1138.

Kawakami, A., Kitsukawa, T., Takagi, S., & Fujisawa, H. (1996). Developmentally regulated expression of a cell surface protein, neuropilin, in the mouse nervous system. *Journal of Neurobiology* **29**, 1–17.

Kawana, E., Sandri, C., & Akert, K. (1971). Ultrastructure of growth cones in the cerebellar cortex of the neonatal rat and cat. *Zeitschrift für Zellforschung* **115**, 284–298.

Kawasaki, Y., Horie, H., & Takenaka, T. (1986). The maturation-dependent change in fibronectin receptor density of mouse dorsal root ganglion neurons. *Brain Research* **397**, 185–188.

Keating, T.J., Peloquin, J.G., Rodionov, V.I., Momcilovic, D., & Borisy, G.G. (1997). Microtubule release from the centrosome. *Proceedings of the National Academy of Sciences U.S.A.* **94**, 5078–5083.

Keino-Masu, K., Masu, M., Hinck, L., Leonardo, E.D., Chan, S.S., Culotti, J.G., & Tessier-Lavigne, M. (1996). Deleted in Colorectal Cancer (DCC) encodes a netrin receptor. *Cell* **87**, 175–185.

Keith, C. (1987). Slow transport of tubulin in the neurites of differentiated PC12 cells. *Journal of Cell Biology* **54**, 1258–1268.

Keith, C.H. (1990). Neurite elongation is blocked if microtubule polymerization is inhibited in PC12 cells. *Cell Motility and the Cytoskeleton* **17**, 95–105.

Keith, C.H., & Farmer, M.A. (1993). Microtubule behaviour in PC12 neurites: Variable results obtained with photobleach technology. *Cell Motility and the Cytoskeleton* **25**, 345–357.

Kempf, M., Clement, A., Faissner, A., Lee, G., & Brandt, R. (1996). Tau binds to the distal axon early in development of polarity in a microtubule- and microfilament-dependent manner. *Journal of Neuroscience* **16**, 5583–5592.

Kennedy, M.B. (1997). The postsynaptic density at glutamatergic synapses. *Trends in the Neurosciences* **6**, 264–268.

Kennedy, T.E., Serafini, T., de la Torre, J.R., & Tessier-Lavigne, M. (1994). Netrins are diffusible chemotropic factors for commissural axons in the embryonic spinal cord. *Cell* **78**, 425–435.

Keshishian, H., & Chiba, A. (1993). Neuromuscular development in *Drosophila*: Insights from single neurons and single genes. *Trends in the Neurosciences* **16**, 278–283.

Keshishian, H., Chiba, A., Chang, T.N., Halfon, M.S., Harkins, E.W., Jarecki, J., Wang, L., Anderson, M., Cash, S., Halpern, M.E., & Johansen, J. (1993). Cellular mechanisms governing synaptic development in *Drosophila melanogaster*. *Journal of Neurobiology* **24**, 757–787.

Keshishian, H., Broadie, K., Chiba, A., & Bate, M. (1996). The *Drosophila* neuromuscular junction: A model system for studying synaptic development and function. *Annual Review of Neuroscience* **19**, 545–575.

Keynes, R.J., & Stern, C.D. (1984). Segmentation in the vertebrate nervous system. *Nature* **310**, 786–789.

Keynes, R.J., & Stern, C.D. (1988). Mechanisms of vertebrate segmentation. *Development* **103**, 413–429.

Keynes, R.J., Tannahill, D., Morgenstern, D.A., Johnson, A.R., Cook, G.M.W., & Pini, A. (1997). Surround repulsion of spinal sensory axons in higher vertebrate embryos. *Neuron* **18**, 889–897.

Khawaja, S., Gundersen, G.F., & Bulinski, J.C. (1988). Enhanced stability of microtubules enriched in detyrosinated alpha-tubulin is not a direct function of detyrosination level. *Journal of Cell Biology* **106**, 141–149.

Kilpatrick, T.J., Brown, A., Lai, C., Gassman, M., Goulding, M., & Lemke, G. (1996). Expression of the *Tyro4/Mek4/Cek4* gene specifically marks a subset of embryonic motor neurons and their muscle targets. *Molecular and Cellular Neuroscience* **7**, 62–74.

Kim, G.J., Shatz, C.J., & McConnell, S.K. (1991). Morphology of pioneer and follower growth cones in the developing cerebral cortex. *Journal of Neurobiology* **22**, 629–642.

Kim, Y.-T., & Wu, C.-F. (1991). Distinctions in growth cone morphology and motility between monopolar and multipolar neurons in *Drosophila* CNS cultures. *Journal of Neurobiology* **22**, 263–275.

Kindt, R.M., & Lander, A.D. (1995). Pertussis toxin specifically inhibits growth cone guidance by a mechanism independent of direct G protein inactivation. *Neuron* **15**, 79–88.

Kira, M., Tanaka, J., & Sobue, K. (1995). Caldesmon and low Mr isoform of tropomyosin are localized in neuronal growth cones. *Journal of Neuroscience Research* **40**, 294–305.

Kirschner, M., & Mitchison, T. (1986a). Beyond self-assembly: From microtubules to morphogenesis. *Cell* **45**, 329–342.

Kirschner, M.W., & Mitchison, T. (1986b). Microtubule dynamics. *Nature* **324**, 621.

Kitsukawa, T., Shimono, A., Kawakami, A., Kondoh, H., & Fujisawa, H. (1995). Overexpression of a membrane protein, neuropilin, in chimeric mice causes anomalies in the cardiovascular system, nervous system and limbs. *Development* **121**, 4309–4318.

Kitsukawa, T., Shimizu, M., Sanbo, M., Hirata, T., Taniguchi, M., Bekku, Y., Yagi, T., & Fujisawa, H. (1997). Neuropilin-semaphorin III/D-mediated chemorepulsion signals play a crucial role in peripheral nerve projection in mice. *Neuron* **19**, 995–1005.

Klämbt, C., Jacobs, J.R., & Goodman, C.S. (1991). The midline of the *Drosophila* central nervous system: A model for the genetic analysis of cell fate, cell migration, and growth cone guidance. *Cell* **64**, 801–815.

Klar, A., Baldassare, M., & Jessell, T.M. (1992). F-spondin: A gene expressed at high levels in the floor plate encodes a secreted protein that promotes neural cell adhesion and neurite extension. *Cell* **69**, 95–110.

Kleitman, N., & Johnson, M.I. (1989). Rapid growth cone translocation on laminin is supported by lamellipodial not filopodial structures. *Cell Motility and the Cytoskeleton* **13**, 288–300.

Kleitman, N., Simon, D., Schachner, M., & Bunge, R. (1988). Growth of embryonic retinal neurites elicited by contact with Schwann cell surfaces is blocked by antibodies to L1. *Experimental Neurology* **102**, 298–306.

Klinz, S.G., Schachner, M., & Maness, P.F. (1995). L1 and NCAM antibodies trigger protein phosphatase activity in growth cone-enriched membranes. *Journal of Neurochemistry* **65**, 84–95.

Klose, M., & Bentley, D. (1989). Transient pioneer neurons are essential for formation of an embryonic peripheral nerve. *Science* **245**, 982–983.

Knops, J., Kosik, K.S., Lee, G., Pardee, J.D., Cohen, G.L., & McConlogue, L. (1991). Overexpression of tau in a non-neuronal cell induces long cellular processes. *Journal of Cell Biology* **114**, 725–732.

Knusel, B., & Hefti, F. (1992). K-252 compounds: modulation of neurotrophin signal transduction. *Journal of Neurochemistry* **59**, 1987–1996.

Knusel, B., Kaplan, D.R., Winslow, J.W., Rosenthal, A., Burton, L.E., Beck, K.D., Rabin, S., Nikolics, K., & Hefti, F. (1992). K-252b selectively potentiates cellular actions and trk tyrosine phosphorylation mediated by neurotrophin-3. *Journal of Neurochemistry* **59**, 715–722.

Kobayashi, H., Watanabe, E., & Murakami, F. (1995). Growth cones of dorsal root ganglion but not retina collapse and avoid oligodendrocytes in culture. *Developmental Biology* **168**, 383–394.

Koch, C.A., Anderson, D., Moran, M.F., Ellis, C., & Pawson, T. (1991). SH2 and SH3 domains: Elements that control interactions of cytoplasmic signaling proteins. *Science* **252**, 668–674.

Koda, L.Y., & Partlow, L.M. (1976). Membrane marker movement on sympathetic axons in tissue culture. *Journal of Neurobiology* **7**, 157–172.

Koenig, E., Kinsman, S., Repasky, E., & Sultz, L. (1985). Rapid mobility of motile varicosities and inclusions containing spectrin, actin and calmodulin in regenerating axons *in vitro*. *Journal of Neuroscience* **5**, 715–729.

Koike, T. (1983). Nerve factor induced neurite outgrowth of rat pheochromocytoma PC12 cells: Dependence upon extracellular Mg^{2+} and Ca^{2+}. *Brain Research* **289**, 293–303.

Kolodkin, A.L. (1996). Semaphorins: Mediators of repulsive growth cone guidance. *Trends in Cell Biology* **6**, 15–22.

Kolodkin, A.L., & Ginty, D.D. (1997). Steering clear of semaphorins: Neuropilins sound the retreat. *Neuron* **19**, 1159–1162.

Kolodkin, A.L., Matthes, D.J., O'Connor, T.P., Patel, N.H., Admon, A., Bentley, D., & Goodman, C.S. (1992). Fasciclin IV: Sequence, expression, and function during growth cone guidance in the grasshopper embryo. *Neuron* **9**, 831–845.

Kolodkin, A.L., Matthes, D.J., & Goodman, C.S. (1993). The *semaphorin* genes encode a family of transmembrane and secreted growth cone guidance molecules. *Cell* **75**, 1389–1399.

Kolodkin, A.L., Levengood, D.V., Rowe, E.G., Tai, Y.-T., Giger, R.J., & Ginty, D.D. (1997). Neuropilin is a semaphorin receptor. *Cell* **90**, 753–762.

Kolodziej, P.A., Timpe, L.C., Mitchell, K.J., Fried, S.R., Goodman, C.S., Jan, L.Y., & Jan, Y.N. (1996). Frazzled encodes a *Drosophila* member of the DCC immunoglobulin subfamily and is required for CNS and motor axon guidance. *Cell* **87**, 197–204.

König, N., Roch, G., & Marty, R. (1975). The onset of synaptogenesis in rat temporal cortex. *Anatomy and Embryology* **148**, 73–87.

Kosik, K.S., & Finch, E.A. (1987). MAP2 and tau segregate into axonal and dendritic domains after the elaboration of morphologically distinct neurites: An immunocytochemical study of cultured rat cerebrum. *Journal of Neuroscience* **7**, 3142–3153.

Kozma, R., Sarner, S., Ahmed, S., & Lim, L. (1997). Rho family GTPases and neuronal growth cone remodelling: Relationships between increased complexity induced by Cdc42, Rac1 and acetylcholine and collapse induced by RhoA and lysophosphatidic acid. *Molecular and Cellular Biology* **17**, 1201–1211.

Krayanek, S., & Goldberg, S. (1981). Oriented extracellular channels and axonal guidance in the embryonic chick retina. *Developmental Biology* **84**, 41–50.

Kreis, T. & Vale, R. (1993). *Guidebook to the Cytoskeletal and Motor Proteins*. Oxford, Oxford University Press, pp. 1–276.

Kreis, T.E. (1986). Microtubules containing detyrosinated tubulin are less dynamic. *EMBO Journal* **6**, 2597–2606.

Kristt, D.A., & Molliver, M.E. (1976). Synapses in newborn rat cerebral cortex. A quantitative ultrastructural study. *Brain Research* **108**, 180–186.

Kröger, S., Horton, S.E., & Honig, L.S. (1996). The developing avian retina expresses agrin isoforms during synaptogenesis. *Journal of Neurobiology* **29**, 165–182.

Kruger, K., Tam, A.S., Lu, C., & Sretavan, D.W. (1998). Retinal ganglion cell axon progression from the optic chiasm to initiate optic tract development requires cell autonomous function of GAP-43. *Journal of Neuroscience* **18**, 5692–5705.

Kruse, J., Kielhauer, G., Faissner, A., Timpl, R., & Schachner, M. (1985). The J1 glycoprotein – a novel nervous system cell adhesion molecule of the L2/HNK-1 family. *Nature* **316**, 146–148.

Krystosek, A., & Seeds, N.W. (1981a). Plasminogen activator release at the neuronal growth cone. *Science* **213**, 1532–1534.

Krystosek, A., & Seeds, N.W. (1981b). Plasminogen-activator secretion by granule neurons in cultures of developing cerebellum. *Proceedings of the National Academy of Sciences U.S.A.* **78**, 7810–7814.

Krystosek, A., & Seeds, N.W. (1984). Peripheral neurons and Schwann cells secrete plasminogen activator. *Journal of Cell Biology* **98**, 773–776.

Kuczmarski, E.R., & Rosenbaum, J.L. (1979). Studies on the organization and localization of actin and myosin in neurons. *Journal of Cell Biology* **80**, 356–371.

Kuhn, L.T., Stoeckli, E.T., Confrau, M.A., Rathjen, F.G., & Sonderegger, P. (1991). Neurite outgrowth on immobilized axonin-1 is mediated by a heterophilic interaction with L1 (G4). *Journal of Cell Biology* **115**, 1113–1126.

Kuhn, T.B., Schmidt, M.F., & Kater, S.B. (1995). Laminin and fibronectin guideposts signal sustained but opposite effects to passing growth cones. *Neuron* **14**, 275–285.

Kuwada, J.Y. (1986). Cell recognition by neuronal growth cones in a simple vertebrate embryo. *Science* **233**, 740–746.

Kuwada, J.Y., Bernhardt, R.R., & Chitnis, A.B. (1990). Pathfinding by identified growth cones in the spinal cord of zebrafish embryos. *Journal of Neuroscience* **10**, 1299–1308.

Laessing, U., Giordano, S., Lottspeich, F., & Stuermer, C.A.O. (1994). Molecular characterization of fish neurolin: A growth associated cell surface protein and

member of the immunoglobulin-superfamily with similarities to chick DM-Grasp/
SC-1/Ben. *Differentiation* **56**, 21–29.

Lagenaur, C., & Lemmon, V. (1987). An L1–like molecule, the 8D9 antigen, is a potent
substrate for neurite extension. *Proceedings of the National Academy of Sciences
U.S.A.* **84**, 7753–7757.

Lamoureux, P., Buxbaum, R.E., & Heidemann, S.R. (1989). Direct evidence that
growth cones pull. *Nature* **340**, 159–162.

Lamoureux, P., Zheng, R.E., Buxbaum, R.E., & Heidemann, S.R. (1992). A cytome-
chanical investigation of neurite growth on different culture surfaces. *Journal of
Cell Biology* **118**, 655–661.

Lamoureux, P., Altun-Glutekin, Z.F., Lin, C., Wagner, J.A., & Heidemann, S.R.
(1997). Rac is required for growth cone function but not neurite assembly.
Journal of Cell Science **110**, 635–641.

Lampert, P.W. (1967). A comparative electron microscopic study of reactive, degener-
ating, regenerating, and dystrophic axons. *Journal of Neuropathology and
Experimental Neurology* **26**, 345–370.

Lance-Jones, C., & Landmesser, L. (1980). Motoneuron projection patterns in the chick
hindlimb following early partial reversals of the spinal cord. *Journal of Physiology
(London)* **302**, 581–602.

Lance-Jones, C., & Landmesser, L. (1981a). Pathway selection by chick lumbrosacral
motoneurons during normal development. *Proceedings of the Royal Society of
London Series B* **214**, 1–18.

Lance-Jones, C., & Landmesser, L. (1981b). Pathway selection of embryonic chick
motoneurons in an experimentally altered environment. *Proceedings of the Royal
Society of London Series B* **214**, 119–152.

Lander, A.D. (1993). Proteoglycans in the nervous system. *Current Opinion in
Neurobiology* **3**, 716–723.

Lander, A.D., Tomaselli, K.J., Calof, A.L., & Reichardt, L.F. (1983). Studies on the
extracellular matrix components that promote neurite outgrowth. *Cold Spring
Harbor Symposium of Quantitative Biology* **48**, 611–624.

Landis, D.M.D., & Reese, T.S. (1977). Structure of the Purkinje cell membrane in
staggerer and weaver mutant mice. *Journal of Comparative Neurology* **171**, 247–260.

Landis, S.C. (1978). Growth cones of cultured sympathetic neurons contain adrenergic
vesicles. *Journal of Cell Biology* **78**, R8–R14

Landis, S.C. (1983). Neuronal growth cones. *Annual Review of Physiology* **45**, 567–580.

Landmesser, L. (1984). The development of specific motor pathways in the chick
embryo. *Trends in Neurosciences* **7**, 336–339.

Landmesser, L.T. (1991). Growth cone guidance in the avian limb: A search for cellular
and molecular mechanisms. In *The Nerve Growth Cone*, ed. P.C. Letourneau, S.B.
Kater & E.R. Macagno, pp. 373–385. New York: Raven Press.

Landmesser, L., Dahm, L., Tang, J.C., & Rutishauser, U. (1990). Polysialic acid as a
regulator of intramuscular nerve branching during embryonic development.
Neuron **4**, 655–667.

Langley, J.N. (1895). Note on the regeneration of pre-ganglionic fibres of the sympathetic. *Journal of Physiology (London)* **18**, 280–284.

Langley, J.N. (1897). On the regeneration of pre-ganglionic and of post-ganglionic
visceral nerve fibres. *Journal of Physiology (London)* **22**, 215–230.

Lankford, K.L., & Letourneau, P.C. (1989). Evidence that calcium may control neurite
outgrowth by regulating the stability of actin filaments. *Journal of Cell Biology* **109**,
1229–1243.

Lankford, K.L., & Letourneau, P.C. (1991). Roles of actin filaments and three second-messenger systems in short-term regulation of chick dorsal root ganglion neurite outgrowth. *Cell Motility and the Cytoskeleton* **20**, 7–29.

Lasek, R.J. (1982). Translocation of the axonal cytoskeleton and axonal locomotion. *Philosophical Transactions of the Royal Society of London Series B* **299**, 313–327.

Lasek, R.J. (1988). Studying the intrinsic determinants of neuronal form and function. In *Intrinsic Determinants of Neuronal Form and Function*, ed. R.J. Lasek & M.M. Black, pp. 1–58. New York: Alan R. Liss, Inc.

Lauderdale, J.D., Davis, N.M., & Kuwada, J.Y. (1997). Axon tracts correlate with netrin-1a expression in the zebrafish embryo. *Molecular and Cellular Neuroscience* **9**, 293–313.

Laurie, D.J., Wisden, W., & Seeberg, P.H. (1992). The distribution of thirteen $GABA_A$ receptor subunit mRNAs in the rat brain. III. Embryonic and postnatal development. *Journal of Neuroscience* **12**, 4151–4172.

LeClerc, N., Baas, P.W., Garner, C.C., & Kosik, K.S. (1996). Juvenile and mature MAP2 isoforms induce distinct patterns of process outgrowth. *Molecular Biology of the Cell* **7**, 443–455.

Lee, G. (1993). Non-motor microtubule-associated proteins. *Current Opinion in Cell Biology* **5**, 88–94.

Lee, G., & Brandt, R. (1992). Microtubule-bundling studies revisited: Is there a role for MAPs? *Trends in Cell Biology* **2**, 286–289.

Lee, G., & Rook, S.L. (1992). Expression of tau protein in non-neuronal cells: Microtubule bundling and stabilization. *Journal of Cell Science* **102**, 227–237.

Lee, M.K., & Cleveland, D.W. (1994). Neurofilament function and dysfunction: Involvement in axonal growth and neuronal disease. *Current Opinion in Cell Biology* **6**, 34–40.

Lefcourt, F., & Bentley, D. (1989). Organization of cytoskeletal elements and organelles preceding growth cone emergence from an identified neuron *in situ*. *Journal of Cell Biology* **108**, 1737–1749.

Lemmon, V., Farr, K.L., & Lagenaur, C. (1989). L1-mediated axon outgrowth occurs via a homophilic binding mechanism. *Neuron* **2**, 1597–1603.

Lemmon, V., Burden, S.M., Payne, H.R., Elmslie, G.J., & Hlavin, M.L. (1992). Neurite growth on different substrates: Permissive versus instructive influences and the role of adhesive strength. *Journal of Neuroscience* **12**, 818–826.

Lentz, T.L. (1967). Fine structure of nerves in the regenerating limb of the newt *Triturus*. *American Journal of Anatomy* **121**, 647–670.

Leonardo, E.D., Hinck, L., Masu, M., Keino-Masu, K., Ackerman, S.L., & Tessier-Lavigne, M. (1997). Vertebrate homologs of *C. elegans* UNC-5 are candidate netrin receptors. *Nature* **386**, 833–838.

Letourneau, P. (1996). The cytoskeleton in nerve growth cone motility and axonal pathfinding. *Perspectives on Developmental Neurobiology* **4**, 111–123.

Letourneau, P.C. (1975a). Possible roles for cell, substrate adhesion in neuronal morphogenesis. *Developmental Biology* **44**, 77–91.

Letourneau, P.C. (1975b). Cell-to-substratum adhesion and guidance of axonal elongation. *Developmental Biology* **44**, 92–101.

Letourneau, P.C. (1978). Chemotactic response of nerve fiber elongation to nerve growth factor. *Developmental Biology* **66**, 183–196.

Letourneau, P.C. (1979). Cell–substratum adhesion of neurite growth cones, and its role in neurite elongation. *Experimental Cell Research* **124**, 127–138.

Letourneau, P.C. (1981). Immunocytochemical evidence for co-localization in neurite growth cones of actin and myosin and their relationship to cell substratum adhesions. *Developmental Biology* **85**, 113–122.

Letourneau, P.C. (1983a). Differences in organization of actin in the growth cones compared with the neurites of cultured neurons from chick embryos. *Journal of Cell Biology* **97**, 963–973.

Letourneau, P.C. (1983b). Axonal growth and guidance. *Trends in Neurosciences* **6**, 451–456.

Letourneau, P.C., & Cypher, C. (1991). Regulation of growth cone motility. *Cell Motility and the Cytoskeleton* **20**, 267–271.

Letourneau, P.C., & Ressler, A.H. (1984). Inhibition of neurite initiation and growth by taxol. *Journal of Cell Biology* **98**, 1355–1362.

Letourneau, P.C., & Shattuck, T.A. (1989). Distribution and possible interactions of actin-associated proteins and cell adhesion molecules of nerve growth cones. *Development* **105**, 505–519.

Letourneau, P.C., & Wessells, N.K. (1974). Migratory cell locomotion versus nerve axon elongation. *Journal of Cell Biology* **61**, 56–69.

Letourneau, P.C., Shattuck, T.A., & Ressler, A.H. (1986). Branching of sensory and sympathetic neurites is inhibited by treatment with taxol. *Journal of Neuroscience* **6**, 1912–1917.

Letourneau, P.C., Shattuck, T.A., & Ressler, A.H. (1987). 'Pull' and 'push' in neurite elongation: Observations on the effects of different concentrations of cytochalasin B and taxol. *Cell Motility and the Cytoskeleton* **8**, 193–209.

Letourneau, P.C., Madsen, A.M., Palm, S.L., & Furcht, L.T. (1988). Immunoreactivity for laminin in the developing ventral longitudinal pathway of the brain. *Developmental Biology* **125**, 135–144.

Letourneau, P.C., Shattuck, T.A., Roche, F.K., Takeichi, M., & Lemmon, V. (1990). Nerve growth cone migration onto Schwann cells involves the calcium-dependent adhesion molecule, N-cadherin. *Developmental Biology* **138**, 430–442.

Letourneau, P.C., Condic, M.L., & Snow, D.M. (1994). Interactions of developing neurons with the extracellular matrix. *Journal of Neuroscience* **14**, 915–928.

Leung-Hagesteijn, C., Spence, A.M., Stern, B.D., Zhou, Y., Su, M.-W., Hedgecock, E.M., & Culotti, J.G. (1992). UNC-5, a transmembrane protein with immunoglobulin and thrombospondin type 1 domains, guides cell and pioneer axon migrations in *C. elegans*. *Cell* **71**, 289–299.

Levi, G. (1917). Connessioni e struttura degli elementi nervosi sviluppati fuori dell'organismo. *Memorie-Accademia Nazionale dei Lincei* **12**, 142–182.

Levi, G. (1926). Ricerche sperimentali sovra elementi nervosi sviluppati *in vitro*. *Archiv für experimentelle Zellforschung* **2**, 244–272.

Levi, G. (1934). Explantation, besonders die Struktur und die biologischen Eigenschaften der in vitro gezüchteten Zellen und Gewebe. *Ergebnisse der Anatomie und Entwicklungsgeschichte* **31**, 125–707.

Levi, G., & Meyer, H. (1945). Reactive, regressive and regenerative processes of neurons cultivated *in vitro* and injured by micromanipulator. *Journal of Experimental Zoology* **99**, 141–181.

Levi-Montalcini, R., & Angeletti, P.U. (1963). Essential role of the nerve growth factor in the survival and maintenance of dissociated sensory and sympathetic nerve cells *in vitro*. *Developmental Biology* **7**, 653–659.

Levi-Montalcini, R., & Angeletti, P.U. (1968). Nerve growth factor. *Physiology Reviews* **48**, 534–569.

Levy, B.T., Sorge, L.K., Meymandi, A., & Maness, P.F. (1984). pp60^{c-src} kinase is in chick and human embryonic tissue. *Developmental Biology* **104**, 9–17.

Lewis, A.K., & Bridgman, P.C. (1992). Nerve growth cone lamellipodia contain two populations of actin filaments that differ in organization and polarity. *Journal of Cell Biology* **119**, 1219–1243.

Lewis, A.K., & Bridgman, P.C. (1996). Mammalian myosin Iα is concentrated near the plasma membrane in nerve growth cones. *Cell Motility and the Cytoskeleton* **33**, 130–150.

Lewis, J., Chevallier, A., Kieny, M., & Wolpert, L. (1981). Muscle nerve branches do not develop in chick wings devoid of muscle. *Journal of Embryology and Experimental Morphology* **64**, 211–232.

Lewis, W.H. (1950). Motion picture of neurons and neuroglia in tissue culture. In *Genetic Neurology*, ed. P. Weiss, pp. 1–21. Chicago: University of Chicago Press.

Lewis, W.H., & Lewis, M.R. (1912). The cultivation of sympathetic nerves from the intestine of chick embryos in saline solutions. *Anatomical Record* **6**, 7–31.

Li, M., Shibata, A., Li, C.M., Braun, P.E., McKerracher, L., Roder, J., Kater, S.B., & David, S. (1996). Myelin-associated glycoprotein inhibits neurite/axon growth and causes growth cone collapse. *Journal of Neuroscience Research* **46**, 404–414.

Li, W., Herman, R.K., & Shaw, J.E. (1992). Analysis of the *Caenorhabditis elegans* axonal guidance and outgrowth gene unc-33. *Genetics* **132**, 675–689.

Liesi, P. (1985). Do neurons in the vertebrate CNS migrate on laminin? *EMBO Journal* **4**, 1163–1170.

Liesi, P., Dahl, D., & Vaheri, A. (1984). Neurons cultured from developing rat brain attach and spread preferentially to laminin. *Journal of Neuroscience Research* **11**, 241–251.

Lilienbaum, A., Reszka, A.A., Horwitz, A.F., & Holt, C.E. (1995). Chimeric integrins expressed in retinal ganglion cells impair process outgrowth in vivo. *Molecular and Cellular Neuroscience* **6**, 139–152.

Lim, S., Sammak, P., & Borisy, G. (1989). Progressive and spatially differentiated stability of microtubules in developing neuronal cells. *Journal of Cell Biology* **109**, 253–263.

Lim, S., Edson, K., Letourneau, P., & Borisy, G. (1990). A test of microtubule translocation during neurite elongation. *Journal of Cell Biology* **111**, 123–130.

Lin, C.-H., & Forscher, P. (1993). Cytoskeletal remodelling during growth cone–target interactions. *Journal of Cell Biology* **121**, 1369–1383.

Lin, C.-H., & Forscher, P. (1995). Growth cone advance is inversely proportional to retrograde F-actin flow. *Neuron* **14**, 763–771.

Lin, C.-H., Thompson, C.A., & Forscher, P. (1994a). Cytoskeletal reorganization underlying growth cone motility. *Current Opinion in Neurobiology* **4**, 640–647.

Lin, C.-H., Espreafico, E.M., Mooseker, M.S., & Forscher, P. (1996). Myosin drives retrograde F-actin flow in neuronal growth cones. *Neuron* **16**, 769–782.

Lin, D.M., & Goodman, C.S. (1994). Ectopic and increased expression of Fasciclin II alters motoneuron growth cone guidance. *Neuron* **13**, 507–523.

Lin, D.M., Fetter, R.D., Kopczynski, C., Grenningloh, G., & Goodman, C.S. (1994b). Genetic analysis of Fasciclin II in *Drosophila:* Defasciculation, refasciculation, and altered fasciculation. *Neuron* **13**, 1055–1069.

Lin, D.M., Auld, V.J., & Goodman, C.S. (1995). Targeted neuronal cell ablation in the *Drosophila* embryo: Pathfinding by follower growth cones in the absence of pioneers. *Neuron* **14**, 707–715.

Lipscombe, D., Madison, D.V., Poenie, M., Reuter, H., Tsien, R.Y., & Tsien, R.W. (1988). Spatial distribution of calcium channels and cytosolic calcium transients in growth cones and cell bodies of sympathetic neurons. *Proceedings of the National Academy of Sciences U.S.A.* **85**, 2398–2402.

Lipton, S.A. (1987). Bursting of calcium-activated cation-selective channels is associated with neurite regeneration in mammalian central neuron. *Neuroscience Letters* **82**, 21–28.

Liu, L., Haines, S., Shew, R., & Akeson, R.A. (1993). Axon growth is enhanced by NCAM lacking the VASE exon when expressed in either the growth substrate or the growing axon. *Journal of Neuroscience Research* **35**, 327–345.

Liu, Y., & Storm, D.R. (1989). Dephosphorylation of neuromodulin by calcineurin. *Journal of Biological Chemistry* **264**, 12800–12804.

Liuzzi, F.J., & Lasek, R.J. (1987). Astrocytes block axonal regeneration in mammals by activating the physiological stop pathway. *Science* **237**, 642–645.

Livesey, F.J., & Hunt, S.P. (1997). Netrin and netrin receptor expression in the embryonic mammalian nervous system suggests roles in retinal, striatal, nigral and cerebellar development. *Molecular and Cellular Neuroscience* **8**, 417–429.

Llinás, R., Hillman, D.E., & Precht, W. (1973). Neuronal circuit reorganization in mammalian agranular cerebellar cortex. *Journal of Neurobiology* **4**, 69–94.

Lochter, A., Vaughan, L., Kaplony, A., Prochianz, A., Schachner, M., & Faissner, A. (1991). J1/tenascin in substrate-bound and soluble form displays contrary effects on neurite outgrowth. *Journal of Cell Biology* **113**, 1159–1171.

Lockerbie, R.O. (1987). The neuronal growth cone: A review of its locomotional, navigational and target recognition capabilities. *Neuroscience* **20**, 719–729.

Lockerbie, R.O. (1990). Biochemical pharmacology of isolated neuronal growth cones: implications for synaptogenesis. *Brain Research Reviews* **15**, 145–165.

Lockerbie, R.O., & Gordon-Weeks, P.R. (1985). γ-Aminobutyric acid-A (GABA$_A$) receptors modulate [^{3}H] GABA release from isolated neuronal growth cones. *Neuroscience Letters* **55**, 273–277.

Lockerbie, R.O., & Gordon-Weeks, P.R. (1986). Further characterization of [^{3}H] gamma-aminobutyric acid release from isolated neuronal growth cones: Role of intracellular Ca^{2+} stores. *Neuroscience* **17**, 1257–1266.

Lockerbie, R.O., Gordon-Weeks, P.R., & Pearce, B. (1985). Growth cones isolated from developing rat forebrain: Uptake and release of GABA and noradrenaline. *Developmental Brain Research* **21**, 265–275.

Lockerbie, R.O., Edde, B., & Prochiantz, A. (1989). Cyclic AMP-dependent protein phosphorylation in isolated neuronal growth cones from developing rat forebrain. *Journal of Neurochemistry* **52**, 786–796.

Lockerbie, R.O., Miller, V.E., & Pfenninger, K.H. (1991). Regulated plasmalemmal expansion in growth cones. *Journal of Cell Biology* **112**, 1215–1227.

Lohof, A. M., Quillan, M., Dan, Y., & Poo, M.-M. (1992). Asymmetric modulation of cytosolic cAMP activity induces growth cone turning. *Journal of Neuroscience* **12**, 1253–1261.

Lohse, K., Helmke, S.M., Wood, M.R., Quiroga, S., Delahoussaye, B.A., Miller, V.E., Negreaminou, P., & Pfenninger, K.H. (1996). Axonal origin and purity of growth cones isolated from fetal rat brain. *Developmental Brain Research* **96**, 83–96.

Lømo, T., & Slater, C.R. (1980). Control of junctional acetycholinesterase by neural and muscular influences in the rat. *Journal of Physiology* **303**, 191–202.

LoPresti, V., Macagno, E.R., & Levinthal, C. (1973). Structure and development of neuronal connection in isogenic organisms: Cellular interactions in the development of the optic lamina of *Daphnia*. *Proceedings of the National Academy of Sciences U.S.A.* **70**, 433–437.

Luckenbill-Edds, L. (1997). Laminin and mechanism of neuronal outgrowth. *Brain Research Reviews* **23**, 1–27.

Luduena, M.A. (1973). The growth of spinal ganglion neurones in serum-free medium. *Developmental Biology* **33**, 470–476.

Luis de Blas, A. (1993). GABA$_A$/benzodiazepine receptors in the developing mammalian brain. In *Receptors in the Developing Nervous System*, ed. I.S. Zagon & P.J. McLaughlin, pp. 105–126. London: Chapman & Hall.

Lumsden, A.G.S., & Davies, A.M. (1983). Earliest sensory nerve fibres are guided to their peripheral targets by attractants other than nerve growth factor. *Nature* **306**, 786–788.

Lumsden, A.G.S., & Davies, A.M. (1986). Chemotropic effect of specific target epithelium in the developing mammalian nervous system. *Nature* **323**, 538–539.

Lumsden, C.E. (1951). Aspects of neurite outgrowth in tissue culture. *Anatomical Record* **110**, 145–168.

Luna, E.J., & Hitt, A.L. (1992). Cytoskeleton–plasma membrane interactions. *Science* **258**, 955–964.

Luo, L., Liao, Y.J., Jan, L.Y., & Jan, Y.N. (1994). Distinct morphogenetic functions of similar small GTPases: *Drosophila* Drac1 is involved in axonal outgrowth and myoblast function. *Genes and Development* **8**, 1787–1802.

Luo, L., Hensch, T.K., Ackerman, L., Barbel, S., Jan, L.Y., & Jan, Y.N. (1996). Differential effects of the Rac GTPase on Purkinje cell axons and dendritic trunks and spines. *Nature* **379**, 837–840.

Luo, L., Jan, L.Y., & Jan, Y.-N. (1997). Rho family small GTP-binding proteins in growth cone signalling. *Current Opinion in Neurobiology* **7**, 81–86.

Luo, Y., & Raper, J.A. (1994). Inhibitory factors controlling growth cone motility and guidance. *Current Opinion in Neurobiology* **4**, 648–654.

Luo, Y., Raible, D., & Raper, J.A. (1993). Collapsin: A protein in brain that induces the collapse and paralysis of neuronal growth cones. *Cell* **75**, 217–227.

Luo, Y., Shepherd, I., Li, J., Renzi, M.J., Chang, S., & Raper, J.A. (1995). A family of molecules related to collapsin in the embryonic chick nervous system. *Neuron* **14**, 1131–1140.

Lyons, W.E., George, E.B., Dawson, T.M., Steiner, J.P., & Snyder, S.H. (1994). Immunosuppressant FK506 promotes neurite outgrowth in cultures of PC-12 cells and sensory ganglia. *Proceedings of the National Academy of Sciences U.S.A.* **91**, 3191–3195.

Macagno, E.R. (1978). Mechanism for the formation of synaptic projections in the arthropod visual system. *Nature* **275**, 318–320.

MacDonald, R., Scholes, J., Strähle, U., Brennan, C., Holder, N., Brand, M., & Wilson, S.W. (1997). The Pax protein Noi is required for commissural axon pathway formation in the rostral forebrain. *Development* **124**, 2397–2408.

Maciver, S.K. (1996). Myosin II function in non-muscle cells. *Bioessays* **18**, 179–182.

Mackay, D.J.G., Nobes, C.D., & Hall, A. (1995). The Rho progress – a potential role during neuritogenesis for the Rho family of GTPases. *Trends in Neurosciences* **18**, 496–501.

MacLennan, A.J., McLaurin, D.L., Marks, L., Vinson, E.N., Pfeifer, M., Szulc, S.V., Heaton, M.B., & Lee, N. (1997). Immunohistochemical localization of netrin-1 in the embryonic chick nervous system. *Journal of Neuroscience* **17**, 5466–5479.

Magill-Solc, C., & McMahan, U.J. (1988). Motor neurons contain agrin-like molecules. *Journal of Cell Biology* 107, 1825–1833.

Mandell, J.W., & Banker, G.A. (1996). A spatial gradient of tau protein phosphorylation in nascent axons. *Journal of Neuroscience* 16, 5727–5740.

Maness, P.F., Sorge, L.K., & Fults, D.W. (1986). An early developmental phase of pp60^{c-src} expression in neural ectoderm. *Developmental Biology* 117, 83–89.

Maness, P.F., Aubry, M., Shores, C.G., Frame, L., & Pfenninger, K.H. (1988). C-src gene product in developing rat brain is enriched in nerve growth cone membranes. *Proceedings of the National Academy of Sciences U.S.A.* 85, 5001–5005.

Maness, P.F., Shores, C.G., & Ignelzi, M. (1990). Localization of the normal cellular src protein to the growth cone of differentiating neurons in brain and retina. *Advances in Experimental Medicine and Biology* 265, 117–125.

Maness, P.F., Beggs, H.E., Klinz, S.G., & Morse, W.R. (1996). Selective neural cell adhesion molecule signaling by *Src* family tyrosine kinases and tyrosine phosphatases. *Perspectives on Developmental Neurobiology* 4, 169–181.

Mansfield, S.G., & Gordon-Weeks, P.R. (1991). Post-translational modification of tubulin in rat cerebral cortical neurons extending neurites in culture: Effects of taxol. *Journal of Neurocytology* 20, 654–666.

Mansfield, S.G., Diaz-Nido, J., Gordon-Weeks, P.R., & Avila, J. (1991). The distribution and phosphorylation of the microtubule-associated protein MAP 1B in growth cones. *Journal of Neurocytology* 21, 1007–1022.

Manthorpe, M., Engvall, E., Ruoslahti, E., Longo, F.M., Davis, G.E., & Varon, S. (1983). Laminin promotes neuritic regeneration from cultured peripheral and central neurons. *Journal of Cell Biology* 97, 1882–1890.

Marcus, R.C., Gale, N.W., Morrison, M.E., Mason, C.A., & Yancopoulos, G.D. (1996). Eph family receptors and their ligands distribute in opposing gradients in the developing mouse retina. *Developmental Biology* 180, 786–789.

Margolis, R.K., Rauch, U., Maurel, P., & Margolis, R.U. (1996). Neurocan and phosphacan: Two major nervous system-specific chondroitin sulfate proteoglycans. *Perspectives on Developmental Neurobiology* 3, 273–290.

Margolis, R.U., & Margolis, R.K. (1997). Chondroitin sulfate proteoglycans as mediators of axon growth and pathfinding. *Cell and Tissue Research* 290, 343–348.

Markham, J.A., & Fifková, E. (1986). Actin filament organization within dendrites and dendritic spines during development. *Developmental Brain Research* 27, 263–269.

Marks, A.R. (1996). Cellular functions of immunophilins. *Physiological Reviews* 76, 631–649.

Marsh, G., & Beams, H.W. (1946a). Orientation of chick nerve fibers by direct electric currents. *Anatomical Record* 94, 370

Marsh, G., & Beams, H.W. (1946b). *In vitro* control of growing chick nerve fibers by applied electric fields. *Journal of Cell and Comparative Physiology* 27, 139–157.

Marsh, L., & Letourneau, P.C. (1984). Growth of neurites without filopodial or lamellipodial activity in the presence of cytochalasin B. *Journal of Cell Biology* 99, 2041–2047.

Martenson, C., Stone, K., Reedy, M., & Sheetz, M. (1993). Fast axonal transport is required for growth cone advance. *Nature* 366, 66–69.

Martin, P., Khan, A., & Lewis, J. (1989). Cutaneous nerves of the embryonic chick wing do not develop in regions denuded of ectoderm. *Development* 106, 335–346.

Martini, R. (1994). Expression and functional roles of neural cell surface molecules and extracellular matrix components during development and regeneration of peripheral nerves. *Journal of Neurocytology* 23, 1–28.

Martini, R., & Schachner, M. (1988). Immunoelectron microscopic localization of neural cell-adhesion molecules (L1, N-CAM, and MAG) and their shared carbohydrate epitope and myelin basic-protein in developing sciatic-nerve. *Journal of Cell Biology* **103**, 2439–2448.

Masiakowski, P., & Carroll, R.D. (1992). A novel family of cell surface receptors with tyrosine kinase-like domain. *Journal of Biological Chemistry* **267**, 26181–26190.

Mason, C.A. (1982). Development of terminal arbors of retinogeniculate axons in the kitten. I. Light microscopical observations. *Neuroscience* **7**, 541–560.

Mason, C.A. (1985a). Growing tips of embryonic cerebellar axons *in vivo*. *Journal of Neuroscience Research* **13**, 55–73.

Mason, C.A. (1985b). How do growth cones grow? *Trends in Neurosciences* **8**, 304–306.

Mason, C.A. (1986). Axon development in mouse cerebellum: Embryonic axon forms and expression of synapsin I. *Neuroscience* **19**, 1319–1333.

Mason, C.A., & Wang, L.C. (1997). Growth cone form is behaviour-specific and, consequently, position-specific along the retinal axon pathway. *Journal of Neuroscience* **17**, 1086–1100.

Matsumoto, T. (1920). The granules, vacuoles and mitochondria in the sympathetic nerve-fibers cultivated *in vitro*. *Bulletin of The Johns Hopkins Hospital* **31**, 91–93.

Matsunaga, M., Hatta, K., Nagafuchi, A., & Takeichi, M. (1988). Guidance of optic nerve fibres by N-cadherin adhesion molecules. *Nature* **334**, 62–64.

Matten, W.T., Aubry, M., West, J., & Maness, P.F. (1990). Tubulin is phosphorylated at serine by pp60^{c-src} in nerve growth cone membranes. *Journal of Cell Biology* **111**, 1959–1970.

Matthes, D.J., Sink, H., Kolodkin, A.L., & Goodman, C.S. (1995). Semaphorin II can function as a selective inhibitor of specific synaptic arborizations. *Cell* **81**, 631–639.

Mattson, M.P., & Kater, S.B. (1987). Calcium regulation of neurite elongation and growth cone motility. *Journal of Neuroscience* **7**, 4034–4043.

Mattson, M.P., & Kater, S.B. (1989). Excitatory and inhibitory neurotransmitters in the generation and degeneration of hippocampal neuroarchitecture. *Brain Research* **478**, 337–348.

Mattson, M.P., Dou, P., & Kater, S.B. (1988a). Outgrowth-regulating actions of glutamate in isolated hippocampal pyramidal neurons. *Journal of Neuroscience* **8**, 2087–2100.

Mattson, M.P., Guthrie, P.B., & Kater, S.B. (1988b). Intracellular messengers in the generation and degeneration of hippocampal neuroarchitecture. *Journal of Neuroscience Research* **21**, 447–464.

Mattson, M.P., Guthrie, P.B., & Kater, S.B. (1988c). Components of neurite outgrowth which determine neuronal architecture: influence of calcium and growth substrate. *Journal of Neuroscience Research* **20**, 331–345.

Mattson, M.P., Lee, R.E., Adams, M.E., Guthrie, P.B., & Kater, S.B. (1988d). Interactions between entorhinal axons and target hippocampal neurons: A role for glutamate in the development of hippocampal circuitry. *Neuron* **1**, 865–876.

Mattson, M.P., Taylor-Hunter, A., & Kater, S.B. (1988e). Neurite outgrowth in individual neurons of a neuronal population is differentially regulated by calcium and cyclic AMP. *Journal of Neuroscience* **8**, 1704–1711.

Mattson, M.P., Guthrie, P.B., & Kater, S.B. (1989). A role for Na$^+$-dependent calcium extrusion in protection against neuronal excitotoxicity. *FASEB Journal* **3**, 2519–2526.

Matus, A. (1988). Microtubule-associated proteins: Their potential role in determining neuronal morphology. *Annual Reviews of Neuroscience* **11**, 29–44.

Matus, A., Bernhardt, R., Bodmer, R., & Alaimo, D. (1986). Microtubule-associated protein 2 and tubulin are differentially distributed in the dendrites of developing neurons. *Neuroscience* 17, 371–389.

Maurel, P., Rauch, U., Flad, M., Margolis, R.K., & Margolis, R.U. (1994). Phosphacan, a chondroitin sulphate proteoglycan of brain that interacts with neurons and neural cell adhesion molecules, is an extracellular variant of a receptor-type protein tyrosine phosphatase. *Proceedings of the National Academy of Sciences U.S.A.* 91, 2512–2516.

McBurney, R.N., & Neering, I.R. (1987). Neuronal calcium homeostasis. *Trends in Neurosciences* 10, 164–169.

McCaig, C.D. (1986). Dynamic aspects of amphibian neurite growth and the effects of an applied electric field. *Journal of Physiology (London)* 375, 55–69.

McCaig, C.D. (1987). Spinal neurite reabsorption and regrowth *in vitro* depend on the polarity of an applied electric field. *Development* 100, 31–41.

McCaig, C.D. (1988). Nerve guidance: A role for bio-electric fields? *Progress in Neurobiology* 30, 449–468.

McCaig, C.D. (1989). Nerve growth in the absence of growth cone filopodia and the effects of a small applied electric field. *Journal of Cell Science* 93, 723–730.

McCaig, C.D. (1990). Nerve branching is induced and oriented by a small applied electric field. *Journal of Cell Science* 95, 605–616.

McCaig, C.D., & Erskine, L. (1996). Nerve growth and nerve guidance in a physiological electric field. In *Nerve Growth and Guidance*, ed. C.D. McCaig, pp. 151–170. London: Portland Press.

McCaig, C.D., & Zhao, M. (1997). Physiological electric fields modify cell behaviour. *Bioessays* 19, 819–825.

McCaig, C.D., Allan, D.W., Erskine, L., Rajnicek, A.M. & Stewart, R. (1994). Growing nerves in an electric field. In *Axon Growth and Guidance in Vitro*, ed. P.R. Gordon-Weeks, pp. 134–141. New York: Academic Press.

McCobb, D.P., Cohan, C.S., Connor, J.A., & Kater, S.B. (1988). Interactive effects of serotonin and acetylcholine on neurite elongation. *Neuron* 1, 377–385.

McConnell, S.K. (1992). The determination of neuronal identity in the mammalian cerebral cortex. In *Determinants of Neuronal Identity*, ed. M. Shankland & E.R. Macagno, pp. 391–432. San Diego, California: Academic Press Inc.

McConnell, S.K., Ghosh, A., & Shatz, C.J. (1989). Subplate neurons pioneer the pathway from cortex to thalamus. *Science* 245, 978–982.

McFarlane, S., & Holt, C.E. (1997). Growth factors: a role in guiding axons? *Trends in Cell Biology* 7, 424–430.

McFarlane, S., McNeill, L., & Holt, C.E. (1995). FGF signaling and target recognition in the developing *Xenopus* visual system. *Neuron* 15, 1017–1028.

McFarlane, S., Cornel, E., Amaya, E., & Holt, C.E. (1996). Inhibition of FGF receptor activity in retinal ganglion cell axons causes errors in target recognition. *Neuron* 17, 245–254.

McGraw, C.F., & McLaughlin, B.J. (1980). Fine structural studies of synaptogenesis in the superficial layers of the chick optic tectum. *Journal of Neurocytology* 9, 79–93.

McGuire, C.B., Snipes, G.J., & Norden, J.J. (1988). Light-microscopic immunolocalization of the growth- and plasticity-associated protein GAP-43 in the developing rat brain. *Developmental Brain Research* 41, 277–291.

McIntire, S.L., Garriga, G., White, J., Jacobson, D., & Horvitz, H.R. (1992). Genes necessary for directed axonal elongation or fasciculation in *C. elegans. Neuron* 8, 307–322.

McKenna, M.P., & Raper, J.A. (1988). Growth cone behaviour on gradients of substratum-bound laminin. *Developmental Biology* 130, 232–236.

McKerracher, L., Chamoux, M., & Arregui, C.O. (1996). Role of laminin and integrin interactions in growth cone guidance. *Molecular Neurobiology* 12, 95–116.

McLoon, S.C., McLoon, L.K., Palm, S.L., & Furcht, L.T. (1988). Transient expression of laminin in the optic nerve of the developing rat. *Journal of Neuroscience* 8, 1981–1990.

McMahan, U.J. (1990). The agrin hypothesis. *Cold Spring Harbor Symposium of Quantitative Biology* 55, 407–418.

Meier, T., Perez, G.M., & Wallace, B.G. (1995). Immobilization of nicotinic acetylcholine receptors in mouse C2 myotubes by agrin-induced protein tyrosine phosphorylation. *Journal of Cell Biology* 131, 441–451.

Meier, T., Hauser, D.M., Chiquet, M., Landmann, L., Ruegg, M.A., & Brenner, H.R. (1997). Neural agrin induces ectopic postsynaptic specializations in innervated muscle fibers. *Journal of Neuroscience* 17, 6534–6544.

Meima, L., Moran, P., Matthews, W., & Caras, I.W. (1997a). Lerk2 (Ephrin-B1) is a collapsing factor for a subset of cortical growth cones and acts by a mechanism different from AL-1 (Ephrin-A5). *Molecular and Cellular Neuroscience* 9, 314–328.

Meima, L., Kljavin, I., Moran, P., Shih, A., Winslow, J.W., & Caras, I.W. (1997b). AL-1–induced growth cone collapse of rat cortical neurons is correlated with REK7 expression and rearrangement of the actin cytoskeleton. *European Journal of Neuroscience* 9, 177–188.

Meiners, S., Powell, E.M., & Geller, H. (1995). A distinct subset of tenascin CS-6–PG-rich astrocytes restricts neuronal growth in vitro. *Journal of Neuroscience* 15, 8096–8108.

Meiri, K.F., & Burdick, D. (1991). Nerve growth factor stimulation of GAP-43 phosphorylation in intact isolated growth cones. *Journal of Neuroscience* 11, 3155–3164.

Meiri, K.F., & Gordon-Weeks, P.R. (1990). GAP-43 in growth cones is associated with areas of membrane that are tightly bound to substrate and is a component of a membrane skeleton subcellular fraction. *Journal of Neuroscience* 10, 256–266.

Meiri, K.F., Pfenninger, K.H., & Willard, M. (1986). Growth-associated protein, GAP-43, a polypeptide that is induced when neurons extend axons, is a component of growth cones and corresponds to pp46, a major polypeptide of a subcellular fraction enriched in growth cones. *Proceedings of the National Academy of Sciences U.S.A.*, 83, 3537–3541.

Meiri, K.F., Bickerstaff, L.E., & Schwob, J.E. (1991). Monoclonal antibodies show that kinase C phosphorylation of GAP-43 during axogenesis is both spatially and temporally restricted in vivo. *Journal of Cell Biology* 112, 991–1005.

Meller, K. (1964). Elektronenmikroskopische Befunde zur Differenzierung der Rezeptorzellen und Bipolarzellen der Retina und ihrer synaptischen Verbindungen. *Zeitschrift für Zellforschung und mikroskopische Anatomie* 64, 733–750.

Menesini-Chen, M.G., Chen, J.S., & Levi-Montalcini, R. (1978). Sympathetic nerve fiber ingrowth in the central nervous system of neonatal rodents upon intracerebral NGF injections. *Archives Italiennes de Biologie* 116, 53–84.

Mercurio, A.M. (1995). Laminin receptors: Achieving specificity through cooperation. *Trends in Neurosciences* 5, 419–423.

Merlie, J.P., & Sanes, J.R. (1985). Concentration of acetylcholine receptor mRNA in synaptic regions of adult muscle fibres. *Nature* 317, 66–68.

Messersmith, E.K., Leonardo, E.D., Shatz, C.J., Tessier-Lavigne, M., Goodman, C.S., & Kolodkin, A.L. (1995). Semaphorin III can function as a selective chemorepellent to pattern sensory projections in the spinal cord. *Neuron* **14**, 949–959.

Métin, C., Deléglise, D., Serafini, T., Kennedy, T.E., & Tessier-Lavigne, M. (1997). A role for netrin-1 in the guidance of cortical efferents. *Development* **124**, 5063–5074.

Metuzals, J., & Tasaki, I. (1978). Subaxolemmal filamentous network in the giant nerve fiber of the squid (*Loligo pealei* L.) and its possible role in excitability. *Journal of Cell Biology* **78**, 597–621.

Meyerson, G., Pfenninger, K.H., & Påhlman, S. (1992). A complex consisting of $pp60^{c\text{-}src}/pp60^{c\text{-}srcN}$ and a 38 kDa protein is highly enriched in growth cones from differentiated SH-SY5Y neuroblastoma cells. *Journal of Cell Science* **103**, 233–243.

Meyerson, G., Parrow, V., Gestblom, C., Johansson, I., & Påhlman, S. (1994). Protein synthesis and mRNA in isolated growth cones from differentiating SH-SY5Y neuroblastoma cells. *Journal of Neuroscience Research* **37**, 303–312.

Michler, A. (1990). Involvement of GABA receptors in the regulation of neurite growth in cultured embryonic chick tectum. *International Journal of Developmental Neuroscience* **8**, 463–472.

Mihailoff, G.A., & Bourell, K.W. (1986). Synapse formation and other ultrastructural features of postnatal development in the basilar pontine nuclei of the rat. *Developmental Brain Research* **28**, 195–212.

Miller, D.R., Lee, G.M., & Maness, P.F. (1993). Increased neurite outgrowth induced by inhibition of protein tyrosine kinase activity in PC12 pheochromocytoma cells. *Journal of Neurochemistry* **60**, 2134–2144.

Miller, M., & Peters, A. (1981). Maturation of rat visual cortex. II. A combined Golgi–electron microscope study of pyramidal neurons. *Journal of Comparative Neurology* **203**, 555–573.

Miller, M., Bower, E., Levitt, P., Deqin, L., & Chantler, P.D. (1992). Myosin II distribution in neurons is consistent with a role in growth cone motility but not in synaptic vesicle mobilization. *Neuron* **8**, 25–44.

Miller, M.W. (1988). Development of projection and local circuit neurons in neocortex. In *Cerebral Cortex. Development and Maturation of Cerebral Cortex*, ed. A. Peters & E.G. Jones, pp. 133–175. New York: Plenum Press.

Mills, L.R., & Kater, S.B. (1990). Neuron-specific and state-specific differences in calcium homeostasis regulate the generation and degeneration of neuronal architecture. *Neuron* **4**, 149–163.

Ming, G.-I., Song, H., Berninger, B., Holt, C.E., Tessier-Lavigne, M., & Poo, M.-M. (1997). cAMP-dependent growth cone guidance by netrin-1. *Neuron* **19**, 1225–1235.

Mitchell, K.J., Doyle, J.L., Serafini, T., Kennedy, T.E., Tessier-Lavigne, M., Goodman, C.S., & Dickson, B.J. (1996). Genetic analysis of netrin genes in *Drosophila*: Nethrins guide CNS commissural axons and peripheral motor axons. *Neuron* **17**, 203–215.

Mitchison, T.J., & Cramer, L.P. (1996). Actin-based cell motility and cell locomotion. *Cell* **84**, 371–379.

Mitchison, T., & Kirschner, M.W. (1984a). Microtubule assembly nucleated by isolated centromeres. *Nature* **312**, 232–237.

Mitchison, T., & Kirschner, M.W. (1984b). Dynamic instability of microtubule growth. *Nature* **312**, 237–242.

Mitchison, T., & Kirschner, M. (1988). Cytoskeletal dynamics and nerve growth. *Neuron* **1**, 761–772.

Molliver, M.E., & Van der Loos, H. (1970). The ontogenesis of cortical circuitry: The spatial distribution of synapses in somesthetic cortex of newborn dog. *Ergebnisse der Anatomie und Entwicklungsgeschichte* 42, 1–54.

Molliver, M.E., Kostovic, I., & Van der Loos, H. (1973). The development of synapses in cerebral cortex of the human fetus. *Brain Research* 50, 403–407.

Molnár, Z., & Blakemore, C. (1991). Lack of regional specificity for connections formed between thalamus and cortex in coculture. *Nature* 351, 475–477.

Molnár, Z., & Blakemore, C. (1995). How do thalamic axons find their way to the cortex? *Trends in Neurosciences* 18, 389–397.

Monschau, B., Kremoser, C., Ohta, K., Tanaka, H., Kaneko, T., Yamada, T., Handwerker, C., Hornberger, M.R., Löschinger, J., Pasquale, E.B., Siever, D.A., Verderame, M.F., Müller, B.K., Bonhoeffer, F., & Drescher, U. (1997). Shared and distinct functions of RAGS and ELF-1 in guiding retinal axons. *EMBO Journal* 16, 1258–1267.

Montell, D.J., & Goodman, C.S. (1989). *Drosophila* laminin: Sequence of B2 subunit and expression of all three subunits during embryogenesis. *Journal of Cell Biology* 109, 2441–2453.

Moorman, S.J., & Hume, R.I. (1990). Growth cones of chick sympathetic preganglionic neurons *in vitro* interact with other neurons in a cell-specific manner. *Journal of Neuroscience* 10, 3158–3163.

Moorman, S.J., & Hume, R.I. (1993). Omega-conotoxin prevents myelin-evoked growth cone collapse in neonatal rat locus coeruleus neurons in vitro. *Journal of Neuroscience* 13, 4727–4736.

Moos, M., Tacke, R., Scherer, H., Teplow, D., Fruh, K., & Schachner, M. (1988). Neural adhesion molecule L1 as a member of the immunoglobulin superfamily with binding domains similar to fibronectin. *Nature* 334, 701–703.

Mooseker, M. (1985). Organization, chemistry, and assembly of the cytoskeletal apparatus of the intestinal brush border. *Annual Review of Cell Biology* 1, 209–241.

Mooseker, M.S., & Cheney, R.E. (1995). Unconventional myosins. *Annual Review of Cell and Developmental Biology* 11, 633–675.

Mooseker, M.S., & Tilney, L.G. (1975). Organization of an actin filament-membrane complex. Filament polarity and membrane attachment in the microvilli of intestinal epithelial cells. *Journal of Cell Biology* 67, 725–743.

Morest, D.K. (1969a). The growth of dendrites in the mammalian brain. *Zeitschrift für Anatomie und Entwicklungsgeschichte* 128, 290–317.

Morest, D.K. (1969b). The differentiation of cerebral dendrites: a study of the post-migratory neuroblast in the medial nucleus of the trapezoid body. *Zeitschrift für Anatomie und Entwicklungsgeschichte* 128, 271–289.

Morissette, N., & Carbonetto, S. (1995). Laminin $\alpha 2$ chain (M chain) is found within the pathway of avian and murine retinal projections. *Journal of Neuroscience* 15, 8067–8082.

Moss, D.J., Fernyhough, P., Chapman, K., Baizer, L., Bray, D., & Allsopp, T. (1990). Chicken growth-associated protein GAP-43 is tightly bound to the actin-rich neuronal membrane skeleton. *Journal of Neurochemistry* 54, 729–736.

Moya, K.L., Benowitz, L.I., Jhaveri, S., & Schneider, G.E. (1988). Changes in rapidly transported proteins in developing hamster retinofugal axons. *Journal of Neuroscience* 8, 4445–4454.

Müller, B.K., Bonhoeffer, F., & Drescher, U. (1996). Novel gene families involved in neural pathfinding. *Current Biology* 6, 469–474.

Muroya, K., Hashimoto, Y., Hattori, S., & Nakamura, S. (1992). Specific inhibition of NGF receptor tyrosine kinase activity by K-252a. *Biochimica et Biophysica Acta* **1135**, 353–356.

Myers, P.Z., & Bastiani, M.J. (1993). Growth cone dynamics during the migration of an identified commissural growth cone. *Journal of Neuroscience* **13**, 127–143.

Myers, P.Z., Eisen, J.S., & Westerfield, M. (1986). Development and axonal outgrowth of identified motoneurons in Zebrafish. *Journal of Neuroscience* **6**, 2278–2289.

Nakai, J. (1956). Dissociated dorsal root ganglia in tissue culture. *American Journal of Anatomy* **99**, 81–129.

Nakai, J. (1960). Studies on the mechanism determining the course of nerve fibres in tissue culture. II. The mechanism of fasciculation. *Zeitschrift für Zellforschung und Mikroskopische Anatomie* **52**, 427–449.

Nakai, J., & Kawasaki, Y. (1959). Studies on the mechanism determining the course of nerve fibers in tissue culture. I. The reaction of the growth cone to various obstructions. *Zeitschrift für Zellforschung und Mikroskopische Anatomie* **51**, 108–122.

Nakamoto, M., Cheng, H.-J., Friedman, G.C., McLaughlin, T., Hansen, M.J., Yoon, C.H., O'Leary, D.D.M., & Flanagan, J.G. (1996). Topographically specific effects of ELF-1 on retinal axon guidance *in vitro* and retinal axon mapping *in vivo*. *Cell* **86**, 755–766.

Neely, M.D. (1993). Role of substrate and calcium in neurite retraction of leech neurons following depolarization. *Journal of Neuroscience* **13**, 1292–1301.

Neely, M.D., & Gesemann, M. (1994). Disruption of microfilaments in growth cones following depolarization and calcium influx. *Journal of Neuroscience* **14**, 7511–7520.

Neubig, R.R., Krodel, E.K., Boyd, N.D., & Cohen, J.B. (1979). Acetylcholine and local anesthetic binding to *Torpedo* nicotinic postsynaptic membranes after removal of nonreceptor peptides. *Proceedings of the National Academy of Sciences U.S.A.* **76**, 690–694.

Neugebauer, K.M., Emmett, C.J., Venstrom, K.A., & Reichardt, L.F. (1991). Vitronectin and thrombospondin promote retinal neurite outgrowth: Developmental regulation and role of integrins. *Neuron* **6**, 345–358.

New, H.V., & Mudge, A.W. (1986). Calcitonin gene-related peptide regulates muscle acetylcholine receptor synthesis. *Nature* **323**, 809–811.

Nishi, R., & Berg, D.K. (1981). Effects of high K^+ concentration on the growth and development of ciliary ganglion neurons in culture. *Developmental Biology* **87**, 301–307.

Nitkin, R.M., Smith, M.A., Magill, C., Fallon, J.R., Yao, Y.-M.M., & McMahan, U.J. (1987). Identification of agrin, a synaptic organizing protein from *Torpedo* electric organ. *Journal of Cell Biology* **105**, 2471–2478.

Noakes, P.G., Gautam, M., Mudd, J., Sanes, J.R., & Merlie, J.P. (1995). Aberrant differentiation of neuromuscular junctions in mice lacking S-laminin/laminin B2. *Nature* **374**, 258–262.

Nordlander, R.H. (1987). Axonal growth cones within the developing amphibian spinal cord. *Journal of Comparative Neurology* **263**, 485–496.

Nordlander, R.H., & Singer, M. (1978). The role of ependyma in regeneration of the spinal cord in the urodele amphibian tail. *Journal of Comparative Neurology* **180**, 349–374.

Nordlander, R.H., & Singer, M. (1982a). Morphology and position of growth cones in the developing *Xenopus* spinal cord. *Developmental Brain Research* **4**, 181–193.

Nordlander, R.H., & Singer, M. (1982b). Spaces preceed axons in *Xenopus* embryonic spinal cord. *Experimental Neurology* **75**, 221–228.

Nordlander, R.H., Gazzerro, J.W., & Cook, H. (1991). Growth cones and axon trajectories of a sensory pathway in the amphibian spinal cord. *Journal of Comparative Neurology* **307**, 539–548.

Nörenberg, U., Wille, H., Wolff, M., Frank, R., & Rathjen, R.G. (1992). The chicken neural extracellular matrix molecule restrictin: Similarity with EGF-, fibronectin type III-, and fibrinogen-like motifs. *Neuron* **8**, 849–863.

Norris, C.R., & Kalil, K. (1990). Morphology and cellular interactions of growth cones in the developing corpus callosum. *Journal of Comparative Neurology* **293**, 268–281.

Nose, A., & Takeichi, M. (1986). A novel cadherin cell adhesion molecule: Its expression pattern is associated with implantation and organogenesis of mouse embryos. *Journal of Cell Biology* **103**, 2649–2658.

Nothias, F., Fischer, I., Murray, M., Mirman, S., & Vincent, J.D. (1996). Expression of a phosphorylated isoform of MAP1B is maintained in adult central nervous system areas that retain capacity for structural plasticity. *Journal of Comparative Neurology* **368**, 317–344.

Nuccitelli, R., & Smart, T. (1989). Extracellular calcium levels strongly influence neural crest cell galvanotaxis. *Biological Bulletin* **176**, 130–135.

Nunez, J. (1986). Differential expression of microtubule components during brain development. *Developmental Neuroscience* **8**, 125–141.

Nuttall, R.P., & Wessells, N.K. (1979). Veils, mounds, and vesicle aggregates in neurons elongating *in vitro*. *Experimental Cell Research* **119**, 163–174.

Oakley, B.R. (1992). γ-Tubulin: the microtubule organizer? *Trends in Cell Biology* **2**, 1–5.

Oakley, C.E., & Oakley, B.R. (1989). Identification of gamma-tubulin, a new member of the tubulin superfamily encoded by *mipA* gene of *Aspergillus nidulans*. *Nature* **338**, 662–664.

Oakley, R.A., & Tosney, K.W. (1993). Contact-mediated mechanisms of motor axon segmentation. *Journal of Neuroscience* **13**, 3773–3792.

Obata, K., Oide, M., & Tanaka, H. (1978). Excitatory and inhibitory actions of GABA and glycine on embryonic chick spinal neurons in culture. *Brain Research* **144**, 179–184.

Oberstar, J.V., Challacombe, J.F., Roche, F.K., & Letourneau, P.C. (1997). Concentration-dependent stimulation and inhibition of growth cone behavior and neurite elongation by protein kinase inhibitors KT5926 and K-252a. *Journal of Neurobiology* **33**, 161–171.

Oblinger, M.M., & Kost, S.A. (1994). Coordinate regulation of tubulin and microtubule associated protein genes during development of hamster brain. *Developmental Brain Research* **77**, 45–54.

Obrietan, K., & van den Pol, A.N. (1995). GABA neurotransmission in the hypothalamus: Developmental reversal from Ca^{2+} elevating to depressing. *Journal of Neuroscience* **15**, 5065–5077.

Obrietan, K., & van den Pol, A.N. (1996). Growth cone calcium rise by GABA. *Journal of Comparative Neurology* **372**, 167–175.

O'Connor, T.P., & Bentley, D. (1993). Accumulation of actin in subsets of pioneer growth cone filopodia in response to neural and epithelial guidance cues in situ. *Journal of Cell Biology* **123**, 935–948.

O'Connor, T.P., Duerr, J.S., & Bentley, D. (1990). Pioneer growth cone steering decisions mediated by single filopodial contacts *in situ*. *Journal of Neuroscience* **10**, 3935–3946.

Oestreicher, A.B., & Gispen, W.H. (1986). Comparison of the immunocytochemical distribution of the phosphoprotein B-50 in the cerebellum and hippocampus of immature and adult rat brain. *Brain Research* **375**, 267–279.

Ohmichi, M., Pang, L., Ribon, V., Gazit, A., Levitzki, A., & Saltiel, A.R. (1993). The tyrosine kinase inhibitor tyrphostin blocks the cellular actions of nerve growth factor. *Biochemistry* **32**, 4650–4658.

Ohta, K., Iwamasa, H., Drescher, U., Terasaki, H., & Tanaka, H. (1997). The inhibitory effect on neurite outgrowth of motoneurons exerted by the ligands ELF-1 and RAGS. *Mechanisms of Development* **64**, 127–135.

Okabe, S., & Hirokawa, N. (1988). Microtubule dynamics in nerve cells: Analysis using microinjection of biotinylated tubulin into PC12 cells. *Journal of Cell Biology* **107**, 651–654.

Okabe, S., & Hirokawa, N. (1989). Turnover of fluorescently labelled tubulin and actin in the axon. *Nature* **343**, 479–482.

Okabe, S., & Hirokawa, N. (1991). Actin dynamics in growth cones. *Journal of Neuroscience* **11**, 1918–1929.

Okabe, S., & Hirokawa, N. (1992). Differential behaviour of photoactivated microtubules in growing axons of mouse and frog neurons. *Journal of Cell Biology* **117**, 105–120.

Okabe, S., & Hirokawa, N. (1993). Do photobleached fluorescent microtubules move? : Re-evaluation of fluorescence laser photobleaching both in vitro and in growing *Xenopus* axon. *Journal of Cell Biology* **120**, 1177–1186.

O'Leary, D.D., Bicknese, A.R., De Carlos, J.A., Heffner, C.D., Koester, S.E., Kutka, L.J., & Terashima, T. (1990). Target selection by cortical axons: alternative mechanisms to establish axonal connections in the developing brain. *Cold Spring Harbor Symposium of Quantitative Biology* **55**, 453–468.

O'Leary, D.D.M., & Terashima, T. (1988). Cortical axons branch to multiple targets by interstitial axon budding: implications for target recognition and waiting periods. *Neuron* **1**, 901–910.

Olinck-Coux, M., & Hollenbeck, P.J. (1996). Localization and active transport of mRNA in axons of sympathetic neurons in culture. *Journal of Neuroscience* **15**, 1346–1358.

Olmsted, J.B. (1986). Microtubule-associated proteins. *Annual Reviews of Neuroscience* **11**, 29–44.

Ono, K., Tomasiewicz, H., Magnuson, T., & Rutishauser, U. (1994). N-CAM mutation inhibits tangential neuronal migration and is phenocopied by enzymatic removal of polysialic acid. *Neuron* **13**, 595–609.

Oohira, A., Matsui, F., & Katoh-Semba, R. (1991). Inhibitory effects of brain chondroitin sulphate proteoglycans on neurite outgrowth from PC12D cells. *Journal of Neuroscience* **11**, 822–827.

Orike, N., & Pini, A. (1996). Axon guidance: Following the Eph plan. *Current Biology* **6**, 108–110.

Orioli, D., & Klein, R. (1997). The Eph receptor family: axonal guidance by contact repulsion. *Trends in Genetics* **13**, 354–359.

Orioli, D., Henkemeyer, M., Lemke, G., Klein, R., & Pawson, T. (1996). Sek4 and Nuk receptors cooperate in guidance of commissural axons and in palate formation. *EMBO Journal* **15**, 6035–6049.

O'Rourke, N.A., & Fraser, S.E. (1990). Dynamic changes in optic fiber terminal arbors lead to retinotopic map formation: An in vivo confocal microscopic study. *Neuron*, 5, 159–171 (Abstract).

O'Rourke, N.A., Cline, H.T., & Fraser, S.E. (1994). Rapid remodeling of retinal arbors in the tectum with and without blockade of synaptic transmission. *Neuron* 12, 921–934 (Abstract).

Osen-Sand, A., Catsica, M., Staple, J.K., Jones, K.A., Ayala, G., Knowles, J., Grenningloh, G., & Catsicas, S. (1993). Inhibition of axonal growth by SNAP-25 antisense oligonucleotides *in vitro* and *in vivo*. *Nature* 364, 445–448.

Osen-Sand, A., Staple, J.K., Naldi, E., Schiavo, G., Rossetto, O., Petitpierre, S., Malgaroli, A., Montecucco, C., & Catsicas, S. (1996). Common and distinct fusion proteins in axonal growth and transmitter release. *Journal of Comparative Neurology* 367, 222–234.

Palka, J., Whitlock, K.E., & Murray, M.A. (1992). Guidepost cells. *Current Opinion in Neurobiology* 2, 48–54.

Paradies, N.E., & Grunwald, G.B. (1993). Purification and characterization of NCAD90, a soluble endogenous form of N-cadherin, which is generated by proteolysis during retinal development and retains adhesive and neurite-promoting function. *Journal of Neuroscience Research* 36, 33–45.

Park, S., Frisen, J., & Barbacid, M. (1996). Aberrant axonal projections in mice lacking EphA8 (Eek) tyrosine kinase receptors. *EMBO Journal* 16, 3106–3114.

Patel, N., & Poo, M.-M. (1982). Orientation of neurite growth by extracellular electric fields. *Journal of Neuroscience* 2, 483–496.

Patel, N.B., Xie, Z.P., Young, S.H., & Poo, M.-M. (1985). Response of growth cones to focal electric currents. *Journal of Neuroscience Research* 13, 245–256.

Patel, N.H., Snow, P.M., & Goodman, C.S. (1987). Characterization and cloning of fasciclin III : A glycoprotein expressed on a subset of neurons and axon pathways in *Drosophila*. *Cell* 48, 975–988.

Patterson, P.H. (1988). On the importance of being inhibited, or saying no to growth cones. *Neuron* 1, 263–267.

Patterson, S.I., & Skene, J.H.P. (1994). Novel inhibitory action of tunicamycin homologs suggests a role for dynamic protein fatty acylation in growth cone-mediated neurite extension. *Journal of Cell Biology* 124, 521–536.

Patton, B.L., Miner, J.H., Chiu, A.Y., & Sanes, J.R. (1997). Distribution and function of laminins in the neuromuscular system of developing, adult, and mutant mice. *Journal of Cell Biology* 139, 1507–1521.

Payne, H.R., Burden, S.M., & Lemmon, V. (1992). Modulation of growth cone morphology by substrate-bound adhesion molecules. *Cell Motility and the Cytoskeleton* 21, 65–73.

Peng, H.B., Chen, Q.M., de Biasi, S., & Zhu, D.L. (1989). Development of calcitonin gene-related peptide (CGRP) immunoreactivity in relationship to the formation of neuromuscular junctions in *Xenopus* myotomal muscle. *Journal of Comparative Neurology* 290, 533–543.

Pepperkok, R., BrÒ, M.H., Davoust, J., & Kreis, T. (1990). Microtubules are stabilized in confluent epithelial cells but not in fibroblasts. *Journal of Cell Biology* 111, 3003–3012.

Perez, R.G., & Halfter, W. (1993). Tenascin in the developing chick visual system: Distribution and potential role as a modulator of retinal axon growth. *Developmental Biology* 156, 278–292.

Perrins, R., & Roberts, A. (1994). Nicotinic and muscarinic Ach receptors in rhythmically active spinal neurons in the *Xenopus laevis* embryo. *Journal of Physiology* **478**, 221–228.

Perrone-Bizzozero, N.I., Benowitz, L.I., Apostolides, P.J., Franck, E.R., Finklestein, S.P., & Bizzozero, O.A. (1989). Protein fatty-acid acylation in developing cortical neurons. *Journal of Neurochemistry* **52**, 1149–1155.

Persohn, E., & Schachner, M. (1990). Immunohistological localization of the neural cell adhesion molecules L1 and N-CAM in the developing hippocampus of the mouse. *Journal of Neurocytology* **19**, 807–819.

Pesheva, P., Spiess, E., & Schachner, M. (1989). J1–160 and J1–180 are oligodendrocyte-secreted nonpermissive substrates for cell adhesion. *Journal of Cell Biology* **109**, 1765–1778.

Pesheva, P., Gennarini, G., Goridis, C., & Schachner, M. (1993). The F111/F11 cell adhesion molecule mediates repulsion of neurons by the extracellular matrix glycoprotein J1–160/180. *Neuron* **10**, 69–82.

Peterson, E.R., & Crain, S.M. (1982). Nerve growth factor attenuates neurotoxic effects of taxol on spinal cord-ganglion explants from fetal mice. *Science* **217**, 377–379.

Pfenninger, K.H. (1979). Subplasmalemmal vesicle clusters: Real or artifact? In *Freeze Fracture: Methods, Artifacts and Interpretations*, ed. J.E. Rash & C.S. Hudson, pp. 71–80. New York: Raven Press.

Pfenninger, K.H., & Friedman, L.B. (1993). Sites of plasmalemmal expansion in growth cones. *Developmental Brain Research* **71**, 181–192.

Pfenninger, K.H., & Maylié-Pfenninger, M.-F. (1981). Lectin labelling of sprouting neurons. II. Relative movement and appearance of glycoconjugates during plasmalemmal expansion. *Journal of Cell Biology* **89**, 547–559.

Pfenninger, K.H., Ellis, L., Johnson, M.P., Friedman, L.B., & Somlo, S. (1983). Nerve growth cones isolated from fetal rat brain. Sub-cellular fractionation and characterization. *Cell* **35**, 573–584.

Pfenninger, K.H., de la Houssaye, B.A., Frame, L., Helmke, S., Lockerbie, R.O., Lohse, K., Miller, V., Negre-Aminou, P., & Wood, M.R. (1992). Biochemical dissection of plasmalemmal expansion at the growth cone. In *The Nerve Growth Cone*, ed. P.C. Letourneau, S.B. Kater & E.R. Macagno, pp. 111–123. New York: Raven Press.

Phelan, P., & Gordon-Weeks, P.R. (1992). Widespread distribution of synaptophysin, a synaptic vesicle glycoprotein, in growing neurites and growth cones. *European Journal of Neuroscience* **4**, 1180–1190.

Phillips, G.R., Edelman, G.M., & Crossin, K.L. (1995). Separate cell binding sites within cytotactin/tenascin differentially promote neurite outgrowth. *Cell Adhesion and Communication* **3**, 257–271.

Phillips, L.L., Autilio-Gambetti, L., & Lasek, R.J. (1983). Bodian's silver method reveals molecular variation in the evolution of neurofilament proteins. *Brain Research* **278**, 219–223.

Phillips, W.D., Kopta, C., Blount, P., Gardner, P.D., Steinbach, J.H., & Merlie, J.P. (1991). ACh receptor-rich membrane domains organized in fibroblasts by recombinant 43-kilodalton protein. *Science* **251**, 568–570.

Pierceall, W.E., Reale, M.A., Candia, A.F., Wright, C.V., Cho, K.R., & Fearon, E.R. (1994). Expression of a homolog of the deleted in colorectal cancer (DCC) gene in the nervous system of developing *Xenopus* embryos. *Developmental Biology* **166**, 654–665.

Pike, S.H., & Eisen, J.S. (1990). Identified primary motoneurons in zebrafish select appropriate pathways in the absence of other primary motoneurons. *Journal of Neuroscience* **10**, 44–49.

Pini, A. (1993). Chemorepulsion of axons in the developing mammalian central nervous system. *Science* **261**, 95–98.

Piperno, G., Ledizet, M., & Chang, X. (1987). Microtubules containing acetylated α-tubulin in mammalian cells in culture. *Journal of Cell Biology* **104**, 289–302.

Pittman, R.N. (1985). Release of plasminogen activator and a calcium-dependent metalloprotease from cultured sympathetic and sensory neurons. *Developmental Biology* **110**, 91–101.

Pittman, R.N., & Williams, A.G. (1989). Neurite penetration into collagen gels requires Ca^{2+}-dependent metalloproteinase activity. *Developmental Neuroscience* **11**, 41–51.

Pittman, R.N., Ivins, J.K., & Buettner, H.M. (1989a). Neuronal plasminogen activators: Cell surface binding sites and involvement in neurite outgrowth. *Journal of Neuroscience* **9**, 4269–4286.

Pittman, R.N., Vos, P., Ivins, J.K., Buettner, H.M., & Repka, A. (1989b). Proteases and inhibitors in the developing nervous system. In *The Assembly of the Nervous System*, ed. L. Landmesser, pp. 109–128. New York: Alan R. Liss, Inc.

Placzek, M., Tessier-Lavigne, M., Jessell, T., & Dodd, J. (1990a). Orientation of commissural axons *in vitro* in response to a floor plate-derived chemoattractant. *Development* **110**, 19–30.

Placzek, M., Tessier-Lavigne, M., Yamada, T., Dodd, J., & Jessell, T.M. (1990b). Guidance of developing axons by diffusible chemoattractants. *Cold Spring Harbor Symposium of Quantitative Biology* **55**, 279–290.

Polak, K.A., Edelman, A.M., Wasley, J.W.F., & Cohan, C.S. (1991). A novel calmodulin antagonist, CGS9343B, modulates calcium dependent changes in neurite outgrowth and growth cone movements. *Journal of Neuroscience* **11**, 534–542.

Pollard, T.D., Doberstein, S.K., & Zot, H.G. (1991). Myosin-I. *Annual Review of Physiology* **53**, 653–681.

Pomerat, C.M., Hendelman, W.J., Raiborn, C.W., Jr. & Massey, J.F. (1967). Dynamic activities of nervous tissue *in vitro*. In *The Neuron*, ed. H. Hydén, pp. 121–178. New York: Elsevier.

Poo, M.-M. (1981). *In situ* electrophoresis of membrane components. *Annual Reviews of Biophysics and Bioengineering* **10**, 245–276.

Popov, S., & Poo, M.-M. (1992). Diffusional transport of macromolecules in developing nerve processes. *Journal of Neuroscience* **12**, 77–85.

Popov, S., Brown, A., & Poo, M.-M. (1993). Forward plasma membrane flow in growing nerve processes. *Science* **259**, 244–246.

Porter, B.E., Weis, J., & Sanes, J.R. (1995). A motoneuron-selective stop signal in the synaptic protein s-laminin. *Neuron* **14**, 549–559.

Poulter, M.O., Barker, J.L., O'Carroll, A.-M., Lolait, S.J., & Mahan, L.C. (1992). Differential and transient expression of $GABA_A$ receptor a-subunit mRNAs in the developing rat CNS. *Journal of Neuroscience* **12**, 2888–2900.

Povlishock, J.T. (1976). The fine structure of the axons and growth cones of the human fetal cerebral cortex. *Brain Research* **114**, 379–389.

Powell, S.K., Rivas, R.J., Rodriguez-Boulan, E., & Hatten, M.E. (1997). Development of polarity in cerebellar granule neurons. *Journal of Neurobiology* **32**, 223–236.

Przywara, D.A., Bhave, S.V., Chowdhury, P.S., Wakade, T.D., & Wakade, A.R. (1993). Sites of transmitter release and relation to intracellular calcium in cultured sympathetic neurons. *Neuroscience* **52**, 973–985.

Purves, D., & Lichtman, J.W. (1985). *Principles of Neural Development.* Sunderland, Massachusetts: Sinauer Associates Inc.

Püschel, A.W., Adams, R.H., & Betz, H. (1995). Murine semaphorin D/collapsin is a member of a diverse gene family and creates domains inhibitory for axonal extension. *Neuron* **14**, 941–948.

Püschel, A.W., Adams, R.H., & Betz, H. (1996). The sensory innervation of the mouse spinal cord may be patterned by differential expression of and differential responsiveness to semaphorins. *Molecular and Cellular Neuroscience* **7**, 419–431.

Rakic, P., & Sidman, R.L. (1973a). Sequence of developmental abnormalities leading to granule cell deficit in cerebellar cortex of weaver mutant mice. *Journal of Comparative Neurology* **152**, 103–132.

Rakic, P., & Sidman, R.L. (1973b). Organization of cerebellar cortex secondary to deficit of granule cells in weaver mutant mice. *Journal of Comparative Neurology* **152**, 133–162.

Rakic, P., & Sidman, R.L. (1973c). Weaver mutant mouse cerebellum: Defective neuronal migration secondary to abnormality of Bergmann glia. *Proceedings of the National Academy of Sciences U.S.A.* **70**, 240–244.

Ramón y Cajal, S. (1890). A quelle époque apparaissent les expansions des cellules nerveuses de la moëlle épinière du poulet? *Anatomischer Anzeiger* **21–22**, 609–613-631-639.

Ramón y Cajal, S. (1892). La rétine des vertébrés. *La Cellule* **9**, 119–258.

Ramón y Cajal, S. (1909). *Histologie du systeme nerveux de l'homme et des vertébrés.* Madrid: CSIC.

Ramón y Cajal, S. (1928). *Degeneration and Regeneration of the Nervous System*, trans. R. M. Kay. Oxford: Oxford University Press.

Ramón y Cajal, S. (1937). *Recollections of My Life*, trans. E. H. Craigie & J. Cano from the third Spanish edition (1923). Philadelphia: American Philosophical Society.

Ramón y Cajal, S. (1960). *Studies on Vertebrate Neurogenesis*, trans. L. Guth. Springfield, Illinois: Charles C. Thomas.

Ramón y Cajal, S. (1989). *Recollections of My Life*, trans. E.H. Craigie. Cambridge, Massachusetts: MIT Press.

Ramón y Cajal, S. (1990). *New Ideas on the Structure of the Nervous System in Man and Vertebrates*, trans. N. Swanson & L. W. Swanson. Cambridge, Massachusetts: MIT Press.

Ranscht, B. (1991). Cadherin cell adhesion molecules in vertebrate neural development. *Seminars in Neuroscience* **3**, 285–296.

Ranscht, B., & Bronner-Fraser, M. (1991). T-Cadherin expression alternates with migrating neural crest cells in the trunk of the avian embryo. *Development* **111**, 15–22.

Raper, J.A., & Grunewald, E.B. (1990). Temporal retinal growth cones collapse on contact with nasal retinal axons. *Experimental Neurobiology* **109**, 71–74.

Raper, J.A., & Kapfhammer, J.P. (1990). The enrichment of a neuronal growth cone collapsing activity from embryonic chick brain. *Neuron* **4**, 21–29.

Raper, J.A., Bastiani, M., & Goodman, C.S. (1983a). Pathfinding by neuronal growth cones in grasshopper embryos. I. Divergent choices made by the growth cones of sibling neurons. *Journal of Neuroscience* **3**, 20–30.

Raper, J.A., Bastiani, M., & Goodman, C.S. (1983b). Pathfinding by neuronal growth cones in grasshopper embryos. II. Selective fasciculation onto specific axonal pathways. *Journal of Neuroscience* **3**, 31–41.

Raper, J.A., Bastiani, M.J., & Goodman, C.S. (1983c). Guidance of neuronal growth cones: Selective fasciculation in the grasshopper embryo. *Cold Spring Harbor Symposium of Quantitative Biology* **48**, 587–598.

Raper, J.A., Bastiani, M.J., & Goodman, C.S. (1984). Pathfinding by neuronal growth cones in grasshopper embryos. IV. The effects of ablating the A and P axons upon the behaviour of the G growth cone. *Journal of Neuroscience* **4**, 2329–2345.

Raper, J.A., Chang, S., & Raible, D.W. (1992). Interactions between growth cones and axons: Selectively distributed extension-promoting and extension-inhibiting components. In *The Nerve Growth Cone*, ed. P.C. Letourneau, S.B. Kater & E.R. Macagno, pp. 207–217. New York: Raven Press.

Rathjen, F.G. (1991). Neural cell contact and axonal growth. *Current Opinion in Cell Biology* **3**, 992–1000.

Rathjen, F.G., Wolff, J.M., Frank, R., Bonhoeffer, F., & Rutishauser, U. (1987a). Membrane glycoproteins involved in neurite fasciculation. *Journal of Cell Biology* **104**, 343–353.

Rathjen, F.G., Wolff, J.M., Chang, S., Bonhoeffer, F., & Raper, J.A. (1987b). Neurofascin: a novel chick cell-surface glycoprotein involved in neurite–neurite interactions. *Cell* **51**, 841–849.

Rauch, U., Karthikeyan, L., Maurel, P., Margolis, R.U., & Margolis, R.K. (1992). Cloning and primary structure of neurocan, a developmentally regulated, aggregating chondroitin sulfate proteoglycan of brain. *Journal of Biological Chemistry* **267**, 19536–19547.

Rausch, D.M., Dickens, G., Doll, S., Fujita, K., Koizumi, S., Rudkin, B.B., Tocco, M., Eiden, L.E., & Guroff, G. (1989). Differentiation of PC12 cells with v-*src*: Comparison with nerve growth factor. *Journal of Neuroscience Research* **24**, 49–58.

Raybin, D., & Flavin, M. (1977). Enzyme which specifically adds tyrosine to the α-chain of tubulin. *Biochemistry* **16**, 2189–2194.

Reber, B.F.X., & Reuter, H. (1991). Dependence of cytosolic calcium in differentiating rat pheochromocytoma cells on calcium channels and intracellular stores. *Journal of Physiology* **435**, 145–162.

Redies, C. (1997). Cadherins and the formation of neural circuitry in the vertebrate CNS. *Cell and Tissue Research* **290**, 405–413.

Redies, C., & Takeichi, M. (1993). N- and R-cadherin expression in the optic nerve of the chicken embryo. *Glia* **8**, 161–171.

Redies, C., Inuzuka, H., & Takeichi, M. (1992). Restricted expression of N- and R-cadherin on neurites of the developing chicken CNS. *Journal of Neuroscience* **12**, 3525–3534.

Redies, C., Engelhart, K., & Takeichi, M. (1993). Differential expression of N- and R-cadherin in functional neuronal systems and other structures of the developing chicken brain. *Journal of Comparative Neurology* **333**, 398–416.

Rees, R.P. (1978). The morphology of interneuronal synaptogenesis: A review. *Federation Proceedings* **37**, 2000–2009.

Rees, R.P., & Reese, T.S. (1981). New structural features of freeze-substituted neuritic growth cones. *Neuroscience* **6**, 247–254.

Rees, R.P., Bunge, M.B., & Bunge, R.P. (1976). Morphological changes in the neuritic growth cone and target neuron during synaptic junction development in culture. *Journal of Cell Biology* **68**, 240–263.

Reh, T.A., & Constantine-Paton, M. (1985). Growth cone–target interactions in the frog retinotectal pathway. *Journal of Neuroscience Research* **15**, 89–100.

Rehder, V., & Kater, S. (1992). Regulation of neuronal growth cone filopodia by intracellular calcium. *Journal of Neuroscience* **12**, 3175–3186.

Reichardt, L.F., & Tomaselli, T.J. (1991). Extracellular matrix molecules and their receptors. *Annual Review of Neuroscience* **14**, 531–570.

Reinsch, S.S., Mitchison, T.J., & Kirschner, M. (1991). Microtubule polymer assembly and transport during axonal elongation. *Journal of Cell Biology* **115**, 365–379.

Reist, N.E., Magill, C., & McMahan, U.J. (1987). Agrin-like molecules at synaptic sites in normal, denervated, and damaged skeletal muscles. *Journal of Cell Biology* **105**, 2457–2469.

Reist, N.E., Werle, M.J., & McMahan, U.J. (1992). Agrin released by motor neurons induces the aggregation of acetylcholine receptors at neuromuscular junctions. *Proceedings of the National Academy of Sciences U.S.A.* **85**, 2825–2829.

Renteria, R.C., & Constantine-Paton, M. (1996). Exogenous nitric oxide causes collapse of retinal ganglion cell axonal growth cone *in vitro*. *Journal of Neurobiology* **29**, 415–428.

Reynolds, M.L., Fitzgerald, M., & Benowitz, L.I. (1991). GAP-43 expression in developing cutaneous and muscle nerves in the rat hindlimb. *Neuroscience* **41**, 201–211.

Ridley, A.J. (1996) Rho: Theme and variations. *Current Biology* **6**, 1256–1264.

Riederer, B., Cohen, R., & Matus, A. (1986). MAP 5: A novel microtubule associated protein under strong developmental regulation. *Journal of Neurocytology* **15**, 763–775.

Riederer, B.M., Guadano-Ferraz, A., & Innocenti, G.M. (1990). Difference in distribution of microtubule-associated protein 5a and 5b during development of cerebral cortex and corpus callosum in cats: Dependence on phosphorylation. *Developmental Brain Research* **56**, 235–243.

Riehl, R., Johnson, K., Bradley, R., Grunwald, G.B., Cornel, E., Lilienbaum, A., & Holt, C.E. (1996). Cadherin function is required for axon outgrowth in retinal ganglion cells in vivo. *Neuron* **17**, 837–848.

Riggott, M.J., & Moody, S.A. (1987). Distribution of laminin and fibronectin along peripheral trigeminal axon pathways in the developing chick. *Journal of Comparative Neurology* **258**, 580–596.

Rivas, R.J., Burmeister, D.W., & Goldberg, D. (1992). Rapid effects of laminin on the growth cone. *Neuron* **8**, 107–115.

Roberts, A. (1976). Neuronal growth cones in amphibian embryo. *Brain Research* **118**, 526–530.

Roberts, A. (1988). The early development of neurons in *Xenopus* embryos revealed by transmitter immunocytochemistry for serotonin, GABA and glycine. In *Developmental Neurobiology of the Frog*, pp. 191–205. New York: Alan R. Liss.

Roberts, A., & Patton, D.J. (1985). Growth cones and the formation of central and peripheral neurites by sensory neurons in amphibian embryos. *Journal of Neuroscience Research* **13**, 23–38.

Roberts, A., & Taylor, J.S.H. (1983). A study of the growth cones of developing embryonic sensory neurites. *Journal of Embryology and Experimental Morphology* **75**, 31–47.

Roberts, A., Dale, N., Ottersen, O.P., & Storm-Mathisen, J. (1987). The early development of interneurons with GABA immunoreactivity in the central nervous system of *Xenopus laevis*. *Journal of Comparative Neurology* **261**, 435–449.

Robson, S.J., & Burgoyne, R.D. (1989a). Differential localisation of tyrosinated, detyrosinated, and acetylated α-tubulins in neurites and growth cones of dorsal root ganglion neurons. *Cell Motility and the Cytoskeleton* **12**, 273–282.

Robson, S.J., & Burgoyne, R.D. (1989b). L-type calcium channels in the regulation of neurite outgrowth from rat dorsal root ganglion neurons in culture. *Neuroscience Letters* **104**, 110–114.

Rocha, M.G., & Avila, J. (1995). Characterization of microtubule-associated protein phosphoisoforms present in isolated growth cones. *Developmental Brain Research* **89**, 47–55.

Rochlin, M.W., Itoh, K., Adelstein, R.S., & Bridgman, P.C. (1995). Localization of myosin IIA and B isoforms in cultured neurons. *Journal of Cell Science* **108**, 3661–3670.

Rochlin, M.W., Wickline, K.M., & Bridgman, P.C. (1996). Microtubule stability decreases axon elongation but not axoplasm production. *Journal of Neuroscience* **16**, 3236–3246.

Rogers, S.L., Letourneau, P.C., Palm, S.L., McCarthy, J.B., & Furcht, L.T. (1983). Neurite extension by peripheral and central nervous system neurons in response to substratum-bound fibronectin and laminin. *Developmental Biology* **98**, 212–220.

Rogers, S.L., Edson, K.J., Letourneau, P.C., & McCloon, S.C. (1986). Distribution of laminin in the developing nervous system of the chick. *Developmental Biology* **113**, 429–435.

Roisen, F., Inczedy-Marcsek, M., Hsu, L., & Yorke, W. (1978). Myosin: Immunofluorescent localization in neuronal and glial cultures. *Science* **199**, 1445–1448.

Roskies, A.L., & O'Leary, D.D. (1994). Control of topographic retinal axon branching by inhibitory membrane-bound molecules. *Science* **265**, 799–803.

Ruchhoeft, M.L., & Harris, W.A. (1997). Myosin functions in *Xenopus* retinal ganglion cell growth cone motility *in vivo*. *Journal of Neurobiology* **32**, 567–578.

Ruegg, M.A. (1996). Agrin, laminin β2 (s-laminin) and ARIA: Their role in neuromuscular development. *Current Opinion in Neurobiology* **6**, 97–103.

Ruegg, M.A., & Bixby, L. (1998). Agrin orchestrates synaptic differentiation at the vertebrate neuromuscular junction. *Trends in the Neurosciences* **21**, 22–27.

Ruegg, M.A., Tsim, K.W.K., Horton, S.E., Kröger, S., Escher, G., Gensch, E.M., & McMahan, U.J. (1992). The agrin gene codes for a family of basal lamina proteins that differ in function and distribution. *Neuron* **8**, 691–699.

Rupp, F., Payan, D.G., Magill-Solc, C., Cowan, D.M., & Scheller, R.H. (1991). Structure and expression of a rat Agrin. *Neuron* **6**, 811–823.

Ruppert, C., Kroschewski, R., & Bähler, M. (1993). Identification, characterization and cloning of myr 1, a mammalian myosin-I. *Journal of Cell Biology* **6**, 1393–1403.

Rutishauser, U. (1991). Pleiotropic biological effects of the neural cell adhesion molecule (NCAM). *Seminars in Neuroscience* **3**, 265–270.

Rutishauser, U. (1993). Adhesion molecules of the nervous system. *Current Opinion in Neurobiology* **3**, 709–715.

Rutishauser, U., Gall, W.E., & Edelman, G.M. (1978). Adhesion among neural cells of the chick embryo. IV. Role of the cell surface molecule CAM in the formation of neurite bundles in cultures of spinal ganglia. *Journal of Cell Biology* **79**, 382–393.

Rutishauser, U., Acheson, A., Hall, A.K., Mann, D.M., & Sunshine, J. (1988). The neural cell adhesion molecule (NCAM) as a regulator of cell–cell interactions. *Science* **240**, 53–57.

Sabry, J., O'Connor, T.P., & Kirschner, M.W. (1995). Axonal transport of tubulin in Ti1 pioneer neurons in situ. *Neuron* **14**, 1247–1256.

Sabry, J.H., O'Connor, T.P., Evans, L., Toroian-Raymond, A., & Kirschner, M. (1991). Microtubule behaviour during guidance of pioneer neuron growth cones in situ. *Journal of Cell Biology* **115**, 381–395.

Sachs, G.M., Jacobson, M., & Caviness, V.S. (1986). Postnatal changes in arborization patterns of murine retinocollicular axons. *Journal of Comparative Neurology* **146**, 395–408.

Safaei, R., & Fischer, I. (1989). Cloning of a cDNA encoding MAP1B in rat brain: Regulation of mRNA levels during development. *Journal of Neurochemistry* **52**, 1871–1879.

Saffell, J.L., Walsh, F.S., & Doherty, P. (1994). Expression of NCAM containing VASE in neurons can account for a developmental loss in their neurite growth response to NCAM in cellular substrata. *Journal of Cell Biology* **125**, 427–436.

Saffell, J.L., Williams, E.J., Mason, I.J., Walsh, F.S., & Doherty, P. (1997). Expression of a dominant negative FGF receptor inhibits axonal growth and FGF receptor phosphorylation stimulated by CAMs. *Neuron* **18**, 231–242.

Saito, S., Fujita, T., Komiya, Y., & Igarashi, M. (1992). Biochemical characterization of nerve growth cones isolated from both fetal and neonatal rat forebrains – the growth cone particle fraction mainly consists of axonal growth cones in both stages. *Developmental Brain Research* **65**, 179–184.

Sakaguchi, D.S., & Murphey, R.K. (1985). Map formation in the developing *Xenopus* retinotectal system: An examination of ganglion cell terminal arborizations. *Journal of Neuroscience* **5**, 3228–3245.

Sanders, M.C., & Wang, Y.-L. (1991). Assembly of actin-containing cortex occurs at distal regions of growing neurites in PC12 cells. *Journal of Cell Science* **100**, 771–780.

Sanes, J.R. (1989). Extracellular matrix molecules that influence neural development. *Annual Review of Neuroscience* **12**, 491–516.

Sanes, J.R. (1993). Basement membrane molecules in vertebrate nervous system. In *Molecular and Cellular Aspects of Basement Membranes*, ed. D.H. Rohrbach & R. Timpl, pp. 67–84. San Diego, California: Academic Press Inc.

Sanes, J.R. (1997). Genetic analysis of postsynaptic differentiation at the vertebrate neuromuscular junction. *Current Opinion in Neurobiology* **7**, 93–100.

Sanes, J.R., Marshall, L.M., & McMahan, U.J. (1978). Reinnervation of muscle fibre basal lamina after removal of myofibres. *Journal of Cell Biology* **78**, 176–198.

Sasaki, Y., Hayashi, K., Shirao, T., Ishikawa, R., & Kohama, K. (1996). Inhibition by drebrin of the actin-bundling activity of brain fascin, a protein localized in filopodia of growth cones. *Journal of Neurochemistry* **66**, 980–988.

Sato, M., Lopez-Mascaraque, L., Heffner, C.D., & O'Leary, D.D. (1994). Action of a diffusible target-derived chemoattractant on cortical axon branch induction and directed growth. *Neuron* **13**, 791–803.

Sawin, K.E., Theriot, J.A., & Mitchison, T.J. (1993). Photoactivation of fluorescence as a probe for cytoskeletal dynamics in mitosis and cell motitlity. In *Fluorescent and Luminescent Probes for Biological Activity*, ed. W.T. Mason, pp. 405–419. San Diego, California: Academic Press.

Scalia, F., & Matsumoto, D.E. (1985). The morphology of growth cones of regenerating optic nerve axons. *Journal of Comparative Neurology* **231**, 323–338.

Schartl, M., & Barnekow, A. (1984). Differential expression of the cellular *src* gene product during vertebrate development. *Developmental Biology* **105**, 415–422.

Scherer, S.S., & Easter, S.S., Jr (1984). Degenerative and regenerative changes in the trochlear nerve of goldfish. *Journal of Neurocytology* **13**, 519–565.

Schiff, P.B., & Horwitz, S.B. (1980). Taxol stabilizes microtubules in mouse fibroblast cells. *Proceedings of the National Academy of Sciences U.S.A.* **77**, 1561–1565.

Schiff, P.B., Fant, J., & Horwitz, S.B. (1979). Promotion of microtubule assembly *in vitro* by taxol. *Nature* **277**, 665–667.

Schlessinger, J., & Ullrich, A. (1992). Growth factor signaling by receptor tyrosine kinases. *Neuron* **9**, 383–391.

Schoenfeld, T.A., McKerracher, L., Obar, R., & Vallee, R.B. (1989). MAP1A and MAP1B are structurally related microtubule associated proteins with distinct developmental patterns in the CNS. *Journal of Neuroscience* **9**, 1712–1730.

Schubert, D., Humphreys, S., Baroni, C., & Cohn, M. (1969). In-vitro differentiation of a mouse neuroblastoma. *Proceedings of the National Academy of Sciences U.S.A.* **64**, 316–323.

Schubert, D., Lacorbiere, M., Whitlock, C., & Stallcup, W. (1978). Alterations in the surface properties of cells responsive to nerve growth factor. *Nature* **273**, 718–721.

Schulze, E., & Kirschner, M. (1986). Microtubule dynamics in interphase cells. *Journal of Cell Biology* **102**, 1020–1031.

Schulze, E., & Kirschner, M. (1987). Dynamic and stable populations of microtubules in cells. *Journal of Cell Biology* **97**, 1249–1254.

Schulze, E., Asai, D.J., Bulinski, J., & Kirschner, M. (1987). Post-translational modification and microtubule stability. *Journal of Cell Biology* **105**, 2167–2177.

Schwab, M.E., & Caroni, P. (1988). Oligodendrocytes and CNS myelin are nonpermissive substrates for neurite growth and fibroblast spreading *in vitro*. *Journal of Neuroscience* **8**, 2381–2393.

Seeds, N.W., Gilman, A.G., Amano, T., & Nirenberg, M.W. (1970). Regulation of axon formation by clonal lines of a neural tumor. *Proceedings of the National Academy of Sciences U.S.A.* **66**, 160–167.

Seeger, M., Tear, G., Ferres-Marco, D., & Goodman, C.S. (1993). Mutations affecting growth cone guidance in *Drosophila*: Genes necessary for guidance toward or away from the midline. *Neuron* **10**, 409–426.

Seeley, P.J., & Greene, L.A. (1983). Short-latency local actions of nerve growth factor at the growth cone. *Proceedings of the National Academy of Sciences U.S.A.* **80**, 2789–2793.

Serafini, T., Kennedy, T.E., Galko, M.J., Mirzayan, C., Jessell, T.M., & Tessier-Lavigne, M. (1994). The netrins define a family of axon outgrowth-promoting proteins homologous to *C. elegans* UNC-6. *Cell* **78**, 409–424.

Serafini, T., Colamarino, S.A., Leonardo, E.D., Wang, H., Beddington, R., Skarnes, W.H., & Tessier-Lavigne, M. (1996). Netrin-1 is required for commissural axon guidance in the developing vertebrate nervous system. *Cell* **87**, 1001–1014.

Serrano, L., & Avila, J. (1990). Structure and function of tubulin regions. In *Microtubule Proteins*, ed. J. Avila, pp. 67–88. Boca Raton, Florida: CRC Press.

Shankland, M., & Bentley, D. (1983). Sensory receptor differentiation and axonal pathfinding in the cercus of the grasshopper embryo. *Developmental Biology* **97**, 468–482.

Shankland, M., & Macagno, E.R. (1992). *Determinants of Neuronal Identity*. San Diego: Academic Press.

Sharma, N., Kress, Y., & Shafit-Zagardo, B. (1994). Antisense MAP2 oligonucleotides induce changes in microtubule assembly and neuritic elongation in pre-existing neurites of rat cortical neurons. *Cell Motility and the Cytoskeleton* **27**, 234–247.

Sharp, D.J., Yu, W., & Baas, P.W. (1995). Transport of dendritic microtubules establishes their nonuniform polarity orientation. *Journal of Cell Biology* **130**, 93–104.

Shatz, C.J., & Luskin, M.B. (1986). The relationship between the geniculocortical afferents and their cortical target cells during the development of the cat's primary visual cortex. *Journal of Neuroscience* **6**, 3655–3668.

Shatz, C.J., Chun, J.M., & Luskin, M.B. (1988). The role of the subplate in the development of the mammalian telencephalon. In *Cerebral Cortex*, ed. A. Peters & E.G. Jones, pp. 35–57. New York: Plenum Press.

Shatz, C.J., Ghosh, A., McConnell, S.K., Allendoerfer, K.L., Friauf, E., & Antonini, A. (1995). Pioneer neurons and target selection in cerebral cortical development. *Cold Spring Harbor Symposium of Quantitative Biology* **55**, 469–480.

Shaw, G., & Bray, D. (1977). Movement and extension of isolated growth cones. *Experimental Cell Research* **104**, 55–62.

Shaw, G., Osborn, M., & Weber, K. (1981). Arrangement of neurofilaments, microtubules and microfilament-associated proteins in cultured dorsal root ganglia cells. *European Journal of Cell Biology* **24**, 20–27.

Shaw, G., Banker, G.A., & Weber, K. (1985). An immunofluorescence study of neurofilament protein expressed in developing hippocampal neurons in tissue culture. *European Journal of Cell Biology* **39**, 205–216.

Shea, T.B. (1994). Delivery of anti-GAP-43 antibodies into neuroblastoma cells reduces growth cone size. *Biochemical and Biophysical Research Communications* **203**, 459–464.

Shea, T.B., & Benowitz, L.I. (1995). Inhibiton of neurite outgrowth following intracellular delivery of anti-GAP-43 antibodies depends upon culture conditions and method of neurite induction. *Journal of Neuroscience Research* **41**, 347–354.

Shea, T.B., Perrone-Bizzozero, N.I., Beerman, M.L., & Benowitz, L.I. (1991). Phospholipid-mediated delivery of anti-GAP-43 antibodies into neuroblastoma cells prevents neuritogenesis. *Journal of Neuroscience* **11**, 1685–1690.

Shea, T.B., Beerman, M.L., Nixon, R.A., & Fischer, I. (1992). Microtubule-associated protein tau is required for axonal neurite elaboration by neuroblastoma cells. *Journal of Neuroscience Research* **32**, 363–374.

Sheetz, M.P., Turney, S., Qian, H., & Elson, E.L. (1989). Nanometre-level analysis demonstrates that lipid flow does not drive membrane glycoprotein movements. *Nature* **340**, 284–288.

Sheetz, M.P., Wayne, D.B., & Pearlman, L. (1992). Extension of filopodia by motor-dependent actin assembly. *Cell Motility and the Cytoskeleton* **22**, 160–169.

Shelden, E., & Wadsworth, P. (1993). Observation and quantification of individual microtubule behaviour *in vivo*: Microtubule dynamics are cell-type specific. *Journal of Cell Biology* **120**, 935–945.

Shelton, E., & Mowczko, W.E. (1978). Membrane blisters. A fixation artifact. A study for scanning electron microscopy. *Scanning* **1**, 166–173.

Shepherd, I., Luo, Y., Raper, J.A., & Chang, S. (1996). The distribution of collapsin mRNA in the developing chick nervous system. *Developmental Biology* **173**, 185–199.

Shepherd, I.T., Luo, Y., Lefcort, F., Reichardt, L.F., & Raper, J.A. (1997). A sensory axon repellent secreted from ventral spinal cord explants is neutralized by antibodies raised against collapsin-1. *Development* **124**, 1377–1385.

Sherrington, C.S. (1947). *The Integrative Action of the Nervous System*. New Haven, Connecticut: Yale University Press.

Shirasaki, R., Tamada, A., Katsumata, R., & Murakami, F. (1995). Guidance of cerebellofugal axons in the rat embryo: Directed growth toward the floor plate and subsequent elongation along the longitudinal axis. *Neuron* **14**, 961–972.

Sidman, R.L., Lane, P.W., & Dickie, M.M. (1962). Staggerer: A new mutation in the mouse affecting the cerebellum. *Science* **137**, 610–612.

Silver, J., & Robb, R.M. (1979). Studies on the development of the eye cup and optic nerve in normal mice and mutants with congenital optic nerve aplasia. *Developmental Biology* **68**, 175–190.

Silver, J., & Rutishauser, U. (1984). Guidance of optic axons in vivo by preformed adhesive pathways and neuroepithelial endfeet. *Developmental Biology* **106**, 486–499.

Silver, J., & Sidman, R.S. (1980). A mechanism for the guidance and topographic patterning of retinal ganglion cell axons. *Journal of Comparative Neurology* **189**, 101–111.

Silver, R.A., Lamb, A.G., & Bolsover, S.R. (1989). Elevated cytosolic calcium in the growth cone inhibits neurite elongation in neuroblastoma cells: Correlation of behavioural states with cytosolic calcium concentration. *Journal of Neuroscience* **9**, 4007–4020.

Silver, R.A., Lamb, A.G., & Bolsover, S.R. (1990). Calcium hotspots caused by L-channel clustering promote morphological changes in neuronal growth cones. *Nature* **343**, 751–754.

Simmons, P.A., Lemmon, V., & Pearlman, A.L. (1982). Afferent and efferent connections of the striate and extrastriate visual cortex of the normal and reeler mouse. *Journal of Comparative Neurology* **211**, 295–308.

Simon, D.K., & O'Leary, D.D. (1992). Responses of retinal axons in vivo and in vitro to position-encoding molecules in the embryonic superior colliculus. *Neuron* **9**, 977–989.

Simons, K., & Ikonen, E. (1997). Functional rafts in cell membranes. *Nature* **387**, 569–572.

Singer, M., Nordlander, R.H., & Egar, M. (1979). Axonal guidance during embryogenesis and regeneration in the spinal cord of the newt: The blueprint hypothesis of neuronal pathway patterning. *Journal of Comparative Neurology* **185**, 1–22.

Skene, J.H.P. (1989). Axonal growth-associated proteins. *Annual Review of Neuroscience* **12**, 127–156.

Skene, J.H.P., & Virag, I. (1989). Posttranslational membrane attachment and dynamic fatty acylation of neuronal growth cone protein. *Journal of Cell Biology* **108**, 613–624.

Skene, J.H.P., & Willard, M. (1981a). Electrophoretic analysis of axonally transported proteins in toad retinal ganglion cells. *Journal of Neurochemistry* **37**, 79–87.

Skene, J.H.P., & Willard, M. (1981b). Axonally transported proteins associated with growth in rabbit central and peripheral nervous system. *Journal of Cell Biology* **89**, 96–103.

Skene, J.H.P., & Willard, M. (1981c). Changes in axonally transported proteins during axon regeneration in toad retinal ganglion cells. *Journal of Cell Biology* **89**, 86–95.

Skene, J.H.P., & Willard, M. (1981d). Characteristics of growth-associated polypeptides in regenerating toad retinal ganglion cell axons. *Journal of Neuroscience* **1**, 419–426.

Skene, J.H.P., Jacobson, R.D., Snipes, G.J., McGuire, C.B., Norden, J.J., & Freeman, J.A. (1986). A protein induced during nerve growth (GAP-43) is a major component of growth-cone membranes. *Science* **233**, 783–786.

Skoff, R.P., & Hamburger, V. (1974). Fine structure of dendritic and axonal growth cones in embryonic chick spinal cord. *Journal of Comparative Neurology* **153**, 107–148.

Slaughter, T., Wang, J., & Black, M.M. (1997). Microtubule transport from the cell body into the axons of growing neurons. *Journal of Neuroscience* **17**, 5807–5819.

Small, R., & Pfenninger, K.H. (1984). Components of the plasma membrane of growing axons. I. Size and distribution of intramembrane particles. *Journal of Cell Biology* **98**, 1422–1433.

Small, R.K., Blank, M., Ghez, R., & Pfenninger, K.H. (1984). Components of the plasma membrane of growing axons. II. Diffusion of membrane protein complexes. *Journal of Cell Biology* **98**, 1434–1443.

Smallheiser, N.R., Crain, S.M., & Reid, L.M. (1984). Laminin as a substrate for retinal axons *in vitro. Developmental Brain Research* **12**, 136–140.

Smith, M.A., Yao, Y.M.M., Reist, N.E., Magill, C., Wallace, B.G., & McMahan, U.J. (1987). Identification of agrin in electric organ extracts and localization of agrin-like molecules in muscle and central nervous system. *Journal of Experimental Biology* **132**, 223–230.

Smith, S.J. (1988). Neuronal cytomechanics: The actin-based motility of growth cones. *Science* **242**, 708–715.

Smolen, A.J. (1981). Postnatal development of ganglionic neurons in the absence of preganglionic input: Morphological observations on synaptic formation. *Developmental Brain Research* **1**, 49–58.

Snow, D.M., & Letourneau, P.C. (1992). Neurite outgrowth on a step gradient of chondroitin sulphate proteoglycan (CS-PG). *Journal of Neurobiology* **23**, 322–336.

Snow, D.M., Steindler, D.A., & Silver, J. (1990). Molecular and cellular characterization of the glial roof plate of the spinal cord and optic tectum: A possible role for a proteoglycan in the development of an axonal barrier. *Developmental Biology* **138**, 359–376.

Snow, D.M., Watanabe, M., Letourneau, P.C., & Silver, J. (1991). A chondroitin sulfate proteoglycan may influence the direction of retinal ganglion cell outgrowth. *Development* **113**, 1473–1485.

Snow, P.M., Bieber, A.J., & Goodman, C.S. (1989). Fasciclin III: A novel homophilic adhesion molecule in *Drosophila. Cell* **59**, 313–323.

Snyder, S.H., & Sabatini, D.M. (1995). Immunophilins and the nervous system. *Nature Medicine* **1**, 32–37.

Sobue, K. (1993). Actin-based cytoskeleton in growth cone activity. *Neuroscience Research* **18**, 91–102.

Sobue, K., & Kanda, K. (1988). Localization of $pp60^{c-src}$ in growth cone of PC12 cells. *Biochemical and Biophysical Research Communications* **157**, 1383–1389.

Sobue, K., & Kanda, K. (1989). α-Actinins, calspectin (brain spectrin or fodrin), and actin participate in adhesion and movement of growth cones. *Neuron* **3**, 311–319.

Solomon, D.J. (1991). Inhibition of MAP2 expression affects both morphological and cell division phenotypes of neuronal differentiation. *Cell* **64**, 817–826.

Solomon, F. (1992). Neuronal cytoskeleton and growth. *Current Opinion in Neurobiology* **2**, 613–617.

Sonderegger, P., & Rathjen, F.G. (1992). Regulation of axonal growth in the vertebrate nervous system by interactions between glycoproteins belonging to two sub groups of the immunoglobulin superfamily. *Journal of Cell Biology* **119**, 1387–1394.

Song, H., Ming, G., & Poo, M.-M. (1997). cAMP-induced switching in turning direction of nerve growth cones. *Nature* **388**, 275–279.

Sorge, L.K., Levy, B.T., & Maness, P.F. (1984). $pp60^{c-src}$ is developmentally regulated in the neural retina. *Cell* **36**, 249–257.

Soriano, P., Montgomery, C., Geske, R., & Bradley, A. (1991). Targeted disruption of the c-*src* proto-oncogene leads to osteopetrosis in mice. *Cell* **64**, 693–702.

Sotello, C. (1973). Permanence and fate of paramembranous synaptic specializations in 'mutants' and experimental animals. *Brain Research* **62**, 345–351.

Sotello, C. (1975). Anatomical, physiological and biochemical studies of the cerebellum from mutant mice. II. Morphological study of cerebellar cortical neurons and circuits in the weaver mouse. *Brain Research* **94**, 19–44.

Sotello, C. (1990). Cerebellar synaptogenesis: What can we learn from mutant mice? *Journal of Experimental Biology* **153**, 225–249.

Sotello, C., & Changeux, J.P. (1974). Trans-synaptic degeneration 'en cascade' in the cerebellar cortex of staggerer mutant mice. *Brain Research* **67**, 519–526.

Sotelo, J., Toh, B.H., Yildiz, A., Osung, O., & Holborrow, E.J. (1979). Immunofluorescence demonstrates the distribution of actin, myosin and intermediate filaments in cultured neuroblastoma cells. *Neuropathology and Applied Neurobiology* **5**, 499–505.

Speidel, C.C. (1933). Studies of living nerves II. Activities of ameboid growth cones, sheath cells, and myelin segments, as revealed by prolonged observation of individual nerve fibers in frog tadpoles. *American Journal of Anatomy* **52**, 1–79.

Speidel, C.C. (1942). Studies of living nerves. IV. Growth adjustments of cutaneous terminal arborizations. *Journal of Comparative Neurology* **76**, 57–69.

Spencer, S.A., Schuh, S.M., Liu, W.-S., & Willard, M.B. (1992). GAP-43, a protein associated with axon growth, is phosphorylated at three sites in cultured neurons and rat brain. *Journal of Biological Chemistry* **267**, 9059–9064.

Sperry, R.W. (1943a). Effect of 180° rotation of the retinal field on visuomotor coordination. *Journal of Experimental Zoology* **92**, 263–279.

Sperry, R.W. (1943b). Visuomotor coordination in the newt *(Triturus viridescens)* after regeneration of the optic nerve. *Journal of Comparative Neurology* **79**, 33–35.

Sperry, R.W. (1963). Chemoaffinity in the orderly growth of nerve fiber patterns and connections. *Proceedings of the National Academy of Sciences U.S.A.* **50**, 703–710.

Spoerri, P. (1988). Neurotrophic effects of GABA in cultures of embryonic chick brain and retina. *Synapse* **2**, 11–22.

Spooner, B.S., & Holladay, C.R. (1981). Distribution of tubulin and actin in neurites and growth cones of differentiating nerve cells. *Cell Motility* **1**, 167–178.

Sretavan, D.W., & Kruger, K. (1998). Randomized retinal ganglion cell axon routing at the optic chiasm of GAP-43-deficient mice: Association with midline recrossing and lack of normal ipsilateral axon turning. *Journal of Neuroscience* **18**, 10502–10513.

Sretavan, D.W., & Reichardt, L.F. (1993). Time-lapse video analysis of retinal ganglion cell axon pathfinding at the mammalian optic chiasm: Growth cone guidance using intrinsic chiasm cues. *Neuron* **10**, 761–777.

Sretavan, D.W., & Shatz, C.J. (1986). Prenatal development of retinal ganglion cell axons: Segregation into eye-specific layers within the cat's lateral geniculate nucleus. *Journal of Neuroscience* **6**, 234–251.

St. Amand, G.S., & Tipton, S.R. (1954). The separation of neuroblasts and other cells from grasshopper embryos. *Science* **119**, 93–94.

Stahl, B., Muller, B., von Boxberg, Y., Cox, E.C., & Bonhoeffer, F. (1990). Biochemical characterization of a putative axonal guidance molecule of the chick visual system. *Neuron* **5**, 735–743.

Stallcup, W.B., & Beasley, L. (1985). Involvement of the nerve growth factor-inducible large external glycoprotein (NILE) in neurite fasciculation in primary cultures of rat brain. *Proceedings of the National Academy of Sciences U.S.A.* **82**, 1276–1280.

Stallcup, W.B., Beasley, L.L., & Levine, J.M. (1985). Antibody against nerve growth factor-inducible large external (NILE) glycoprotein labels nerve fiber tracts in the developing rat nervous system. *Journal of Neuroscience* **5**, 1090–1101.

Steedman, J.G., & Landreth, G.E. (1989). Expression of pp60$^{c\text{-}src}$ in adult and developing rat central nervous system. *Developmental Brain Research* **45**, 161–167.

Stefanelli, A. (1951). The Mauthnerian apparatus in the ichthyopsida; its nature and function and correlated problems in neurohistogenesis. *Quarterly Review of Biology* **26**, 17–34.

Steindler, D.A., Cooper, N.G.F., Faissner, A., & Schachner, M. (1989). Boundaries defined by adhesion molecules during development of the cerebral cortex: The J1/tenascin glycoprotein in the mouse somatosensory cortical barrel field. *Developmental Biology* **131**, 243–260.

Stelzner, D.J., Martin, A.H., & Scott, G.L. (1973). Early stages of synaptogenesis in the cervical spinal cord of the chick embryo. *Zeitschrift für Zellforschung und mikroskopische Anatomie* **138**, 475–488.

Stevens, J.K., Trogadis, J., & Jacobs, J.R. (1988). Developmental control of axial neurite form: A serial electronmicroscopic analysis. In *Intrinsic Determinants of Neuronal Form and Function*, ed. R.J. Lasek & M.M. Black, pp. 115–145. New York: Alan R. Liss Inc.

Steward, O. (1997). mRNA localization in neurons: A multipurpose mechanism? *Neuron* **18**, 9–12.

Steward, O., & Banker, G.A. (1992). Getting the message from the gene to the synapse: Sorting and intracellular transport of RNA in neurons. *Trends in Neurosciences* **15**, 180–186.

Stewart, R., Erskine, L., & McCaig, C.D. (1995). Calcium channel subtypes and intracellular calcium stores modulate electric field-stimulated and orientated nerve growth. *Developmental Biology* **171**, 340–351.

Stirling, R.V., & Dunlop, S.A. (1995). The dance of the growth cones – where to next? *Trends in Neurosciences* **18**, 111–115.

Stoeckli, E.T., & Landmesser, L.T. (1995). Axonin-1, Nr-CAM, and Ng-CAM play different roles in the in vivo guidance of chick commissural neurons. *Neuron* **14**, 1165–1179.

Stoeckli, E.T., Sonderegger, P., Pollerberg, G.E., & Landmesser, L.T. (1997). Interference with axonin-1 and NrCAM interactions unmasks a floor-plate activity inhibitory for commissural axons. *Neuron* **18**, 209–221.

Strähle, U., Fischer, N., & Blader, P. (1997). Expression and regulation of a netrin homolog in the zebrafish embryo. *Mechanisms of Development* **62**, 147–160.

Strittmatter, S.M., & Fishman, M.C. (1991). The neuronal growth cone as a specialized transduction system. *Bioessays* **13**, 127–134.

Strittmatter, S.M., Valenzuela, D., Kennedy, T.E., Neer, E.J., & Fishman, M.C. (1990). G_o is a major growth cone protein subject to regulation by GAP-43. *Nature* **344**, 836–841.

Strittmatter, S.M., Valenzuela, D., Sudo, Y., Linder, M.E., & Fishman, M.C. (1991). An intracellular guanine-nucleotide release protein for G_0 – GAP-43 stimulates isolated alpha-subunits by a novel mechanism. *Journal of Biological Chemistry* **266**, 22465–22471.

Strittmatter, S.M., Vartanian, T., & Fishman, M.C. (1992). GAP-43 as a plasticity protein in neuronal form and repair. *Journal of Neurobiology* **23**, 507–520.

Strittmatter, S.M., Cannon, S.C., Ross, E.M., Higashijima, T., & Fishman, M.C. (1993). GAP-43 augments G protein-coupled receptor transduction in *X. laevis* oocytes. *Proceedings of the National Academy of Sciences U.S.A.* **90**, 5327–5331.

Strittmatter, S.M., Valenzuela, D., & Fishman, M.C. (1994a). An amino terminal domain of the growth-associated protein GAP-43 mediates its effects on filopodial formation and cell spreading. *Journal of Cell Science* **107**, 195–204.

Strittmatter, S.M., Igarashi, M., & Fishman, M.C. (1994b). GAP-43 amino terminal peptides modulate growth cone morphology and neurite outgrowth. *Journal of Neuroscience* **14**, 5503–5513.

Strittmatter, S.M., Fankhauser, C., Huang, P.L., Mashimo, H., & Fishman, M.C. (1995). Neuronal pathfinding is abnormal in mice lacking the neuronal growth cone protein GAP-43. *Cell* **80**, 445–452.

Suarez-Isla, B.A., Pelto, D.J., Thompson, J.M., & Rapoport, S.I. (1984). Blockers of calcium permeability inhibit neurite extension and formation of neuromuscular synapses in cell culture. *Developmental Brain Research* **14**, 263–270.

Sugisaki, N., Hirata, T., Naruse, I., Kawakami, A., Kitsukawa, T., & Fujisawa, H. (1996). Positional cues that are strictly localized in the telencephalon induce preferential growth of mitral cell axons. *Journal of Neurobiology* **29**, 127–137.

Suidan, H.S., Stone, S.R., Hemmings, B.A., & Monard, D. (1992). Thrombin causes neurite retraction in neuronal cells through activation of cell surface receptors. *Neuron* **8**, 363–375.

Sun, Y., & Poo, M.-M. (1987). Evoked release of acetylcholine from the growing embryonic neuron. *Proceedings of the National Academy of Sciences U.S.A.* **84**, 2540–2544.

Sydor, A.M., Su, A.L., Wang, F.-S., Xu, A., & Jay, D.G. (1995). Talin and vinculin play distinct roles in filopodial motility in the neuronal growth cone. *Journal of Cell Biology* **134**, 1197–1207.

Symons, M. (1996). Rho-family GTPases – the cytoskeleton and beyond. *Trends in Biochemistry* **21**, 178–181.

Taghert, P.H., Bastiani, M.J., Ho, R.K., & Goodman, C.S. (1982). Guidance of pioneer growth cones: Filopodial contacts and coupling revealed with an antibody to lucifer yellow. *Developmental Biology* **94**, 391–399.

Takai, Y., Sasaki, T., Tanaka, K., & Nakanishi, H. (1995). Rho as a regulator of the cytoskeleton. *Trends in Biochemical Sciences* **20**, 227–231.

Takagi, S., Tsuji, T., Amagai, T., Takamatsu, T., & Fujisawa, H. (1987). Specific cell surface labels in the visual centers of *Xenopus laevis* tadpole identified using monoclonal antibodies. *Developmental Biology* **122**, 90–100.

Takagi, S., Hirata, T., Agata, K., Mochii, M., Eguchi, G., & Fujisawa, H. (1991). The A5 antigen, a candidate for the neuronal recognition molecule, has homologies to complement component and coagulation factors. *Neuron* **7**, 295–307.

Takagi, S., Kasuya, Y., Shimizu, M., Matsuura, T., Tsuboi, M., Kawakami, A., & Fujisawa, H. (1995). Expression of a cell adhesion molecule, neuropilin, in the developing chick nervous system. *Developmental Biology* **170**, 207–222.

Takeda, S., Funakoshi, T., & Hirokawa, N. (1995). Tubulin dynamics in neuronal axons of living zebrafish embryos. *Neuron* **14**, 1257–1264.

Takei, Y., Kondo, S., Harada, A., Inomata, S., Noda, T., & Hirokawa, N. (1997). Delayed development of nervous system in mice homozygous for disrupted microtubule-associated protein 1B (MAP1B) gene. *Journal of Cell Biology* **137**, 1615–1626.

Takeichi, M. (1988). The cadherins: Cell–cell adhesion molecules controlling animal morphogenesis. *Development* **102**, 639–655.

Takemura, R., Okabe, S., Umeyama, T., Kanai, Y., Cowan, N.J., & Hirokawa, N. (1992). Increased microtubule stability and alpha tubulin acetylation in cells transfected with microtubule-associated proteins MAP-1B, MAP2 and tau. *Journal of Cell Science* **103**, 953–964.

Tamada, A., Shirasaki, R., & Murakami, F. (1995). Floor plate chemoattracts crossed axons and chemorepels uncrossed axons in the vertebrate brain. *Neuron* **14**, 1083–1093.

Tanaka, E., & Sabry, J. (1995). Making the connection: Cytoskeletal rearrangements during growth cone guidance. *Cell* **83**, 171–176.

Tanaka, E., Ho, T., & Kirschner, M.W. (1995). The role of microtubule dynamics in growth cone motility and axonal growth. *Journal of Cell Biology* **128**, 139–155.

Tanaka, E.M., & Kirschner, M.W. (1991). Microtubule behaviour in the growth cones of living neurons during axon elongation. *Journal of Cell Biology* **115**, 345–363.

Tanaka, J., Kira, M., & Sobue, K. (1993). Gelsolin is localized in neuronal growth cones. *Developmental Brain Research* **76**, 268–271.

Tanaka, Y., Kawahata, K., Nakata, T., & Hirokawa, N. (1992). Chronological expression of microtubule-associated proteins (MAPs) in EC cell P19 after neuronal induction by retinoic acid. *Brain Research* **596**, 269–278.

Tang, J.C., Landmesser, L., & Rutishauser, U. (1992). Polysialic influences specific pathfinding by avian motoneurons. *Neuron* **8**, 1031–1044.

Taniguchi, M., Yuasa, S., Fujisawa, H., Naruse, I., Saga, S., Mishina, M., & Yagi, T. (1997). Disruption of *Semaphorin III/D* gene causes severe abnormality in peripheral nerve projection. *Neuron* **19**, 519–530.

Tannahill, D., Cook, G.M.W., & Keynes, R.J. (1997). Axon guidance and somites. *Cell and Tissue Research* **290**, 275–283.

Tapon, N., & Hall, A. (1997). Rho, Rac and Cdc42 GTPases regulate the organization of the actin cytoskeleton. *Current Opinion in Cell Biology* **9**, 86–92.

Tawil, N.J., Wilson, P., & Carbonetto, S. (1993). Integrins in point contacts mediate cell spreading: factors that regulate integrin accumulation in point contacts vs. focal contacts. *Journal of Cell Biology* **120**, 261–271.

Taylor, J., & Gordon-Weeks, P.R. (1989). Developmental changes in the calcium dependency of gamma-aminobutyric acid release from isolated growth cones: correlation with growth cone morphology. *Journal of Neurochemistry* **53**, 834–843.

Taylor, J., & Gordon-Weeks, P.R. (1991). Calcium-independent GABA release from GABAergic growth cones: Role of GABA transport. *Journal of Neurochemistry* **56**, 273–280.

Taylor, J., Docherty, M., & Gordon-Weeks, P.R. (1990). GABAergic growth cones: Release of endogenous GABA precedes the expression of synaptic vesicle antigens. *Journal of Neurochemistry* **54**, 1689–1699.

Taylor, J., Pesheva, P., & Schachner, M. (1993). The influence of janusin and tenascin on growth cone behaviour *in vitro*. *Journal of Neuroscience Research* **35**, 347–363.

Tear, G., Seeger, M.A., & Goodman, C.S. (1993). To cross or not to cross: A genetic analysis of guidance at the midline. *Perspectives on Developmental Neurobiology* **1**, 183–194.

Tear, G., Harris, R., Sutaria, S., Kilomanski, K., & Goodman, C.S. (1996). *Commissureless* controls growth cone guidance across the CNS midline in *Drosophila* and encodes a novel membrane protein. *Neuron* **16**, 501–514.

Teichman-Weinberg, A., Littauer, U.Z., & Ginzburg, I. (1988). The inhibition of neurite outgrowth in PC12 cells by tubulin antisense oligodeoxynucleotides. *Gene* **72**, 297–307.

Tennyson, V.M. (1970). The fine structure of the axon and growth cone of the dorsal root neuroblast of the rabbit embryo. *Journal of Cell Biology* **44**, 62–79.

Terasaki, M. (1993). Probes for the endoplasmic reticulum. In *Fluorescent and Luminescent Probes for Biological Activity*, ed. W.T. Mason, pp. 120–123. London: Academic Press.

Tessier-Lavigne, M. (1992). Axon guidance by molecular gradients. *Current Opinion in Neurobiology* **2**, 60–65.

Tessier-Lavigne, M. (1994). Axon guidance by diffusible repellants and attractants. *Current Opinion in Genetics and Development* **4**, 596–601.

Tessier-Lavigne, M. (1995). Eph receptor tyrosine kinases, axon repulsion, and the development of topographic maps. *Cell* **82**, 345–348.

Tessier-Lavigne, M., & Placzek, M. (1991). Target attraction: Are developing axons guided by chemotropism? *Trends in Neurosciences*, **14**, 303–310.

Tessier-Lavigne, M., Placzek, M., Lumsden, A.G.S., Dodd, J., & Jessell, T.M. (1988). Chemotropic guidance of developing axons in the mammalian nervous system. *Nature* **336**, 775–778.

Tessler, A.L., Autilio-Gambetti, L., & Gambetti, P. (1980). Axonal growth during regeneration: A quantitative autoradiographic study. *Journal of Cell Biology* **87**, 197–203.

Thanos, S., Bonhoeffer, F., & Rutishauser, U. (1984). Fiber–fiber interaction and tectal cues influence the development of chick retinotectal projection. *Proceedings of the National Academy of Sciences U.S.A.* **81**, 1906–1910.

Theriot, J.A., & Mitchison, T.J. (1992). The nucleation–release model of actin filament dynamics in cell motility. *Trends in Cell Biology* **2**, 219–222.

Thomas, W.S., O'Dowd, D.K., & Smith, M.A. (1993). Developmental expression and alternative splicing of chick agrin RNA. *Developmental Biology* **158**, 525–535.

Thompson, C., Lin, C.-H., & Forscher, P. (1996). An *Aplysia* cell adhesion molecule associated with site-directed actin filament assembly in neuronal growth cones. *Journal of Cell Science* **109**, 2843–2854.

Thompson, J.M., & Pelto, D.J. (1982). Attachment, survival and neurite extension of chick embryo retinal neurons on various culture substrates. *Developmental Neuroscience* **5**, 447–457.

Timpl, R., & Brown, J.C. (1994). The laminins. *Matrix Biology* **14**, 275–281.

Titus, M.A. (1997). Unconventional myosins: New frontiers in actin-based motors. *Trends in Cell Biology* **7**, 119–123.

Tolkovsky, A.M., Walker, A.E., Murrell, R.D., & Suidan, H.S. (1990). Ca^{2+} transients are not required as signals for long-term neurite outgrowth from cultured sympathetic neurons. *Journal of Cell Biology* **110**, 1295–1306.

Tomaselli, K.J., & Reichardt, L.F. (1988). Peripheral motoneuron interactions with laminin and Schwann cell-derived neurite-promoting molecules: developmental regulation of laminin receptor function. *Journal of Neuroscience Research* **21**, 275–285.

Tomaselli, K.J., Reichardt, L.F., & Bixby, J.L. (1986). Distinct molecular interactions mediate neuronal process outgrowth on non-neuronal cell surfaces and extracellular matrices. *Journal of Cell Biology* **103**, 2659–2672.

Tomaselli, K.J., Damsky, C.H., & Reichardt, L.F. (1987). Interactions of a neuronal cell line (PC12) with laminin, collagen IV, and fibronectin: Identification of integrin-related glycoproteins involved in attachment and process outgrowth. *Journal of Cell Biology* **105**, 2347–2358.

Tomaselli, K.J., Neugebauer, K.M., Bixby, J.L., Lilien, J., & Reichardt, L.F. (1988). N-Cadherin and integrins: Two receptor systems that mediate neuronal process outgrowth on astrocyte surfaces. *Neuron* **1**, 33–43.

Tomaselli, K., Hall, D.E., Flier, L.A., Gehlsen, K.R., Turner, D.C., Carbonetto, S., & Reichardt, L.F. (1990). A neuronal cell line (PC12) expresses two β1-class integrins $\alpha 1\beta 1$ and $\alpha 3\beta 1$ that recognize different outgrowth-promoting domains in laminin. *Neuron* **5**, 651–662.

Tomasiewicz, H., Ono, K., Yee, D., Thompson, C., Goridis, C., Rutishauser, U., & Magnuson, T. (1993). Genetic deletion of a neural cell adhesion molecule variant (N-CAM-180) produces distinct defects in the central nervous system. *Neuron* **11**, 1163–1174.

Tosney, K. (1987). Proximal tissues and patterned neurite outgrowth at the lumbosacral level of the chick embryo: Deletion of the dermamyotome. *Developmental Biology* **122**, 540–558.

Tosney, K.W., & Landmesser, L.T. (1985a). Growth cone morphology and trajectory in the lumbosacral region of the chick embryo. *Journal of Neuroscience* **5**, 2345–2358.

Tosney, K.W., & Landmesser, L.T. (1985b). Specificity of early motoneuron growth cone outgrowth in the chick embryo. *Journal of Neuroscience* **5**, 2336–2344.

Tosney, K.W., & Landmesser, L.T. (1985c). Development of the major pathways for neurite outgrowth in the chick hindlimb. *Developmental Biology* **109**, 193–214.

Tosney, K.W., & Wessells, N.K. (1983). Neuronal motility: The ultrastructure of veils and microspikes corrrelates with their motile activities. *Journal of Cell Science* **61**, 389–411.

Tsien, R.Y. (1988). Fluorescence measurement and photochemical manipulation of cytosolic free calcium. *Trends in Neurosciences* **11**, 419–424.

Tsui, H.T., Ris, H., & Klein, L. (1983). Ultrastructural networks in growth cones and neurites of cultured central nervous system neurones. *Proceedings of the National Academy of Sciences U.S.A.* **80**, 5779–5783.

Tsui, H.T., Lanford, K.L., Ris, H., & Klein, W.L. (1984). Novel organization of microtubules in culture central nervous system neurons: Formation of hair-pin loops at ends of maturing neurites. *Journal of Neuroscience* **4**, 3002–3013.

Tsukada, Y., Chiba, K., Yamazaki, M., & Mohri, T. (1994). Inhibition of the nerve growth factor-induced outgrowth by specific tyrosine kinase and phospholipase inhibitors. *Biological Pharmacology Bulletin* **17**, 370–375.

Tsukita, S., Kobayashi, T., & Matsumoto, G. (1986). Subaxolemmal cytoskeleton in squid giant axon. II. Morphological identification of microtubule- and microfilament-associated domains of axolemma. *Journal of Cell Biology* **102**, 1710–1725.

Tucker, J. (1992). The microtubule-organizing centre. *Bioessays* **14**, 861–867.

Tucker, R.P. (1990). The roles of microtubule-associated proteins in brain morphogenesis: a review. *Brain Research Review* **15**, 101–120.

Tucker, R.P., & Matus, A. (1988). Microtubule-associated proteins characteristic of embryonic brain are found in the adult mammalian retina. *Developmental Biology* **130**, 423–434.

Tucker, R.P., Binder, L.I., & Matus, A. (1988). Neuronal microtubule-associated proteins in the embryonic avian spinal cord. *Journal of Comparative Neurology* **271**, 44–55.

Turner, D.C., & Flier, L.A. (1989). Receptor-mediated active adhesion to the substratum is required for neurite outgrowth. *Developmental Neuroscience* **11**, 300–312.

Ulloa, L., Díez-Guerra, J., Avila, J., & Díaz-Nido, J. (1997). Localization of differentially phosphorylated isoforms of microtubule associated protein 1B (MAP1B) in cultured rat hippocampal neurons. *Neuroscience* **61**, 211–223.

VanBerkum, M.F.A., & Goodman, C.S. (1995). Targeted disruption of Ca^{2+}-calmodulin signaling in *Drosophila* growth cones leads to stalls in axon extension and errors in axon guidance. *Neuron* **14**, 43–56.

Vancura, K.L., & Jay, D.G. (1998). G proteins and axon growth. *Seminars in Neuroscience* **9**, 209–219.

van den Pol, A.N. (1997). GABA immunoreactivity in hypothalamic neurons and growth cones in early development in vitro before synapse formation. *Journal of Comparative Neurology* **383**, 178–188.

Van der Loos, H. (1965). The 'improperly' oriented pyramidal cell in the cerebral cortex and its possible bearing on problems of growth and cell orientation. *Bulletin of Johns Hopkins Hospital* **117**, 228–250.

Van Hooff, C.O.M., DeGraan, P.N.E., Oestreicher, A.B., & Gispen, W.H. (1988). B-50 phosphorylation and phosphoinositide metabolism in nerve growth cone membranes. *Journal of Neuroscience* **8**, 1789–1795.

Van Lookeren Campagne, M., Beate Oestreicher, A., Van Bergen en Henegouwen, P.M.P., & Gispen, W.H. (1989). Ultrastructural immunocytochemical localization of B-50/GAP43, a protein kinase C substrate, in isolated presynaptic nerve terminals and neuronal growth cones. *Journal of Neurocytology* **18**, 479–489.

van Mier, P., Joosten, H.W.J., van Rheden, R., & ten Donkelaar, H.J. (1986). The development of serotonergic raphespinal projections in *Xenopus laevis*. *International Journal of Developmental Neuroscience* **4**, 465–476.

Vanselow, J., Thanos, S., Godement, P., Henke-Fahle, S., & Bonhoeffer, F. (1989). Spatial arragement of glia and ingrowing retinal axons in the chick optic tectum during development. *Developmental Brain Research* **45**, 15–27.

Varnum-Finney, B., & Reichardt, L.F. (1994). Vinculin-deficient PC12 cell lines extend unstable lamellipodia and filopodia and have reduced rate of neurite outgrowth. *Journal of Cell Biology* **127**, 1071–1084.

Varnum-Finney, B., Venstrom, K., Muller, U., Kypta, R., Backus, C., Chiquet, M., & Reichardt, L.F. (1995). The integrin receptor $\alpha8\beta1$ mediates interactions of embryonic chick motor and sensory neurons with tenascin-C. *Neuron* **14**, 1213–1222.

Varon, S., & Raiborn, C.W. (1969). Dissociation, fractionation and culture of embryonic brain cells. *Brain Research* **12**, 180

Vaughan, L., Weber, P., D'Alessandri, L., Zisch, A.H., & Winterhalter, K.H. (1994). Tenascin-contactin/F11 interactions: A clue for a developmental role? *Perspectives on Developmental Neurobiology* **2**, 43–52.

Vaughn, J.E. (1989). Review: Fine structure of synaptogenesis in the vertebrate central nervous system. *Synapse* **3**, 255–285.

Vaughn, J.E., & Sims, T.J. (1978). Axonal growth cones and developing axonal collaterals form synaptic junctions in developing mouse spinal cord. *Journal of Neurocytology* **7**, 337–363.

Vaughn, J.E., Henrikson, C.K., & Grieshaber, J.A. (1974). A quantitative study of synapses on motor neuron dendritic growth cones in developing mouse spinal cord. *Journal of Cell Biology* **60**, 664–672.

Vestal, D.J., & Ranscht, B. (1992). Phosphatidylinositol-anchored T-cadherin mediates calcium-dependent, homophilic adhesion. *Journal of Cell Biology* **119**, 451–461.

Viereck, C., & Matus, A. (1990). The expression of phosphorylated and nonphosphorylated forms of MAP5 in the amphibian CNS. *Brain Research* **508**, 257–264.

Viereck, C., Tucker, R.P., & Matus, A. (1989). The adult rat olfactory system expresses microtubule-associated proteins found in the developing brain. *Journal of Neuroscience* **9**, 3547–3557.

Vigers, A.J., & Pfenninger, K.H. (1991). N type and L type calcium channels are present in nerve growth cones. Numbers increase on synaptogenesis. *Developmental Brain Research* **60**, 197–203.

Vitkovic, L., Steisslinger, H.W., Aloyo, V.J., & Mersel, M. (1988). The 43-kDa neuronal growth-associated protein (GAP-43) is present in plasma membranes of rat astrocytes. *Proceedings of the National Academy of Sciences U.S.A.* **85**, 8296–8300.

Volkmer, H., Leuschner, R., Zacharias, U., & Rathjen, F.G. (1996). Neurofascin induces neurites by heterophilic interactions with axonal NrCAM while NrCAM requires F11 on the axonal surface to extend neurites. *Journal of Cell Biology* **135**, 1059–1069.

Wadsworth, W.G., Bhatt, H., & Hedgecock, E.M. (1996). Neuroglia and pioneer neurons express UNC-6 to provide global and local netrin cues for guiding migrations in *C. elegans. Neuron* **16**, 35–46.

Wagner, M.C., Barylko, B., & Albanesi, J.P. (1992). Tissue distribution and subcellular localization of mammalian myosin I. *Journal of Cell Biology* **119**, 163–170.

Walbot, V., & Holder, N. (1987). *Developmental Biology.* New York: Random House.

Wallace, B.G. (1989). Agrin-induced specializations contain cytoplasmic, membrane, and extracellular matrix-associated components of the postsynaptic apparatus. *Journal of Neuroscience* **9**, 1294–1302.

Wallace, B.G. (1994). Staurosporine inhibits agrin-induced acetylcholine receptor phosphorylation and aggregation. *Journal of Cell Biology* **125**, 661–668.

Wallace, B.G. (1996). Signalling mechanisms mediating synapse formation. *Bioessays* **18**, 777–780.

Wallace, B.G., Qu, Z., & Huganir, R.L. (1991). Agrin induces phosphorylation of the nicotinic acetylcholine receptor. *Neuron* **6**, 869–878.

Walsh, F.S., & Doherty, P. (1993). Factors regulating the expression and function of calcium-independent cell adhesion molecules. *Current Opinion in Cell Biology* **5**, 791–796.

Walsh, F.S., & Doherty, P. (1997). Neural cell adhesion molecules of the immunoglobulin superfamily: Role in axon growth and guidance. *Annual Reviews of Cell and Developmental Biology* **13**, 425–456.

Walter, J., Henke-Fahle, S., & Bonhoeffer, F. (1987a). Avoidance of posterior tectal membranes by temporal retinal axons. *Development* **101**, 909–913.

Walter, J., Kern-Veits, B., Huf, J., Stolze, B., & Bonhoeffer, F. (1987b). Recognition of position-specific properties of tectal cell membranes by retinal axons *in vitro. Development* **101**, 685–696.

Walter, J., Allsopp, T.E., & Bonhoeffer, F. (1990). A common denominator of growth cone guidance and collapse? *Trends in Neurosciences* **13**, 447–452.

Walton, M.K., Schaffner, A.E., & Barker, J.L. (1993). Sodium channels, $GABA_A$ receptors, and glutamate receptors develop sequentially on embryonic rat spinal cord cells. *Journal of Neuroscience* **13**, 2068–2084.

Wandosell, F., Serrano, L., & Avila, J. (1994). Phosphorylation of α-tubulin carboxyl-terminal tyrosine prevents its incorporation into microtubules. *Journal of Biological Chemistry* **262**, 8268–8273.

Wang, F.-S., Wolenski, J.S., Cheney, R.E., Mooseker, M.S., & Jay, D.G. (1996a). Function of myosin-V in filopodial extension of neuronal growth cones. *Science* **273**, 660–663.

Wang, L.-C., Rachel, R.A., Marcus, R.C., & Mason, C.A. (1996b). Chemosuppression of retinal axon growth by the mouse optic chiasm. *Neuron* **17**, 849–862.

Wang, S., & Hazelrigg, T. (1994). Implications for *bcd* mRNA localization from spatial distribution of *exu* protein in *Drosophila* oogenesis. *Nature* **369**, 400–403.

Weber, A. (1938). Croissance des fibres nerveuses commissurales lors de lésions de la moelle épiniere chez de jeunes embryons de Poulet. *Biomorphosis* **1**, 30–35.

Weeds, A. (1982). Actin-binding proteins – regulators of cell architecture and motility. *Nature* **296**, 811–816.

Wehrle, B., & Chiquet, M. (1990). Tenascin is accumulated along developing peripheral nerves and allows neurite outgrowth in vitro. *Development* **110**, 401–415.

Wehrle-Haller, B., & Chiquet, M. (1993). Dual function of tenascin: Simultaneous promotion of neurite growth and inhibition of glial migration. *Journal of Cell Science* **106**, 597–610.

Wehrle-Haller, B., Koch, M., Baumgartner, S., Spring, J., & Chiquet, M. (1991). Nerve-dependent and independent tenascin expression in the developing chick limb bud. *Development* **112**, 627–637.

Weiss, P. (1924). Die Funktion transplantierter Amphibienextremitaten. Aufstellung einer Resonanztheorie der motorischen Nerventätigkeit auf Grund abgestimmter Endorgane. *Archiv für Mikroskopische Anatomie und Entwicklungsmechanik* **102**, 635–672.

Weiss, P. (1928). Erregungspecifität and Erregungsresonanz. *Ergebnisse der Biologie* **3**, 1–115.

Weiss, P. (1934). *In vitro* experiments on the factors determining the course of the outgrowing nerve fiber. *Journal of Experimental Zoology* **68**, 393–448.

Weiss, P. (1941). Nerve patterns: The mechanics of nerve growth. *Third Growth Symposium* **5**, 163–203.

Weisshaar, B., Doll, T., & Matus, A. (1992). Reorganisation of the microtubular cytoskeleton by embryonic microtubule-associated protein 2 (MAP2c). *Development* **116**, 1151–1161.

Welch, M.D., Mallavarapu, A., Rosenblatt, J., & Mitchison, T.J. (1997). Actin dynamics *in vivo*. *Current Opinion in Cell Biology* **9**, 54–61.

Weldon, P.R. (1975). Pinocytotic uptake and intracellular distribution of colloidal thorium dioxide by cultured sensory neurites. *Journal of Neurocytology* **4**, 341–356.

Wells, D.G., & Fallon, J.R. (1996). Neuromuscular junctions: The state of the union. *Current Biology* **6**, 1073–1075.

Wessells, N.K., & Ludueña, M.A. (1974). Thorotrast uptake and transit in embryonic glia, heart fibroblasts and neurons *in vitro*. *Tissue and Cell* **6**, 757–776.

Wessells, N.K., & Nuttall, R.P. (1978). Normal branching, induced branching, and steering of cultured parasympathetic motor neurons. *Experimental Cell Research* **115**, 111–122.

Wessells, N.K., Spooner, B.S., Ash, J.F., Ludueña, M.A., & Wrenn, J.T. (1971). Cytochalasin B: Microfilaments and 'contractile' processes. *Science* **173**, 356–359.

Wessells, N.K., Letourneau, P.C., Nuttall, R.P., Ludueña-Anderson, M., & Geiduschek, J.M. (1980). Responses to cell contacts between growth cones, neurites, and ganglionic non-neuronal cells. *Journal of Neurocytology* **9**, 647–665.

Wessells, N.K., Letourneau, P.C., Nuttall, R.P., Ludueña-Anderson, M., & Geiduschek, J.M. (1981). Responses to cell contacts between growth cones, neurites and ganglionic non-neuronal cells. *Journal of Neurocytology* **9**, 647–664.

Westerfield, M. (1987). Substrate interactions affecting motor growth cone guidance during development and regeneration. *Journal of Experimental Biology* **132**, 161–175.

Westrum, L. (1975). Electron microscopy of synaptic structures in olfactory cortex of early postnatal rats. *Journal of Neurocytology* **4**, 713–732.

Wettstein, R., & Sotelo, J.R. (1963). Electron-microscope study on the regenerative process of peripheral nerves of mice. *Zeitschrift für Zellforschung und mikroskopische Anatomie* **59**, 708–723.

Wiche, G., Oberkanins, C., & Himmler, A. (1991). Molecular structure and function of microtubule-associated proteins. *International Review of Cytology* **124**, 217–273.

Widmer, F., & Caroni, P. (1993). Phosphorylation site mutagenesis of the growth-associated protein GAP-43 modulates its effects on cell spreading and morphology. *Journal of Cell Biology* **120**, 503–512.

Wiestler, O.D., & Walter, G. (1988). Developmental expression of two forms of pp60[c-Src]. *Molecular and Cellular Biology* **8**, 502–504.

Williams, C.V., Davenport, R.W., Dou, P., & Kater, S.B. (1995). Developmental regulation of plasticity along neurite shafts. *Journal of Neurobiology* **27**, 127–140.

Williams, E.J., Doherty, P., Turner, G., Reid, R.A., Hemperly, J.J., & Walsh, F.S. (1992). Calcium influx into neurons can solely account for cell contact-dependent neurite outgrowth stimulated by transfected L1. *Journal of Cell Biology* **119**, 883–892.

Williams, E.J., Walsh, F.S., & Doherty, P. (1994a). Tyrosine kinase inhibitors can differentially inhibit integrin dependent and CAM stimulated neurite outgrowth. *Journal of Cell Biology* **124**, 1029–1037.

Williams, E.J., Walsh, F.S., & Doherty, P. (1994b). The production of arachidonic acid can account for calcium channel activation in the second messenger pathway underlying neurite outgrowth stimulated by NCAM, N-cadherin and L1. *Journal of Neurochemistry* **62**, 1231–1234.

Williams, E.J., Furness, J., Walsh, F.S., & Doherty, P. (1994c). Activation of the FGF receptor underlies neurite outgrowth stimulated by L1, NCAM and N-cadherin. *Neuron* **13**, 583–594.

Williams, E.J., Mittal, B., Walsh, F.S., & Doherty, P. (1995a). A Ca^{2+}/calmodulin kinase inhibitor, KN-62, inhibits neurite outgrowth stimulated by CAMs and FGF. *Molecular and Cellular Neuroscience* **6**, 69–79.

Williams, E.J., Mittal, B., Walsh, F.S., & Doherty, P. (1995b). FGF inhibits neurite outgrowth over monolayers of astrocytes and fibroblasts expressing transfected cell adhesion molecules. *Journal of Cell Science* **108**, 3523–3530.

Williams, R.W., Bastiani, M.J., Lia, B., & Chalupa, L.M. (1986). Growth cones, dying axons, and developmental fluctuations in the fiber population of the cat's optic nerve. *Journal of Comparative Neurology* **246**, 32–69.

Williamson, T., Gordon-Weeks, P.R., Schachner, M., & Taylor, J. (1996). Microtubule reorganisation is obligatory for growth cone turning. *Proceedings of the National Academy of Sciences U.S.A.* **93**, 15221–15226.

Wilson, L. & Jordan, M.A. (1994). Pharmacological probes of microtubule function. In *Microtubules*, ed. J. Hyams & C.W. Lloyd, pp. 59–83. New York: Wiley–Liss, Inc.

Wilson, S.W., & Easter, S.S., Jr (1991). A pioneering growth cone in the embryonic zebrafish brain. *Proceedings of the National Academy of Sciences U.S.A.* **88**, 2293–2296.

Wong, E.V., Kenwrick, S., Willems, P., & Lemmon, V. (1995). Mutations in the cell adhesion molecule L1 cause mental retardation. *Trends in Neurosciences* **18**, 168–172.

Wong, E.V., Schaefer, A.W., Landreth, G., & Lemmon, V. (1996). Involvement of p90rsk in neurite outgrowth mediated by the cell adhesion molecule L1. *Journal of Biological Chemistry* **271**, 18217–18223.

Wong, J.T.W., Yu, W.T.C., & O'Connor, T.P. (1997). Transmembrane grasshopper semaphorin I promotes axon outgrowth in vivo. *Development* **124**, 3597–3607.

Woolf, C.J., Reynolds, M.L., Chong, M.S., Emson, P., Irwin, N., & Benowitz, L.I. (1992). Denervation of motor endplate results in the rapid expression by terminal Schwann cells of the growth-associated protein GAP-43. *Journal of Neuroscience* **12**, 3999–4010.

Worley, T., & Holt, C. (1996). Inhibition of protein tyrosine kinases impairs axon extension in the embryonic optic tract. *Journal of Neuroscience* **16**, 2294–2306.

Wright, D.E., White, F.A., Gerfen, R.W., Silos-Santiago, I., & Snider, W.D. (1995). The guidance molecule semaphorin III is expressed in regions of spinal cord and periphery avoided by growing sensory axons. *Journal of Comparative Neurology* **361**, 321–333.

Wu, D.-Y., & Goldberg, D.J. (1993). Regulated tyrosine phosphorylation at the tips of growth cone filopodia. *Journal of Cell Biology* **123**, 653–664.

Wu, D.-Y., Wang, L.-C., Mason, C.A., & Goldberg, D.J. (1996). Association of β1 integrin with phosphotyrosine in growth cone filopodia. *Journal of Neuroscience* **16**, 1470–1478.

Xie, Z., & Poo, M.-M. (1986). Initial events in the formation of neuromuscular synapse. *Proceedings of the National Academy of Sciences U.S.A.* **83**, 7069–7073.

Yaginuma, H., Homma, S., Künzi, R., & Oppenheim, R.W. (1991). Pathfinding by growth cones of commissural interneurons in the chick embryo spinal cord: A light and electron microscopic study. *Journal of Comparative Neurology* **304**, 78–102.

Yamada, K., Spooner, B.S., & Wessells, N.K. (1970). Axon growth: Roles of microfilaments and microtubules. *Proceedings of the National Academy of Sciences U.S.A.* **66**, 1206–1212.

Yamada, K.M., Spooner, B.S., & Wessells, N.K. (1971). Ultrastructure and function of growth cones and axons of cultured nerve cells. *Journal of Cell Biology* **49**, 614–635.

Yamamoto, N., Yamada, K., Kurotani, T., & Toyama, K. (1992). Laminar specificity of extrinsic cortical connections studied in coculture preparations. *Neuron* **9**, 217–228.

Yamasaki, H., Itakura, C., & Mizutani, M. (1991). Hereditary hypotrophic axonopathy with neurofilament deficiency in a mutant strain of Japanese quail. *Acta Neuropathologica* **82**, 427–434.

Yamasaki, H., Bennett, G.S., Itakura, C., & Mizutani, M. (1992). Defective expression of neurofilament protein subunits in hereditary hypotrophic axonopathy of quail. *Laboratory Investigation* **66**, 734–743.

Yanker, B.A., Benowitz, L.I., Villa-Komaroff, L., & Neve, R.L. (1990). Transfection of PC12 cells with the human GAP-43 gene: Effects on neurite outgrowth and regeneration. *Molecular Brain Research* **7**, 39–44.

Yip, J.W., & Yip, Y.P.L. (1992). Laminin-developmental expression and role in axonal outgrowth in the peripheral nervous system of the chick. *Developmental Brain Research* **68**, 23–33.

Young, H.S., & Poo, M.-M. (1983). Spontaneous release of transmitter from growth cones of embryonic neurons. *Nature* **305**, 634–637.

Yu, W., & Baas, P.W. (1995). The growth of the axon is not dependent upon net microtubule assembly at its distal tip. *Journal of Neuroscience* **15**, 6827–6833.

Yu, W., Centonze, V.E., Ahmad, F.J., & Baas, P.W. (1993). Microtubule nucleation and release from the neuronal centrosome. *Journal of Cell Biology* **122**, 349–359.

Yu, W., Schwei, M.J., & Baas, P.W. (1996). Microtubule transport and assembly during axon growth. *Journal of Cell Biology* **133**, 151–157.

Zagon, I.S. & McLaughlin, B.J. (1993). *Receptors in the Developing Nervous System.* London: Chapman & Hall.

Zauner, W., Kratz, J., Staunton, J., Feick, P., & Wiche, G. (1992). Identification of two distinct microtubule binding domains on recombinant rat MAP 1B. *European Journal of Cell Biology* **57**, 66–74.

Zhang, J.-H., Cerretti, D.P., Yu, T., Flanagan, J.G., & Zhou, R. (1996). Detection of ligands in regions anatomically connected to neurons expressing the Eph receptor BSK: Potential roles in neuron–target interaction. *Journal of Neuroscience* **16**, 7182–7192.

Zheng, J., Buxbaum, R.E., & Heidemann, S.R. (1993). Investigation of microtubule assembly and organization accompanying tension-induced neurite initiation. *Journal of Cell Science* **104**, 1239–1250.

Zheng, J.Q., Felder, M., Connor, J.A., & Poo, M.-M. (1994). Turning of nerve growth cones induced by neurotransmitters. *Nature* **368**, 140–144.

Zheng, J.Q., Wan, J., & Poo, M.-M. (1996). Essential role of filopodia in chemotropic turning of nerve growth cone induced by a glutamate gradient. *Journal of Neuroscience* **16**, 1140–1149.

Zhou, F.C. (1990). Four patterns of laminin-immunoreactive structure in developing rat brain. *Developmental Brain Research* **55**, 191–201.

Zimprich, F., & Bolsover, S.R. (1996). Calcium channels in neuroblastoma cell growth cones. *European Journal of Neuroscience* **8**, 467–475.

Zipkin, I.D., Kindt, R.M., & Kenyon, C.J. (1997). Role of a new Rho family member in cell migration and axon guidance in *C. elegans. Cell* **90**, 883–894.

Zisch, A.H., D'Alessandri, L., Ranscht, B., Falchetto, R., Winterhalter, K.H., & Vaughan, L. (1992). Neuronal cell adhesion molecule contactin/F11 binds to tenascin via its immunoglobulin-like domains. *Journal of Cell Biology* **119**, 203–213.

Zoran, M.J., Doyle, R.T., & Haydon, P.G. (1990). Target-dependent induction of secretory capabilities in an identified motoneuron during synaptogenesis. *Developmental Biology* **138**, 202–213.

Zoran, M.J., Doyle, R.T., & Haydon, P.G. (1991). Target contact regulates the calcium responsiveness of the secretory machinery during synaptogenesis. *Neuron* **6**, 145–151.

Zoran, M.J., Funte, L.R., Kater, S.B., & Haydon, P.G. (1993). Neuron-muscle contact changes presynaptic resting calcium set-point. *Developmental Biology* **158**, 163–171.

Zuber, M.X., Goodman, D.W., Karns, L.R., & Fishman, M.C. (1989). The neuronal growth associated protein GAP-43 induces filopodial formation in non-neuronal cells. *Science* **244**, 1193–1195.

Zuber, M.X., Strittmatter, S.M., & Fishman, M.C. (1989). A membrane-targeting signal in the amino terminus of the neuronal protein GAP-43. *Nature* **341**, 345–348.

Zwiers, H., Veldhuis, D., Schotman, P., & Gispen, W.H. (1976). ACTH, cyclic nucleotides, and brain protein phosphorylation in vitro. *Neurochemistry Research* **1**, 699–677.

Zwiers, H., Schotman, P., & Gispen, W.H. (1980). Purification and some characteristics of an ACTH-sensitive protein kinase and its substrate protein in rat brain membranes. *Journal of Neurochemistry* **34**, 1689–1699.

Zerner, H. [...]

Zucker, P. and J. Gibson [...] 1869.

Index